Elementary Induction on Abstract Structures

ELEMENTARY INDUCTION ON ABSTRACT STRUCTURES

Yiannis N. Moschovakis

Professor of Mathematics
University of California, Los Angeles
and
Emeritus Professor of Mathematics
University of Athens

DOVER PUBLICATIONS, INC.
Mineola, New York

Bibliographical Note

This Dover edition, first published in 2008, is an unabridged republication of the work first published by North-Holland Publishing Company, Amsterdam, in 1974 as Volume 77 of the series Studies in Logic and the Foundations of Mathematics.

Library of Congress Cataloging-in-Publication Data

Moschovakis, Yiannis N.
Elementary induction on abstract structures / Yiannis N. Moschovakis. — Dover ed.
p. cm.
Originally published: Amsterdam : North-Holland Pub. ; New York : American Elsevier Pub., 1974; in series: Studies in logic and the foundations of mathematics ; v. 77.
ISBN-13: 978-0-486-46678-1
ISBN-10: 0-486-46678-7
1. Recursive functions. 2. Induction (Mathematics) I. Title.

QA9.615.M67 2008
511.3'52—dc22

2007053068

Manufactured in the United States of America
Dover Publications, Inc., 31 East 2nd Street, Mineola, N.Y. 11501

For Joanie

PREFACE

This monograph originated with a seminar I gave at UCLA in Winter 1972. I wrote an extended first draft of the first seven chapters in Spring 1972, while on leave from teaching on a Sloan Fellowship, and in Summer 1972, while on a research grant from the National Science Foundation. The project was mostly finished in the fall of that year, while I was teaching a course on Inductive Definability at the University of Wisconsin, Madison.

I am grateful to the Sloan Foundation and the National Science Foundations for their support and to the University of Wisconsin for its hospitality.

I am also greatly indebted to P. Aczel, K. J. Barwise, A. S. Kechris, K. Kunen and J. R. Moschovakis for innumerable corrections, comments and suggestions which altered substantially my original conception of the book and helped shape its final form. Several other mathematicians made useful remarks which clarified and illuminated the text in places, particularly H. Enderton, H. J. Keisler, A. Nyberg and J. Schlipf. Anne Beate Nyberg typed the manuscript beautifully.

Finally, I am deeply grateful to my wife Joan. She was the only person to attend both the original seminar at UCLA and the course in Madison and she made numerous specific helpful suggestions—some of them on points of detail in the exercises, the technical parts of proofs or the style, that no one else would take the time to comment on. She also tolerated me this past year. One of the non-mathematical things I learned from writing this book is that the traditional thanks that authors give their spouses in prefaces are probably sincere and certainly deserved.

I am grateful to Dr. Perry Smith for correcting proofs and making up the index.

Palaion Phaliron, Greece January 1973

CONTENTS

INTRODUCTION

One of the chief concerns of logic is the study of those relations on an abstract structure $\mathfrak{A} = \langle A, R_1, \ldots, R_l \rangle$ which are explicitly definable in the first order language of $\mathfrak{A}$. We study here the relations on $\mathfrak{A}$ which are *inductively definable* in the same language.

Consider first a typical example of the kind of inductive definition we have in mind.

Let $\langle G, \cdot \rangle$ be a group and $b_1, \ldots, b_k$ fixed members of G, and take

$$H = [b_1, \ldots, b_k] = \textit{the subgroup generated by } b_1, \ldots, b_k.$$

There are two traditional ways of defining this notion in an algebra course.

One is to say that H is the least subset of G which satisfies

(1) $\quad b_1, \ldots, b_k \in H,$

(2) $\quad$ if $y, z \in H$, then $y \cdot z^{-1} \in H$.

Putting

$$(3) \qquad \varphi(x, S) \equiv x = b_1 \vee x = b_2 \vee \ldots \vee x = b_k \vee (\exists y)(\exists z)[y \in S \,\&\, z \in S \,\&\, x = y \cdot z^{-1}],$$

we have the explicit definition

$$x \in H \Leftrightarrow (\forall S)\{(\forall x')[\varphi(x', S) \Rightarrow x' \in S] \Rightarrow x \in S\}.$$

The other method is to define by induction the sets I_φ^n,

$$(4) \qquad x \in I_\varphi^n \Leftrightarrow \varphi(x, \textstyle\bigcup_{j<n} I_\varphi^j)$$

and put

$$(5) \qquad H = \textstyle\bigcup_n I_\varphi^n.$$

It is an easy exercise to show that both definitions yield the same set. The advantage of the first method is that it yields an explicit definition for H—but notice that this is in the second order language over the group structure $\langle G, \cdot \rangle$. The second approach makes clear that there is an induction involved in the definition and appears to be more constructive.

From our point of view the significant observation is that with either explication *the clauses* (1), (2) *of the induction* are in the first order language over $\langle G, \cdot \rangle$. Equivalently, the formula $\varphi(x, S)$ is elementary over $\langle G, \cdot \rangle$. Moreover, the relation variable S *occurs positively* in $\varphi(x, S)$; it is not hard to see that whenever we have clauses like (1), (2) then the formula that combines them will be positive in the relation variable of the induction.

In the most general case we will study, there will be a formula

$$(6) \qquad \varphi(\bar{x}, S) \equiv \varphi(x_1, \ldots, x_n, S)$$

in the first order language of a structure $\mathfrak{A} = \langle A, R_1, \ldots, R_l \rangle$, with n free variables and only positive occurrences of the n-ary relation variable S. The *set built up by* φ is defined explicitly by

$$(7) \qquad \bar{x} \in I_\varphi \Leftrightarrow (\forall S)\{(\forall \bar{x}')[\varphi(\bar{x}', S) \Rightarrow \bar{x}' \in S] \Rightarrow \bar{x} \in S\}.$$

The second approach of the example may lead to a transfinite induction in the general case,

$$(8) \qquad \bar{x} \in I_\varphi^\xi \Leftrightarrow \varphi(\bar{x}, \textstyle\bigcup_{\eta<\xi} I_\varphi^\eta),$$

but as in the example, the two methods lead to the same set,

$$(9) \qquad I_\varphi = \textstyle\bigcup_\xi I_\varphi^\xi.$$

These sets of the form I_φ that come directly from inductions are the *fixed points* of the structure $\mathfrak{A}$. We will call a relation R on A *inductive on* $\mathfrak{A}$ if there is a fixed point I_φ and constants $\bar{a} = a_1, \ldots, a_k$ in A such that

$$R(\bar{x}) \Leftrightarrow (\bar{a}, \bar{x}) \in I_\varphi,$$

i.e., the inductive relations are those *reducible* to fixed points. Finally the *hyperelementary* relations will be those which are inductive and have inductive complements.

One meets inductive and hyperelementary relations in practically every field of mathematics. The examples in algebra are rather obvious, e.g. the *algebraically closed subfield of F generated by $b_1, \ldots, b_k$* is inductive in the field structure of an algebraically closed field F.

In logic, perhaps the typical example is the *truth set* for arithmetic,

$T = \{e$: *e is the Gödel number of a true sentence of the structure of arithmetic*$\}$,

which is hyperelementary.

In set theory we can cast all transfinite recursions in this form; for example, if

$$F: \lambda \twoheadrightarrow L_\lambda$$

is the Gödel function enumerating the constructible sets of order less than the infinite cardinal λ, then the relation

$$P(\eta, \xi) \Leftrightarrow F(\eta) \in F(\xi)$$

is hyperelementary on the structure $\langle \lambda, \in \restriction \lambda \rangle$.

Finally in recursion theory, the basic definitions in the theories of constructive ordinals and recursion in higher types are the most obvious important examples, but of course inductive definitions of various types pervade the whole subject.

It is perhaps amusing that a notion which appears to be fundamental and widely applicable has not been explicitly isolated in any published paper that we can find. The very recent papers Grilliot [1971], Barwise–Gandy–Moschovakis [1971], Moschovakis [1970] and [1971a] come close to an abstract approach, but they only study rather special structures. In fact, in Barwise–Gandy–Moschovakis [1971] we read "given a set A equipped with some recursion theoretic structure, one can attempt to formulate . . .". Before these, the most general attack on the problem is Spector's fundamental [1961] where he defines and studies extensively the inductive relations on the structure of arithmetic.

But the preceding paragraph is very bad history. Specifically for the structure of arithmetic, long before Spector's [1961] Kleene had obtained all the key results in the pioneering papers [1944], [1955a], [1955b], [1955c]. In these, Kleene is consciously studying inductive definitions, even though he explicitly draws back from considering *all of them* in [1944]. But the "special cases" he studies are so general, that Spector credits to Kleene the fact that the inductive sets on the integers are precisely the Π^1_1 sets. This means that the hyperelementary sets are the Δ^1_1 sets which Kleene had already identified with the "hyperarithmetical" sets and which he had studied exhaustively.

Thus for the special case of the structure of the integers our subject specializes to the fully developed and justly acclaimed theory of Π^1_1 and hyperarithmetical sets. The approach to this theory through inductive definitions was not made explicit until Spector's [1961] and was not followed up very much after that, simply because Kleene and his students and followers looked at the subject as recursion theorists and chose to formulate their results in recursion theoretic rather than model theoretic terms.

The theory was extended to almost arbitrary structures in Moschovakis [1969a], [1969b], [1969c]. Here again the approach was entirely recursion theoretic in form, so much so that the identification of "semihyperprojective" relations with the inductive relations of the appropriate structure was only added as an afterthought in Remark 21 of the revised version of [1969b]. Nevertheless, the introduction to [1969a] says "the main technical contribution of this paper is the introduction and systematic exploitation of existential, nondeterministic clauses in inductive definitions", i.e. again the results were mostly about inductive definability even though they were cast in recursion theoretic terms.

During UCLA's Logic Year 1967–1968, Gandy argued repeatedly and forcefully that the key notion of abstract recursion theory should be that of an inductive definition. There is a counterargument to this, that recursion theory should have something to say about "computations". Nevertheless, it was obvious that it would be useful to have a development of the theory in Moschovakis [1969a], [1969b], [1969c] from the point of view of inductive definability, as many of the recursion theoretic arguments and methods of these papers seemed somehow irrelevant to the main results. To do this is one of our aims here.

One of the important tools for an exposition of these results from a model theoretic, inductive definability approach is the introduction and systematic exploitation of *open games*. There was a glimpse of this idea in the last section of Moschovakis [1969b] titled "A game theoretic characterization of semi-hyperprojective sets", but some of the missing tricks did not become available until Moschovakis [1970] and [1971a]. This allowed for a neat exposition of most of the first two papers [1969a] and [1969b].

The present work was motivated by some further extensions of these game theoretic ideas which yielded fairly comprehensible proofs of the rather delicate theorems in the third paper [1969c]—this was where we generalized Kleene's deepest results on the hyperarithmetical sets in Kleene [1959a].

A byproduct of attempting to lecture on these matters in a seminar was the somewhat surprising discovery that a very substantial part of the theory of inductive and hyperelementary sets goes through for completely arbitrary structures. This comprises the first four chapters here. Afterwards we specialize to "acceptable" structures and then in Chapter 8 to countable acceptable structures, but even then the flavor is decidedly model theoretic and there are no explicit references to recursion theory. In Chapter 9 we prove a very general version of the main result of Barwise–Gandy–Moschovakis [1971] which is the key to many applications of the present methods to abstract recursion theory.

The exercises at the end of each chapter are an integral part of the text. They give a stock of examples to keep in mind as we develop the general theory, they outline extensions of this theory and they also establish a link between the abstract approach here and the more familiar development of the theory of hyperarithmetical and Π_1^1 sets of integers.

The text is technically accessible to a student who is familiar with the basic notions of logic, model theory and set theory, the material usually covered in the first semester of a graduate or advanced undergraduate logic course. Some of the exercises require a deeper knowledge of set theory. However, the motivation and some of the implications of the results will be better understood by those students who have some acquaintance with the classical theory of recursive and hyperarithmetical sets, e.g. as developed in Rogers [1967] and Shoenfield [1967].

I have tried hard to attribute all results and ideas to the mathematicians who first discovered them, but the task is difficult and I am sure that there are both errors and omissions.

Much of the exciting current research in abstract recursion theory is concerned with very general inductions—nonmonotone inductive definitions and definitions in very restricted or very rich languages. We hope that this work will provide a point of reference by giving a neat exposition of the simplest and most developed part of the theory of inductive definability. We also have hopes that the model theorists may find something to interest them here, both in the results and in some of the methods.

CHAPTER 1

POSITIVE ELEMENTARY INDUCTIVE DEFINITIONS

In this first chapter we introduce the classes of *inductive* and *hyper-elementary* relations on a structure, we prove some of their simple properties and we discuss briefly some important examples of the theory.

1A. Monotone operators

Let A be an infinite set. We use $a, b, c, \ldots, x, y, z, \ldots$ to denote members of A and $P, Q, R, \ldots$ to denote relations on A of any (finite) number of arguments. Barred letters will denote *finite sequences* from A, e.g.

$$\bar{x} = x_1, \ldots, x_n, \qquad \bar{y}_1 = y_{11}, y_{12}, \ldots, y_{1m}.$$

If $P \subseteq A^n$ is an n-ary relation and $\bar{x}$ an n-tuple, we write interchangeably

$$\bar{x} \in P \Leftrightarrow P(\bar{x}).$$

The cardinal number of a set X is $|X|$, so in particular for every $n \geqslant 1$,

$$|A^n| = |A|,$$

since A is infinite. If λ is an ordinal number, then

$$\lambda^+ = \textit{least cardinal number greater than } \lambda.$$

An operator

$$\Gamma\colon \textit{Power}\,(A^n) \to \textit{Power}\,(A^n)$$

is *monotone* if it preserves inclusion, i.e.

$$S \subseteq S' \Rightarrow \Gamma(S) \subseteq \Gamma(S').$$

For each such monotone Γ and each ordinal ξ we define the set I_Γ^ξ by the transfinite recursion

$$I_\Gamma^\xi = \Gamma(\textstyle\bigcup_{\eta<\xi} I_\Gamma^\eta).$$

Of course each I_Γ^ξ is an n-ary relation on A. We let

$$I_\Gamma = \textstyle\bigcup_\xi I_\Gamma^\xi$$

be the *relation defined inductively by* Γ, or simply the set *built up by* Γ.

It is also convenient to put

$$I_\Gamma^{<\xi} = \bigcup_{\eta<\xi} I_\Gamma^\eta,$$

so that for each ξ,

$$I_\Gamma^\xi = \Gamma(I_\Gamma^{<\xi}).$$

1A.1. THEOREM. *Let A be an infinite set, let Γ be a monotone operator on the n-ary relations on A, let I_Γ^ξ, I_Γ be defined as above.*

(i) *If $\zeta \leqslant \xi$, then $I_\Gamma^\zeta \subseteq I_\Gamma^\xi$.*

(ii) *For some ordinal κ of cardinality $\leqslant |A|$,*

$$I_\Gamma = I_\Gamma^\kappa = I_\Gamma^{<\kappa};$$

we call the least such κ the closure ordinal *of Γ, $\kappa = \|\Gamma\|$.*

(iii) *The set built up by Γ is the smallest fixed point of Γ, i.e.*

$$\Gamma(I_\Gamma) = I_\Gamma,$$
$$I_\Gamma = \bigcap\{S: \Gamma(S) = S\}.$$

PROOF. (i) follows directly from the monotonicity of Γ, since for $\zeta \leqslant \xi$,

$$I_\Gamma^\zeta = \Gamma(I_\Gamma^{<\zeta}) \subseteq \Gamma(I_\Gamma^{<\xi}) = I_\Gamma^\xi.$$

To prove (ii) notice that if we had

$$I_\Gamma^{<\xi} \subsetneqq I_\Gamma^\xi$$

for every $\xi < |A|^+$, then we could choose some

$$\bar{x}_\xi \in I_\Gamma^\xi - I_\Gamma^{<\xi}$$

for each $\xi < |A|^+$ and then the set

$$X = \{\bar{x}_\xi : \xi < |A|^+\}$$

would be a subset of A^n of cardinality $|A|^+$, which is absurd. Hence for some $\kappa < |A|^+$ we have

$$I_\Gamma^\kappa = I_\Gamma^{<\kappa}$$

from which an immediate transfinite induction shows that for every $\xi \geqslant \kappa$, $I_\Gamma^\xi = I_\Gamma^\kappa$, so that $I_\Gamma = I_\Gamma^\kappa$.

This argument also proves part of (iii), that I_Γ is a fixed point of Γ, since if κ is the closure ordinal we have

$$\Gamma(I_\Gamma) = \Gamma(I_\Gamma^\kappa) = \Gamma(I_\Gamma^{<\kappa}) = I_\Gamma^\kappa = I_\Gamma.$$

On the other hand, if P is any fixed point, i.e.

$$\Gamma(P) = P,$$

we can show by transfinite induction on ξ that

$$I_\Gamma^\xi \subseteq P,$$

because

$$I_\Gamma^\xi = \Gamma(I_\Gamma^{<\xi}) \subseteq \Gamma(P) = P,$$

using the induction hypothesis and the monotonicity of Γ. Hence $I_\Gamma = \bigcup_\xi I_\Gamma^\xi \subseteq P$. ⊣

1B. Relative positive inductive definability

Again let A be an arbitrary set. We will be working with formulas of the *lower predicate calculus with individual and relation constants from A*, call it $\mathscr{L}^A$. More precisely, the language $\mathscr{L}^A$ has an infinite list of individual variables $x, y, z, \ldots$, a constant c for each element c of A, an infinite list $S, T, U, \ldots$ of n-ary relation variables for each $n \geqslant 1$, a constant relation symbol P for each relation P on A, in particular the identity symbol $=$, and the usual logical symbols $\neg$, &, $\vee$, $\rightarrow$, $\exists$, $\forall$. *Formulas* are defined as usual, with the quantifiers $\exists$, $\forall$ applied only to the individual variables—this is a first order language. *Individual terms* are the individual variables and constants and *relation symbols* are the relation variables and constants. Relation variables are always *free*. Formulas of $\mathscr{L}^A$ with no free variables of either kind are called *sentences*; they are either *true* or *false* under the natural interpretation of this language.

We write

$$\varphi \equiv \psi$$

to indicate that "φ" and "ψ" are names of the same formula. This metamathematical convention is useful in defining formulas and establishing notation.

Let S be a relation symbol. The class $\mathscr{P}(S)$ of *formulas in which S occurs positively*, briefly *S-positive formulas*, is the smallest collection $\mathscr{F}$ of formulas with the following properties:

(i) All formulas in which S does not occur are in $\mathscr{F}$.

(ii) If S is n-ary and $\bar{t} = t_1, \ldots, t_n$ is an n-tuple of individual terms, then the formula $S(\bar{t})$ is in $\mathscr{F}$.

(iii) If φ, ψ are in $\mathscr{F}$ and x is any variable, then $\varphi \mathbin{\&} \psi$, $\varphi \vee \psi$, $(\exists x)\varphi$, $(\forall x)\varphi$ are all in $\mathscr{F}$.

An easy induction on the length of formulas proves the following.

1B.1. MONOTONICITY PROPERTY OF POSITIVE FORMULAS. *Suppose S is an n-ary variable which occurs positively in $\varphi(S)$ and suppose no other variable of either kind occurs free in $\varphi(S)$. If P, P' are n-ary relations on A, then*

$$\text{if } P \subseteq P' \text{ and } \varphi(P), \text{ then } \varphi(P') \qquad \dashv$$

Here of course $\varphi(P)$ is the result of substituting the relation constant P for the relation variable S in $\varphi(S)$ and similarly for P'.

Suppose now that A is infinite. An operator

$$\Gamma: Power(A^n) \to Power(A^n)$$

is *positive elementary in* $Q_1, \ldots, Q_m$ if there is a formula

$$\varphi \equiv \varphi(\bar{x}, S) \equiv \varphi(x_1, \ldots, x_n, S, =, Q_1, \ldots, Q_m)$$

such that the following conditions hold:

(i) The relation constants which occur in $\varphi(\bar{x}, S)$ are among $=, Q_1, \ldots, Q_m$ and the only relation variable of $\varphi(\bar{x}, S)$ is the n-ary variable S.

(ii) The symbols $S, Q_1, \ldots, Q_m$ all occur positively in $\varphi(\bar{x}, S)$.

(iii) The free individual variables of $\varphi(\bar{x}, S)$ are among $x_1, \ldots, x_n$.

(iv) The formula $\varphi(\bar{x}, S)$ *defines* Γ, i.e. for each $S \subseteq A^n$,

$$\Gamma(S) = \{\bar{x}: \varphi(\bar{x}, S)\}.$$

Notice that we allow individual constants in $\varphi(\bar{x}, S)$ as well as arbitrary occurrences of $=$, but we insist that all the other relation symbols occur positively.

It is immediate from the monotonicity property of positive formulas that if Γ is positive elementary, then Γ is monotone. If $\varphi \equiv \varphi(\bar{x}, S)$ defines Γ in the sense of (i)–(iv) above, it is convenient to put

$$I_\varphi^\xi = I_\Gamma^\xi, \qquad I_\varphi^{<\xi} = \bigcup_{\eta<\xi} I_\varphi^\eta, \qquad I_\varphi = I_\Gamma;$$

we then have for each ordinal ξ,

$$\bar{x} \in I_\varphi^\xi \Leftrightarrow \varphi(\bar{x}, \bigcup_{\eta<\xi} I_\varphi^\eta) \Leftrightarrow \varphi(\bar{x}, I_\varphi^{<\xi}),$$

$$I_\varphi = \bigcup_\xi I_\varphi^\xi$$

and

$$\bar{x} \in I_\varphi \Leftrightarrow \varphi(\bar{x}, I_\varphi).$$

We call I_φ the *set built up by* φ.

An n-ary relation R on A is *positive elementary inductively definable in* $Q_1, \ldots, Q_m$, or simply *inductive in* $Q_1, \ldots, Q_m$, if there are constants $\bar{a} = a_1, \ldots, a_k$ in A and an operator

$$\Gamma: Power(A^{k+n}) \to Power(A^{k+n})$$

which is positive elementary in $Q_1, \ldots, Q_m$, such that

$$R(\bar{x}) \Leftrightarrow (\bar{a}, \bar{x}) \in I_\Gamma.$$

By the definition above, we then have a formula

$$\varphi \equiv \varphi(\bar{u}, \bar{x}, S) \equiv \varphi(\bar{u}, \bar{x}, S, =, Q_1, \ldots, Q_m)$$

in which the only relation symbols that occur are those that show, and except for $=$ they all occur positively, such that

$$R(\bar{x}) \Leftrightarrow (\bar{a}, \bar{x}) \in I_\varphi.$$

The constants $\bar{a}$ are called the *parameters of the induction*, but recall that there may be other individual constants occurring in φ.

We can picture the construction of the fixed point I_φ in the $\bar{u} \times \bar{x}$ plane and the definition of R from I_φ by projection along the $\bar{u}$-axis as shown in Fig. 1.1.

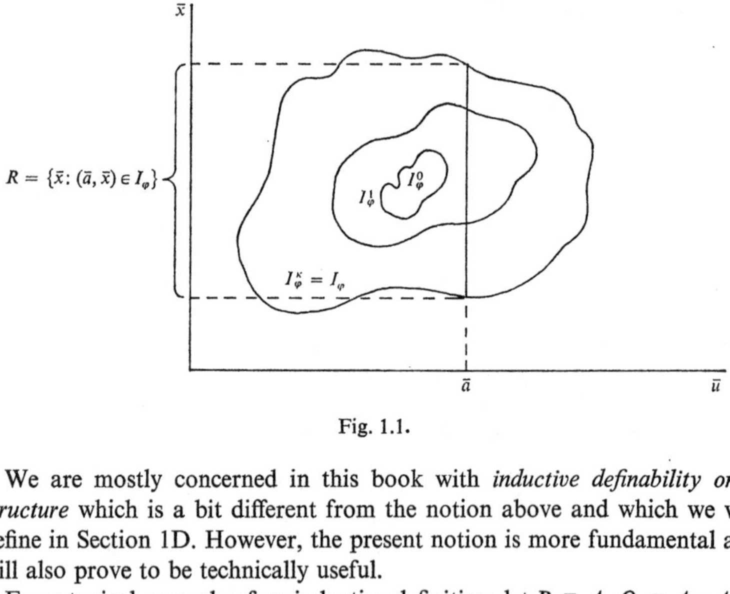

Fig. 1.1.

We are mostly concerned in this book with *inductive definability on a structure* which is a bit different from the notion above and which we will define in Section 1D. However, the present notion is more fundamental and will also prove to be technically useful.

For a typical example of an inductive definition, let $P \subseteq A$, $Q \subseteq A \times A$ be given and consider the *transitive closure of P relative to Q*,

$R(x) \Leftrightarrow$ *there is some sequence* $y_1, y_2, \ldots, y_n$ *such that* $P(y_1)$, *and* $Q(y_1, y_2), Q(y_2, y_3), \ldots, Q(y_{n-1}, y_n)$, *and* $x = y_1$ *or* $x = y_n$.

Put

$$\varphi(x, S) \equiv P(x) \vee (\exists y)[S(y) \mathbin{\&} Q(y, x)].$$

It is easy to verify that $R = I_\varphi$ by showing by induction on ξ that

$$x \in I^\xi_\varphi \Rightarrow R(x)$$

and then by induction on $n \geqslant 1$ that

$$P(y_1) \mathbin{\&} Q(y_1, y_2) \mathbin{\&} \ldots \mathbin{\&} Q(y_{n-1}, y_n) \Rightarrow y_n \in I^{<\omega}_\varphi.$$

Thus R is inductive in P, Q. Notice that this induction has no parameters, and that *it closes at* ω, i.e.

$$I_\varphi = I^\omega_\varphi = I^{<\omega}_\varphi.$$

For a typical example of an induction that does not close in ω steps, let $\leqslant$ be a linear ordering on A and take the *wellordered initial segment* of $\leqslant$,

$$W(x) \Leftrightarrow \textit{there is no infinite sequence } x > x_1 > x_2 > \ldots.$$

Put

$$\psi(x, S) \equiv (\forall u)[u < x \Rightarrow u \in S].$$

Again it is easy to show by induction on ξ that

$$x \in I^\xi_\psi \Rightarrow W(x);$$

because if $x \in I^\xi_\psi$, then by definition $(\forall u)[u < x \Rightarrow u \in I^{<\xi}_\psi]$, so by induction hypothesis $(\forall u)[u < x \Rightarrow W(u)]$, which immediately implies $W(x)$. On the other hand,

$$x \notin I_\psi \Rightarrow (\exists u_1)[u_1 < x \mathbin{\&} u_1 \notin I_\psi]$$

and using the same implication on u_1 we find some $u_2 < u_1$ such that $u_2 \notin I_\psi$, etc., so that an infinite sequence starts with x and $\neg\, W(x)$. Thus

$$W(x) \Leftrightarrow x \in I_\psi.$$

It is not hard to verify that the closure ordinal of this induction is precisely the ordinal of the largest wellordered initial segment of $\leqslant$.

We collect in one theorem some of the trivial properties of relative inductive definability.

1B.2. THEOREM. *Let A be an infinite set, and R, Q_1, Q_2, ... relations on A.*

(i) *The relations $x = y$, $x \neq y$, $x = c$ (c a fixed element of A), $x \neq c$ are inductive (in the empty list of relations).*

(ii) *If R is inductive in $Q_1, \ldots, Q_m$ and each Q_i occurs in the list $Q'_1, \ldots, Q'_{m'}$, then R is inductive in $Q'_1, \ldots, Q'_{m'}$.*

(iii) *R is inductive in R.*

(iv) *If*

$$P(\bar{x}) \Leftrightarrow R_1(\bar{x}) \mathbin{\&} R_2(\bar{x}),$$
$$Q(\bar{x}) \Leftrightarrow R_1(\bar{x}) \vee R_2(\bar{x}),$$

then both P and Q are inductive in R_1, R_2.

(v) *If*

$$P(\bar{x}) \Leftrightarrow (\forall y)R(y, \bar{x}),$$

$$Q(\bar{x}) \Leftrightarrow (\exists y)R(y, \bar{x}),$$

then both P and Q are inductive in R.

PROOF. All of the verifications are completely trivial. For example, if

$$\varphi(x, y, S) \equiv x = y,$$

then easily

$$I_\varphi^\xi = \{(x, y): x = y\}$$

for all ξ, which immediately shows that the identity relation is inductive. ⊣

1C. Combining inductions

The first nontrivial result of the book is the Transitivity Theorem 1C.3 which is the key to the closure properties of the class of inductive relations. Let us establish first a useful but much easier result whose proof is somewhat similar.

We will be using the following convenient and natural notation. If $\varphi(S)$ is a formula in which occurs the n-ary relation symbol S (among others), if U is an $m+n$-ary relation symbol and $\bar{t}$ is an m-tuple of terms, then

$$\varphi(\{\bar{u}: U(\bar{t}, \bar{u})\})$$

is the result of replacing each occurrence of $S(\bar{v})$ in $\varphi(S)$ by $U(\bar{t}, \bar{v})$.

1C.1. SIMULTANEOUS INDUCTION LEMMA. *Suppose*

$$\psi(\bar{y}, S, T) \equiv \psi(\bar{y}, =, Q_1, \ldots, Q_m, S, T),$$

$$\varphi(\bar{x}, S, T) \equiv \varphi(\bar{x}, =, Q_1, \ldots, Q_m, S, T)$$

are formulas in the language over an infinite set A in which the only relation symbols that occur are those that show and except for =, they all occur positively. Define J_0^ξ, J_1^ξ by the simultaneous induction

$$\bar{y} \in J_0^\xi \Leftrightarrow \psi(\bar{y}, J_0^{<\xi}, J_1^{<\xi}),$$

$$\bar{x} \in J_1^\xi \Leftrightarrow \varphi(\bar{x}, J_0^{<\xi}, J_1^{<\xi}),$$

where

$$J_0^{<\xi} = \bigcup_{\eta<\xi} J_0^\eta, \qquad J_1^{<\xi} = \bigcup_{\eta<\xi} J_1^\eta.$$

Then both $J_0 = \bigcup_\xi J_0^\xi$ and $J_1 = \bigcup_\xi J_1^\xi$ are inductive in $Q_1, \ldots, Q_m$.

PROOF. Let c_0, c_1 be distinct members of A and choose $\bar{y}^*$, $\bar{x}^*$ to be sequences of elements of A of the same length as the sequences of variables $\bar{y}$, $\bar{x}$. Put

$$\chi(t, \bar{y}, \bar{x}, U) \Leftrightarrow [t = c_0 \ \&\ \psi(\bar{y}, \{\bar{y}': U(c_0, \bar{y}', \bar{x}^*)\}, \{\bar{x}': U(c_1, \bar{y}^*, \bar{x}')\})]$$
$$\vee\ [t = c_1 \ \&\ \varphi(\bar{x}, \{\bar{y}': U(c_0, \bar{y}', \bar{x}^*)\}, \{\bar{x}': U(c_1, \bar{y}^*, \bar{x}')\})].$$

We claim that for each ξ,

$$\bar{y} \in J_0^\xi \Leftrightarrow (c_0, \bar{y}, \bar{x}^*) \in I_\chi^\xi,$$
$$\bar{x} \in J_1^\xi \Leftrightarrow (c_1, \bar{y}^*, \bar{x}) \in I_\chi^\xi.$$

Proof is by a transfinite induction on ξ simultaneously for both equivalences, e.g.

$$\bar{y} \in J_0^\xi \Leftrightarrow \psi(\bar{y}, J_0^{<\xi}, J_1^{<\xi}) \qquad \text{(by definition)}$$
$$\Leftrightarrow \chi(c_0, \bar{y}, \bar{x}^*, \{(c_0, \bar{y}', \bar{x}^*): \bar{y}' \in J_0^{<\xi}\} \cup \{(c_1, \bar{y}^*, \bar{x}'): \bar{x}' \in J_1^{<\xi}\}) \qquad \text{(by def. of } \chi)$$
$$\Leftrightarrow \chi(c_0, \bar{y}, \bar{x}^*, I_\chi^{<\xi}) \qquad \text{(by ind. hyp.)}$$
$$\Leftrightarrow (c_0, \bar{y}, \bar{x}^*) \in I_\chi^\xi) \qquad \text{(by def. of } I_\chi^\xi).$$

Proof of the equivalence for J_1^ξ is similar. It follows that

$$\bar{y} \in J_0 \Leftrightarrow (c_0, \bar{y}, \bar{x}^*) \in I_\chi,$$
$$\bar{x} \in J_1 \Leftrightarrow (c_1, \bar{y}^*, \bar{x}) \in I_\chi,$$

so that both J_0 and J_1 are inductive in $Q_1, \ldots, Q_m$, since evidently $Q_1, \ldots, Q_m$ occur positively in χ. ⊣

The proof of 1C.1 illustrates the use of the parameters of induction in defining inductive relations.

The Transitivity Theorem deals with a more complicated kind of combination of inductions. It will be useful to codify in a fairly messy lemma the precise combinatory principle involved.

1C.2. COMBINATION LEMMA. *Suppose*

$$\psi \equiv \psi(\bar{u}, \bar{y}, S) \equiv \psi(u_1, \ldots, u_k, y_1, \ldots, y_m, S)$$

is S-positive in the language over an infinite set A, let $\bar{a} = a_1, \ldots, a_k$ *be constants from A and put*

$$Q(\bar{y}) \Leftrightarrow (\bar{a}, \bar{y}) \in I_\psi.$$

Suppose

$$\varphi \equiv \varphi(\bar{x}, Q, T) \equiv \varphi(x_1, \ldots, x_n, Q, T)$$

is T-positive in the language over A and Q occurs positively in φ. *Let* $c_0 \neq c_1$ *and* $\bar{u}^* = u_1^*, \ldots, u_k^*$, $\bar{y}^* = y_1^*, \ldots, y_m^*$, $\bar{x}^* = x_1^*, \ldots, x_n^*$ *be constants from A and put*

$$\chi(t, \bar{u}, \bar{y}, \bar{x}, U) \equiv [t = c_0 \ \& \ \psi(\bar{u}, \bar{y}, \{(\bar{u}', \bar{y}') : U(c_0, \bar{u}', \bar{y}', \bar{x}^*)\})]$$
$$\vee \ [t = c_1 \ \& \ \varphi(\bar{x}, \{\bar{y}' : U(c_0, \bar{a}, \bar{y}', \bar{x}^*)\}, \{\bar{x}' : U(c_1, \bar{u}^*, \bar{y}^*, \bar{x}')\})].$$

Then for every ordinal ξ *we have*

(1) $$(\bar{u}, \bar{y}) \in I_\psi^\xi \Leftrightarrow (c_0, \bar{u}, \bar{y}, \bar{x}^*) \in I_\chi^\xi,$$

(2) $$(c_1, \bar{u}^*, \bar{y}^*, \bar{x}) \in I_\chi^\xi \Rightarrow \bar{x} \in I_\varphi^\xi,$$

(3) $$\bar{x} \in I_\varphi^\xi \Rightarrow (c_1, \bar{u}^*, \bar{y}^*, \bar{x}) \in I_\chi;$$

hence, in particular,

(4) $$(\bar{u}, \bar{y}) \in I_\psi \Leftrightarrow (c_0, \bar{u}, \bar{y}, \bar{x}^*) \in I_\chi,$$

(5) $$\bar{x} \in I_\varphi \Leftrightarrow (c_1, \bar{u}^*, \bar{y}^*, \bar{x}) \in I_\chi.$$

PROOF. To simplify notation, put

$$J_0^\xi = \{(\bar{u}, \bar{y}) : (c_0, \bar{u}, \bar{y}, \bar{x}^*) \in I_\chi^\xi\},$$
$$J_1^\xi = \{\bar{x} : (c_1, \bar{u}^*, \bar{y}^*, \bar{x}) \in I_\chi^\xi\}$$

and as usual

$$J_0^{<\xi} = \bigcup_{\eta<\xi} J_0^\eta, \qquad J_1^{<\xi} = \bigcup_{\eta<\xi} J_1^\eta.$$

Now equivalence (1) asserts that

$$I_\psi^\xi = J_0^\xi$$

and is immediate by induction on ξ:

$$\begin{aligned} (\bar{u}, \bar{y}) \in I_\psi^\xi &\Leftrightarrow \psi(\bar{u}, \bar{y}, I_\psi^{<\xi}) \\ &\Leftrightarrow \psi(\bar{u}, \bar{y}, J_0^{<\xi}) && \text{(by ind. hyp.)} \\ &\Leftrightarrow \chi(c_0, \bar{u}, \bar{y}, \bar{x}^*, I_\chi^{<\xi}) && \text{(by def. of } \chi) \\ &\Leftrightarrow (c_0, \bar{u}, \bar{y}, \bar{x}^*) \in I_\chi^\xi \\ &\Leftrightarrow (\bar{u}, \bar{y}) \in J_0^\xi. \end{aligned}$$

Thus the induction determined by χ imitates step-by-step in its first component the induction determined by ψ. In particular, (1) immediately implies (4).

The combined induction χ is not equally faithful in its second component which defines the stages J_1^ξ, but still in the limit,

$$\bigcup_\xi I_\varphi^\xi = \bigcup_\xi J_1^\xi.$$

To see this, compute first the transfinite recursion which determines J_1^ξ:

$$\begin{aligned} \bar{x} \in J_1^\xi &\Leftrightarrow (c_1, \bar{u}^*, \bar{y}^*, \bar{x}) \in I_\chi^\xi \\ &\Leftrightarrow \chi(c_1, \bar{u}^*, \bar{y}^*, \bar{x}, I_\chi^{<\xi}) \\ &\Leftrightarrow \varphi(\bar{x}, \{\bar{y}' : (c_0, \bar{a}, \bar{y}', \bar{x}^*) \in I_\chi^{<\xi}\}, J_1^{<\xi}) && \text{(by def. of } \chi) \\ &\Leftrightarrow \varphi(\bar{x}, \{\bar{y}' : (\bar{a}, \bar{y}') \in I_\psi^{<\xi}\}, J_1^{<\xi}) && \text{(by (1)).} \end{aligned}$$

Letting for each ξ, Q^ξ be the part of Q that is determined by I_ψ^ξ,

$$\bar{y} \in Q^\xi \Leftrightarrow (\bar{a}, \bar{y}) \in I_\psi^\xi,$$
$$Q^{<\xi} = \bigcup_{\eta<\xi} Q^\eta,$$

we have

(6) $$\bar{x} \in J_1^\xi \Leftrightarrow \varphi(\bar{x}, Q^{<\xi}, J_1^{<\xi}).$$

It is useful to compare (6) with the transfinite recursion which determines the stages I_φ^ξ:

$$\bar{x} \in I_\varphi^\xi \Leftrightarrow \varphi(\bar{x}, Q, I_\varphi^{<\xi}).$$

The only difference is that in defining I_φ^ξ from the preceding stages $I_\varphi^{<\xi}$ we can use the whole relation Q, while in the definition of J_1^ξ from $J_1^{<\xi}$ we can only use the piece $Q^{<\xi}$ of Q. (See Fig. 1.2.)

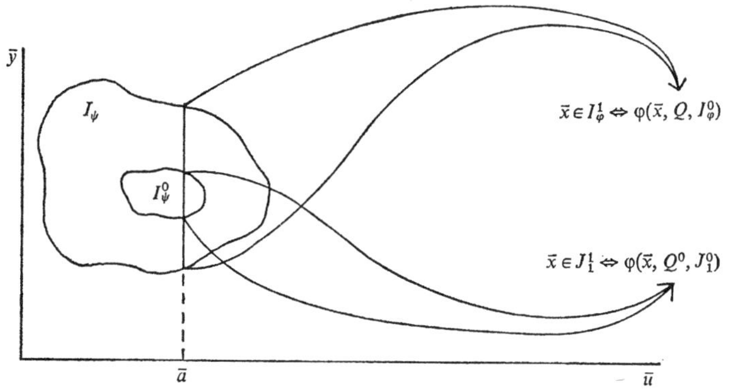

Fig. 1.2.

Implication (2) is also proved by induction on ξ:

$$\begin{aligned} (c_1, \bar{u}^*, \bar{y}^*, \bar{x}) \in I_\chi^\xi &\Rightarrow \bar{x} \in J_1^\xi \\ &\Rightarrow \varphi(\bar{x}, Q^{<\xi}, J_1^{<\xi}) && \text{(by (6))} \\ &\Rightarrow \varphi(\bar{x}, Q, J_1^{<\xi}) && \text{(by monotonicity of } \varphi \text{ in } Q) \\ &\Rightarrow \varphi(\bar{x}, Q, I_\varphi^{<\xi}) && \text{(by ind. hyp. and monotonicity)} \\ &\Rightarrow \bar{x} \in I_\varphi^\xi. \end{aligned}$$

Finally, (3) holds because eventually all of Q becomes available in the construction of J_1^ξ. Formally we show (3) again by induction on ξ:

$$\begin{aligned}
\bar{x} \in I_\varphi^\xi &\Rightarrow \varphi(\bar{x}, Q, I_\varphi^{<\xi}) \\
&\Rightarrow \varphi(\bar{x}, Q, \{\bar{x}' : (c_1, \bar{u}^*, \bar{y}^*, \bar{x}') \in I_\chi\}) && \text{(by ind. hyp. and monotonicity)} \\
&\Rightarrow \varphi(\bar{x}, \{\bar{y}' : (c_0, \bar{a}, \bar{y}', \bar{x}^*) \in I_\chi\}, \{\bar{x}' : (c_1, \bar{u}^*, \bar{y}^*, \bar{x}') \in I_\chi\}) && \text{(by (4))} \\
&\Rightarrow \chi(c_1, \bar{u}^*, \bar{y}^*, \bar{x}, I_\chi) && \text{(by def. of } \chi) \\
&\Rightarrow (c_1, \bar{u}^*, \bar{y}^*, \bar{x}) \in I_\chi && \text{(since } I_\chi \text{ is a fixed point of } \chi).
\end{aligned}$$

Again (2) and (3) immediately imply (5). ⊣

1C.3. Transitivity Theorem. *Let A be an infinite set, let $R, Q, Q_1, \ldots, Q_m$ be relations on A. If R is inductive in $Q, Q_1, \ldots, Q_m$ and Q is inductive in $Q_1, \ldots, Q_m$, then R is inductive in $Q_1, \ldots, Q_m$.*

Proof. By hypothesis there are formulas

$$\psi \equiv \psi(\bar{u}, \bar{y}, S) \equiv \psi(\bar{u}, \bar{y}, Q_1, \ldots, Q_m, S),$$

$$\varphi \equiv \varphi(\bar{v}, \bar{z}, Q, T) \equiv \varphi(\bar{v}, \bar{z}, Q, Q_1, \ldots, Q_m, T)$$

with only the indicated relation symbols (and $=$) and all those occurring positively and constants $\bar{a}, \bar{b}$ such that

$$Q(\bar{y}) \Leftrightarrow (\bar{a}, \bar{y}) \in I_\psi,$$

$$R(\bar{z}) \Leftrightarrow (\bar{b}, \bar{z}) \in I_\varphi.$$

Let

$$\chi \equiv \chi(t, \bar{u}, \bar{y}, \bar{v}, \bar{z}, U) \equiv \chi(t, \bar{u}, \bar{y}, \bar{v}, \bar{z}, Q_1, \ldots, Q_m, U)$$

be the combination formula that is assigned to ψ and φ by the Combination Lemma 1C.2. It has various other constants $c_0, c_1, \bar{u}^*, \bar{y}^*, \bar{x}^* = \bar{v}^*, \bar{z}^*$ in it, but it is obvious from its definition that except for $=$, the only relation constants in χ are $Q_1, \ldots, Q_m$ and they all occur positively. By (5) of 1C.2 we then have

$$(\bar{v}, \bar{z}) \in I_\varphi \Leftrightarrow (c_1, \bar{u}^*, \bar{y}^*, \bar{v}, \bar{z}) \in I_\chi,$$

so that

$$\bar{z} \in R \Leftrightarrow (c_1, \bar{u}^*, \bar{y}^*, \bar{b}, \bar{z}) \in I_\chi$$

and R is inductive in $Q_1, \ldots, Q_m$. ⊣

1C.4. COROLLARY. *The class of relations on some infinite set A which are inductive in fixed relations $Q_1, \ldots, Q_m$ is closed under the positive operations* $\&, \vee, \exists, \forall$.

PROOF is immediate from the Transitivity Theorem and Theorem 1B.2. For example, if

$$P(\bar{x}) \Leftrightarrow (\forall y)R(y, \bar{x})$$

and R is inductive in $Q_1, \ldots, Q_m$, then P is inductive in $R, Q_1, \ldots, Q_m$, hence P is inductive in $Q_1, \ldots, Q_m$. ⊣

1D. Inductive definability on a structure

Suppose

$$\mathfrak{A} = \langle A, R_1, \ldots, R_l \rangle$$

is an infinite structure. The (first order) *language of* $\mathfrak{A}$ consists of all formulas of the language $\mathscr{L}^A$ whose only relation constants are $=, R_1, \ldots, R_l$. The *elementary* relations on $\mathfrak{A}$ are those which can be defined by formulas of the language of $\mathfrak{A}$. (Recall that we allow arbitrary constants from the domain A of the structure.)

A relation P on $\mathfrak{A}$ is a *fixed point* if there is an S-positive formula $\varphi \equiv \varphi(\bar{x}, S)$ in the language of $\mathfrak{A}$ such that $P = I_\varphi$. A relation R is *positive elementary inductively definable* or simply *inductive* on $\mathfrak{A}$ if there is a fixed point I_φ and constants $\bar{a} = a_1, \ldots, a_k$ in A such that

$$R(\bar{x}) \Leftrightarrow (\bar{a}, \bar{x}) \in I_\varphi.$$

Tracing the definitions, this means that R is inductive on $\mathfrak{A}$ if and only if R is inductive in the relations $R_1, \neg R_1, \ldots, R_l, \neg R_l$.

Finally, we call R *coinductive* on $\mathfrak{A}$ if $\neg R$ is inductive on $\mathfrak{A}$ and we call R *hyperelementary* on $\mathfrak{A}$ if R is both inductive and coinductive.

A function $f: A^n \to A^m$ is *elementary* or *hyperelementary* if its representing relation (the *graph*)

$$G_f(\bar{x}, \bar{y}) \Leftrightarrow f(\bar{x}) = \bar{y}$$

is elementary or hyperelementary respectively.

We are mostly interested in studying the inductive and hyperelementary relations on a structure $\mathfrak{A}$, but it is clear that in order to do that we will also have to look at the fixed points. These relations are interesting in their own right since they can be defined by the simplest inductions, without parameters.

Before going on to prove the simple closure properties of these classes of relations we list some of the most important examples of the theory.

The original and standard example is the ordinary *structure of arithmetic*

$$\mathbb{N} = \langle \omega, S, P \rangle,$$

where

$$S(x, y, z) \Leftrightarrow z = x+y,$$

$$P(x, y, z) \Leftrightarrow z = x \cdot y.$$

It is customary to call the elementary relations here *arithmetical* and the hyperelementary relations *hyperarithmetical.* One of the main efforts in the theory of positive inductive definability is to generalize the theory of hyperarithmetical relations to arbitrary (or almost arbitrary) structures. The original definitions of hyperarithemetical relations in Davis [1950], Mostowski [1951], Kleene [1955b] were not from the point of view of inductive definability but rather came as direct attempts to extend the hierarchy of arithmetical relations on $\mathbb{N}$. It was apparently Spector in [1961] who first realized the significance of inductive definability.

Another very important example is second order arithmetic or *analysis.* This is often considered as a two-sorted structure, but we will describe it here as a structure with a single domain so that it is covered by the general theory;

$$\mathbb{R} = \langle \omega \cup {}^{\omega}\omega, N, S, P, A \rangle,$$

where

$$N(x) \Leftrightarrow x \in \omega,$$

$$A(y, x, z) \Leftrightarrow y \in {}^{\omega}\omega \ \& \ x \in \omega \ \& \ y(x) = z$$

and S, P are the sum and product relations, taken as false if any of the arguments are not in ω. The elementary relations of this structure are called *projective* and the hyperelementary relations *hyperprojective.*

The special case of the structure $\mathbb{R}$ was the chief motivation for the work reported in Moschovakis [1969a], [1969b], [1969c] which was the first systematic study of inductive and hyperelementary relations on (almost) arbitrary structures. Because of this we used there the term "hyperprojective" for the hyperelementary relations of an abstract structure, but we now think that "hyperelementary" is more appropriate for the general case.

As with the original development of the theory for $\mathbb{N}$, the approach in Moschovakis [1969a], [1969b], [1969c] was not directly from the point of view of inductive definability but in terms of (search) *computability in functionals.* The characterization (for most structures) of "semihyperprojective" relations as those inductively definable by positive formulas is proved in Remark 21 of [1969b] but it is stated there in a particularly obscure notation.

With each ordinal λ, there is naturally associated the structure

$$\boldsymbol{\lambda} = \langle \lambda, \leqslant \rangle$$

of λ with its ordering. These are particularly interesting examples in studying abstract inductive definability, since they have very "few" elementary relations but (as we will see) a very rich collection of hyperelementary and inductive relations.

These structures are special cases of structures of the form

$$\boldsymbol{A} = \langle A, \in \restriction A \rangle,$$

where A is any transitive, infinite set. Other interesting special cases of these are

$$A = V_\lambda,$$

the set of sets of *rank* less than λ with λ a limit ordinal, and

$$A = L_\lambda,$$

the set of sets *constructible* before λ, with λ again a limit ordinal.

More examples can be constructed by adding relations to the structures above.

The results of 1C give immediately the simple closure properties of the classes of inductive and hyperelementary relations. A relation $P(\bar{x})$ is defined from a relation $R(y_1, \ldots, y_m)$ by *hyperelementary substitution* if there are hyperelementary functions $f_1(\bar{x}), \ldots, f_m(\bar{x})$ such that

$$P(\bar{x}) \Leftrightarrow R(f_1(\bar{x}), \ldots, f_m(\bar{x})).$$

1D.1. THEOREM. *The class of inductive relations on an infinite structure $\mathfrak{A}$ is closed under the positive operations* &, $\vee$, $\exists$, $\forall$ *and hyperelementary substitution.*

The class of hyperelementary relations on $\mathfrak{A}$ includes all elementary relations and is closed under all the elementary operations $\neg$, &, $\vee$, $\rightarrow$, $\exists$, $\forall$ *and hyperelementary substitution.* ⊣

There is another very useful corollary of the Transitivity Theorem which is worth stating explicitly.

1D.2. THEOREM. *Let $\mathfrak{A} = \langle A, R_1, \ldots, R_l \rangle$ be an infinite structure, suppose $Q_1, \ldots, Q_m$ are hyperelementary on $\mathfrak{A}$ and R is inductive on*

$$\mathfrak{A}' = (\mathfrak{A}, Q_1, \ldots, Q_m) = \langle A, R_1, \ldots, R_l, Q_1, \ldots, Q_m \rangle.$$

Then R is inductive on $\mathfrak{A}$.

Proof. It is enough to prove the result for $m = 1$, since we can then apply this m times for the general case. If R is inductive on $\langle A, R_1, \ldots, R_l, Q_1\rangle$, it is inductive in the relations $R_1, \neg R_1, \ldots, R_l, \neg R_l, Q_1, \neg Q_1$ by definition and since both $\neg Q_1, Q_1$ are inductive in $R_1, \neg R_1, \ldots, R_l, \neg R_l$, it must be that R is inductive in $R_1, \neg R_1, \ldots, R_l, \neg R_l$ by two applications of the Transitivity Theorem 1C.2. ⊣

From this follows trivially that if Q is hyperelementary on $\mathfrak{A}$ and R is hyperelementary on $(\mathfrak{A}, Q)$, then R is hyperelementary on $\mathfrak{A}$.

The *second order language over a set* A, $\mathcal{L}_2^A$ is obtained by allowing quantification of the relation variables in the language $\mathcal{L}^A$. The *second order language* $\mathcal{L}_2^{\mathfrak{A}}$ *for a structure* $\mathfrak{A} = \langle A, R, \ldots, R_l\rangle$ consists of those formulas in $\mathcal{L}_2^A$ whose relation constants are among $=, R_1, \ldots, R_l$. The relations on A (even those with relation arguments) which are definable in $\mathcal{L}_2^{\mathfrak{A}}$ are naturally called *second order definable.* According to the usual classification of these relations, R is Π_1^1 if there is an elementary formula $\varphi(S_1, \ldots, S_k, \bar{x})$ in the language of $\mathfrak{A}$ such that

$$R(\bar{x}) \Leftrightarrow (\forall S_1) \ldots (\forall S_k)\varphi(S_1, \ldots, S_k, \bar{x}),$$

i.e. if we can define R using only a block of universal relation quantifiers applied directly to an elementary formula. Similarly, R is Σ_1^1 if $\neg R$ is Π_1^1 and R is Δ_1^1 if R is both Π_1^1 and Σ_1^1.

It is well known that in the structure $\mathbb{N}$ of arithmetic the inductive relations coincide with the Π_1^1 relations—we will prove a generalization of this basic result in Chapter 8. Here we show the trivial half of this equivalence which holds for arbitrary structures.

1D.3. Theorem. *Every inductive relation on an infinite structure is* Π_1^1 *and hence every hyperelementary relation is* Δ_1^1.

Proof. If R is inductive, then for some $\varphi(\bar{u}, \bar{x}, S)$ and constants $\bar{a}$ we have

$$R(\bar{x}) \Leftrightarrow (\bar{a}, \bar{x}) \in I_\varphi;$$

now the characterization of I_φ as the least fixed point of the operator defined by φ, given in 1A, yields

$$R(\bar{x}) \Leftrightarrow (\forall S)[(\forall \bar{u})(\forall \bar{x}')[S(\bar{u}, \bar{x}') \Leftrightarrow \varphi(\bar{u}, \bar{x}', S)] \rightarrow S(\bar{a}, \bar{x})]. \quad ⊣$$

Exercises for Chapter 1

1.1. Prove that if the graph of a function $f\colon A^n \rightarrow A^m$ is inductive on $\mathfrak{A} = \langle A, R_1, \ldots, R_l\rangle$, then f is hyperelementary on $\mathfrak{A}$. ⊣

1.2. Suppose $\varphi(S)$ is a formula in the language over a set A. Prove that there is a formula $\varphi^{\dagger}(S_1, S_2)$ in which both S_1 and S_2 occur positively such that

$$\begin{aligned}\varphi(S) &\Leftrightarrow \varphi^{\dagger}(S, \neg S)\\ &\Leftrightarrow \varphi^{\dagger}(S, \{\bar{x} : \bar{x} \notin S\}).\end{aligned}$$ ⊣

A *wellfounded relation with field* $B \subseteq A^n$ is a subset $\prec$ of A^{2n} such that

$$\bar{x} \in B \Leftrightarrow (\exists \bar{y})\{(\bar{x}, \bar{y}) \in \prec \vee (\bar{y}, \bar{x}) \in \prec\},$$

$$S \subseteq B \;\&\; S \neq \emptyset \Rightarrow (\exists \bar{x} \in S)(\forall \bar{y} \in S) \neg [(\bar{y}, \bar{x}) \in \prec].$$

We write $\bar{x} \prec \bar{y}$ for $(\bar{x}, \bar{y}) \in \prec$.

1.3. Suppose $\prec$ is a wellfounded relation with field $B \subseteq A^n$ which is hyperelementary on $\mathfrak{A}$ and suppose

$$f: B \times A^k \to A^m$$

is a function such that for some formula $\varphi(S, \bar{x}, \bar{z}, \bar{y})$ in the language of $\mathfrak{A}$ and all $\bar{x} \in B$, $\bar{z} \in A^k$,

$$f(\bar{x}, \bar{z}) = \bar{y} \Leftrightarrow \varphi(\{(\bar{x}', \bar{z}', f(\bar{x}', \bar{z}')) : \bar{x}' \prec \bar{x} \;\&\; \bar{z}' \in A^k\}, \bar{x}, \bar{z}, \bar{y}).$$

Prove that f is hyperelementary on $\mathfrak{A}$.

HINT: Use Exercise 1.2 and the Transitivity Theorem. ⊣

This problem shows that ordinary transfinite induction along a wellfounded relation is a special case of the general inductions we have been studying.

A *copy of* $\boldsymbol{\omega}$ is any structure $\langle N, \leqslant \rangle$ which is isomorphic to $\boldsymbol{\omega} = \langle \omega, \leqslant \rangle$. Relative to a fixed structure $\mathfrak{A} = \langle A, R_1, \ldots, R_l \rangle$, an *elementary* (or *hyperelementary*) *copy of* $\boldsymbol{\omega}$ *in* $\mathfrak{A}$ is a copy of $\boldsymbol{\omega}$ $\langle N, \leqslant \rangle$ such that $N \subseteq A$, $\leqslant \subseteq A \times A$ and both N and $\leqslant$ are elementary (or hyperelementary) on $\mathfrak{A}$. We often label the members of a copy of $\boldsymbol{\omega}$ by $0, 1, \ldots$, where 0 is the $\leqslant$-least member, 1 is the $\leqslant$-next member, etc.

1.4. Let $\langle N, \leqslant \rangle$ be a hyperelementary copy of $\boldsymbol{\omega}$ in $\mathfrak{A} = \langle A, R_1, \ldots, R_l \rangle$, let $s: N \to N$ be the successor function of N,

$$s(i) = j \Leftrightarrow i, j \in N \;\&\; i \leqslant j \;\&\; i \neq j \;\&\; (\forall k)[(k \leqslant j \;\&\; k \neq j) \Rightarrow k \leqslant i],$$

let $g: A^n \to A^m$, $h: A^m \times N \times A^n \to A^m$ be hyperelementary functions and assume $f: N \times A^n \to A^m$ satisfies

$$\begin{aligned}f(0, \bar{x}) &= g(\bar{x}),\\ f(s(i), \bar{x}) &= h(f(i, \bar{x}), i, \bar{x}).\end{aligned}$$

Prove that f is hyperelementary on $\mathfrak{A}$. (*Primitive recursion*). ⊣

1.5. Prove that $\mathbb{N} = \langle \omega, S, P \rangle$ and $\omega = \langle \omega, \leqslant \rangle$ have the same inductive relations. ⊣

An *elementary* (*hyperelementary*) *pair* on a structure $\mathfrak{A} = \langle A, R_1, \ldots, R_l \rangle$ is an *elementary* (*hyperelementary*) *one-to-one* function

$$f: A \times A \to A.$$

1.6. Prove that the structure $\mathbb{R}$ of analysis admits an elementary (projective) pair. ⊣

A *coding scheme* on a structure $\mathfrak{A}$ consists of a copy $\langle N, \leqslant \rangle$ of ω in $\mathfrak{A}$ together with a mapping

$$(x_1, \ldots, x_n) \mapsto \langle x_1, \ldots, x_n \rangle$$

which assigns a member of A to each finite sequence from A and which is one-to-one, i.e.

$$\langle x_1, \ldots, x_n \rangle = \langle x'_1, \ldots, x'_m \rangle \Rightarrow [n = m \;\&\; x_1 = x'_1 \;\&\; \ldots \;\&\; x_n = x'_n].$$

(The empty sequence is in the domain of $\langle\,\rangle$ by the notational convention

$$(x_1, \ldots, x_n) = \emptyset \text{ if } n = 0.)$$

With each coding scheme we associate the *decoding relations and functions*

$$Seq(x) \Leftrightarrow x = \langle \emptyset \rangle \textit{ or for some } x_1, \ldots, x_n,\ x = \langle x_1, \ldots, x_n \rangle,$$

$$lh(x) = \begin{cases} 0 & \textit{if} \neg Seq(x), \\ n & \textit{if } Seq(x) \textit{ and } x = \langle x_1, \ldots, x_n \rangle, \end{cases}$$

$$q(x, i) = \begin{cases} x_i & \textit{if for some } x_1, \ldots, x_n,\ 1 \leqslant i \leqslant n \textit{ and } x = \langle x_1, \ldots, x_n \rangle, \\ 0 & \textit{otherwise.} \end{cases}$$

Notice that q is assumed defined on all of $A \times A$ even though only its values on $A \times N$ matter. It is common to use the notation

$$(x)_i = q(x, i).$$

A coding scheme is *elementary* (or *hyperelementary*) if $N, \leqslant, Seq, lh, q$ are all elementary (hyperelementary). A structure $\mathfrak{A}$ is *acceptable* if it admits an elementary coding scheme.

1.7. Prove that if there is a hyperelementary copy of ω in $\mathfrak{A}$ and if $\mathfrak{A}$ admits a hyperelementary pair, then $\mathfrak{A}$ admits a hyperelementary coding scheme. Moreover, there exists an expansion

$$\mathfrak{A}' = (\mathfrak{A}, Q_1, \ldots, Q_m)$$

of $\mathfrak{A}$ which is acceptable and has the same inductive relations as $\mathfrak{A}$. ⊣

Problem 1.7 implies that $\mathbb{N}$ and $\mathbb{R}$ admit hyperelementary coding schemes. Actually both are acceptable structures—this is almost trivial for $\mathbb{R}$ but requires use of Gödel's β-function for $\mathbb{N}$, e.g. see Kleene [1952]. We will assume that both $\mathbb{N}$ and $\mathbb{R}$ are acceptable, but one who does not want to go through the computations with the β-function might as well substitute acceptable $\mathbb{N}'$, $\mathbb{R}'$ with the same inductive relations for them throughout this book.

It is convenient and useful to introduce the customary *two-sorted language* $\mathscr{L}^*$ for the structure $\mathbb{R}$ of analysis. This has variables $x, y, z, \ldots$ varying over ω and another sort of variables, $\alpha, \beta, \gamma, \ldots$, varying over ${}^{\omega}\omega$. The prime formulas are those of the form

$$S(x, y, z) \Leftrightarrow x+y = z,$$

$$P(x, y, z) \Leftrightarrow x \cdot y = z,$$

$$\alpha(x) = y, \qquad x = y$$

and more complicated formulas are constructed by applying the logical operations including quantification on both sorts of variables, $\exists x$, $\forall x$, $\exists \alpha$, $\forall \alpha$.

Let us observe first that this language has the same expressive power (in the proper meaning of this) as the elementary language on $\mathbb{R}$.

1.8. Let

$$R(\bar{x}, \bar{\alpha}) \Leftrightarrow R(x_1, \ldots, x_n, \alpha_1, \ldots, \alpha_m)$$

be a relation on $\omega^n \times ({}^{\omega}\omega)^m$. Prove that R is elementary on $\mathbb{R}$ (projective) if and only if R is definable in $\mathscr{L}^*$.

Similarly, R is definable by a formula of the language of $\mathbb{R}$ with no constants from ${}^{\omega}\omega$ if and only if R is definable by a formula of $\mathscr{L}^*$ with no constants from ${}^{\omega}\omega$. $\dashv$

For each n, m we can extend the language $\mathscr{L}^*$ to the language $\mathscr{L}^*(S)$ by adding a relation variable S varying over subsets of $\omega^n \times ({}^{\omega}\omega)^m$. If $\varphi(\bar{x}, \bar{\alpha}, S)$ is in $\mathscr{L}^*(S)$ with only positive occurrences of S, the sets I^{ξ}_{φ}, I_{φ} are defined exactly as in 1B.

It is obvious that we could study induction using many sorted or higher order languages. This is a simple case for a two-sorted language which can be reduced to the case of positive, elementary inductive definability on which we concentrate in this book.

1.9. Let $R \subseteq \omega^n \times ({}^{\omega}\omega)^m$. Prove that R is a fixed point of the structure $\mathbb{R}$ if and only if there is a formula $\varphi(\bar{x}, \bar{\alpha}, S)$ in the language $\mathscr{L}^*(S)$ such that

$$R(\bar{x}, \bar{\alpha}) \Leftrightarrow (\bar{x}, \bar{\alpha}) \in I_{\varphi}. \qquad \dashv$$

In the next three problems we outline a proof that on $\mathbb{N}$, the inductive relations are precisely the Π^1_1 relations. The result holds for all countable acceptable structures and we will prove it in that generality in Chapter 8. The classical proof here works only for $\mathbb{N}$ (and structures very much like $\mathbb{N}$), but it is good to know and understand this fact before we go on to the more general theory.

1.10. Prove that every Π^1_1 relation on $\mathbb{N}$ satisfies

$$R(\bar{x}) \Leftrightarrow (\forall\alpha)\varphi(\alpha, \bar{x}),$$

where $\varphi(\alpha, \bar{x})$ is a formula of $\mathscr{L}^*$ with no constants from ${}^\omega\omega$.
HINT: Use the elementary coding scheme on $\mathbb{N}$. ⊣

If $\alpha \in {}^\omega\omega$ and $t \in \omega$, put

$$\bar{\alpha}(t) = \langle\emptyset\rangle \qquad \textit{if } t = 0,$$
$$\bar{\alpha}(t) = \langle\alpha(0), \ldots, \alpha(t-1)\rangle \quad \textit{if } t > 0,$$

where the sequence codes are computed relative to some fixed elementary coding scheme on $\mathbb{N}$.

1.11. Prove that for each formula $\varphi(x_1, \ldots, x_n, \alpha_1, \ldots, \alpha_m)$ of $\mathscr{L}^*$ which has no quantifiers of the form $\exists\alpha$, $\forall\alpha$ and no constants from ${}^\omega\omega$, there are formulas $\psi(v, x_1, \ldots, x_n, u_1, \ldots, u_m)$, $\chi(v, x_1, \ldots, x_n, u_1, \ldots, u_m)$ of the language of $\mathbb{N}$ such that

$$\varphi(x_1, \ldots, x_n, \alpha_1, \ldots, \alpha_m) \Leftrightarrow (\forall\beta)(\exists t)\psi(\bar{\beta}(t), x_1, \ldots, x_n, \bar{\alpha}_1(t), \ldots, \bar{\alpha}_m(t))$$
$$\Leftrightarrow (\exists\beta)(\forall t)\chi(\bar{\beta}(t), x_1, \ldots, x_n, \bar{\alpha}_1(t), \ldots, \bar{\alpha}_m(t)).$$

HINT: Use induction on the construction of φ. ⊣

1.12. Prove that every Π^1_1 relation on $\mathbb{N}$ is of the form

$$R(\bar{x}) \Leftrightarrow (\forall\beta)(\exists t)\psi(\bar{\beta}(t), \bar{x}),$$

where $\psi(v, \bar{x})$ is a formula in the language of $\mathbb{N}$ and for every β, t, s,

$$\psi(\bar{\beta}(t), \bar{x}) \;\&\; t < s \Rightarrow \psi(\bar{\beta}(s), \bar{x}).$$ ⊣

1.13. Prove that every Π^1_1 relation on $\mathbb{N}$ is inductive. (Kleene [1955a], Spector [1961].)
HINT: Use the representation of Exercise 1.12 and put

$$\varphi(u, \bar{x}, S) \Leftrightarrow Seq(u) \;\&\; [\psi(u, \bar{x}) \vee (\forall t)S(u^\frown\langle t\rangle, \bar{x})],$$

where if $u = \langle u_1, \ldots, u_m\rangle$, then $u^\frown\langle t\rangle = \langle u_1, \ldots, u_m, t\rangle$. Show then that

$$R(\bar{x}) \Leftrightarrow (\langle\emptyset\rangle, \bar{x}) \in I_\varphi.$$ ⊣

These last two problems are the basic facts about Π_1^1 and inductive relations on $\mathbb{N}$. In addition to showing the identity of these two notions, Exercise 1.12 often yields very simple proofs for the case of $\mathbb{N}$ of results that are quite hard to establish for arbitrary structures, or even for countable acceptable structures. We will suggest some of these easier proofs in the exercises as we go along, but the reader should always keep the example of $\mathbb{N}$ in mind and attempt to obtain easier proofs of the general results for this special case.

1.14. Let

$$\Gamma : Power(\omega^n) \to Power(\omega^n)$$

be an operator which is *monotone*, i.e.

$$S \subseteq S' \Rightarrow \Gamma(S) \subseteq \Gamma(S'),$$

and Π_1^1, i.e.

$$\bar{x} \in \Gamma(S) \Leftrightarrow (\forall S_1) \ldots (\forall S_k)\varphi(S_1, \ldots, S_k, \bar{x}, S),$$

where φ is some elementary formula in the language of $\mathbb{N}$. Prove that the set I_Γ built up by Γ is Π_1^1 on $\mathbb{N}$, hence inductive by Exercise 1.13. (Spector [1961].) ⊣

The next problem shows that we can define all inductive relations using very simple positive formulas. Call a formula φ *simple existential* if it is of the form $(\exists t)\psi(t)$, where $\psi(t)$ is quantifier free. Similarly, call φ *simple universal* if it is of the form $(\forall t)\psi(t)$ with a quantifier free $\psi(t)$.

1.15. Show that if $R(\bar{x})$ is inductive on the infinite structure $\mathfrak{A}$, then there exists an S-positive formula $\varphi(\bar{u}, \bar{x}, S)$ which is a finite disjunction of simple existential and simple universal formulas such that

$$R(\bar{x}) \Leftrightarrow (\bar{a}, \bar{x}) \in I_\varphi$$

with suitable constants $\bar{a}$.

Hint: Take the case that $R = I_\psi$ with

$$\psi \equiv (\forall s)(\exists t)\chi(s, t, \bar{x}, S)$$

and χ quantifier free. Define by a simultaneous recursion sets J_0^ξ, J_1^ξ such that

$$(s, \bar{x}) \in J_0^\xi \Leftrightarrow (\exists t)\chi(s, t, \bar{x}, J_1^{<\xi}),$$
$$\bar{x} \in J_1^\xi \Leftrightarrow (\forall s)(s, \bar{x}) \in J_0^{<\xi},$$

and prove that

$$\bar{x} \in I_\psi \Leftrightarrow \bar{x} \in \bigcup_\xi J_1^\xi.$$

Now reduce the simultaneous recursion giving J_0^ξ, J_1^ξ to one induction using Lemma 1C.1.

For the general case, reduce first to prenex normal form. ⊣

It is not clear who proved this result first, but the proof outlined in the hint is due to P. Aczel.

The last problem of this chapter gives an example as far away from $\mathbb{N}$ as possible—here the theory is trivial.

1.16. Let $\mathfrak{A} = \langle A \rangle$ be the structure with no relations on an infinite set. Prove that every inductive relation on $\mathfrak{A}$ is elementary.

HINT: Prove that for each $a_1, \ldots, a_n \in A$ there are only finitely many sets definable by formulas whose individual constants are among $a_1, \ldots, a_n$. ⊣

CHAPTER 2

THE STAGES OF AN INDUCTIVE DEFINITION

The most prominent feature of an inductive definition determined by an S-positive formula $\varphi(\bar{x}, S)$ in the language over some set A is the natural resolution of the fixed point I_φ into *the stages of the induction*, the sets I^ξ_φ. This assigns ordinals to the members of I_φ in the obvious way,

$$|\bar{x}|_\varphi = \textit{least } \xi \textit{ such that } \bar{x} \in I^\xi_\varphi, \qquad (\bar{x} \in I_\varphi).$$

We prove in 2A a basic regularity property of the ordinal assignment $|\ |_\varphi$ and then we reap some of the consequences in 2B.

2A. The Stage Comparison Theorem

The result of this section is in many ways the central result of the theory of inductive relations. Several versions and corollaries of it have played an important part in the development of the classical theory of hyperarithmetical and Π^1_1 relations on ω and were certainly known to Kleene and Spector—in particular see Spector [1955]. A version of it for almost arbitrary structures is Theorem 7 of Moschovakis [1969b], where it is billed as "the main tool for establishing the basic properties of the hyperprojective (hyperelementary) hierarchy".

My original proof of the present very general version used the methods of Chapter 4 and was quite complicated. The simpler argument given below was discovered by P. Aczel and K. Kunen, independently.

First a simple lemma which will also be useful in Chapter 4.

2A.1. LEMMA. *Suppose $\varphi(\bar{x}, S)$, $\psi(\bar{x}, S)$ are formulas in the language over a set A such that the n-ary relation variable S occurs positively in both of them and such that if $S \subsetneqq A^n$, then*

$$\varphi(\bar{x}, S) \Leftrightarrow \psi(\bar{x}, S).$$

If I^ξ_φ, I^ξ_ψ are defined as in Section 1B, *then for each ξ,*

$$I^\xi_\varphi = I^\xi_\psi,$$

so that both φ and ψ build up the same set $I_\varphi = I_\psi$.

PROOF is immediate by induction on ξ, taking cases on whether

$$I_\varphi^{<\xi} = A^n \quad or \quad I_\varphi^{<\xi} \subsetneq A^n. \qquad \dashv$$

The lemma implies that in studying the induction determined by some $\varphi(\bar{x}, S)$ we may always assume that for all $\bar{x}$, $\varphi(\bar{x}, A^n)$ is true; if not take

$$\psi(\bar{x}, S) \equiv \varphi(\bar{x}, S) \vee (\forall \bar{x}')[\bar{x}' \in S],$$

and then by Lemma 2A.1 for each ordinal ξ,

$$I_\varphi^\xi = I_\psi^\xi,$$

so that ψ determines the same induction as φ.

2A.2. STAGE COMPARISON THEOREM. *Let $\varphi(\bar{x}, S)$, $\psi(\bar{y}, T)$ be formulas in the language of an infinite structure $\mathfrak{A}$, respectively positive in S, T. Define the relations $\leqslant^*_{\varphi,\psi}$, $<^*_{\varphi,\psi}$ by*

$$\bar{x} \leqslant^*_{\varphi,\psi} \bar{y} \Leftrightarrow \bar{x} \in I_\varphi \,\&\, [\bar{y} \notin I_\psi \vee |\bar{x}|_\varphi \leqslant |\bar{y}|_\psi],$$

$$\bar{x} <^*_{\varphi,\psi} \bar{y} \Leftrightarrow \bar{x} \in I_\varphi \,\&\, [\bar{y} \notin I_\psi \vee |\bar{x}|_\varphi < |\bar{y}|_\psi].$$

Then both $\leqslant^*_{\varphi,\psi}$ and $<^*_{\varphi,\psi}$ are fixed points of the structure $\mathfrak{A}$.

PROOF. It is convenient to extend the stage assigning functions by setting

$$|\bar{x}|_\varphi = |A|^+ \quad \text{if } \bar{x} \notin I_\varphi,$$

$$|\bar{y}|_\psi = |A|^+ \quad \text{if } \bar{y} \notin I_\psi,$$

so that

$$\bar{x} \in I_\varphi \Leftrightarrow |\bar{x}|_\varphi < |A|^+; \qquad \bar{y} \in I_\psi \Leftrightarrow |\bar{y}|_\psi < |A|^+.$$

We then have very simple definitions for the relations $\leqslant^* = \leqslant^*_{\varphi,\psi}$, $<^* = <^*_{\varphi,\psi}$,

$$\bar{x} \leqslant^* \bar{y} \Leftrightarrow \bar{x} \in I_\varphi \,\&\, |\bar{x}|_\varphi \leqslant |\bar{y}|_\psi,$$

$$\bar{x} <^* \bar{y} \Leftrightarrow |\bar{x}|_\varphi < |\bar{y}|_\psi.$$

Assuming that S is n-ary and T is m-ary, we may assume without loss of generality that for all $\bar{x}$, $\bar{y}$,

$$\varphi(\bar{x}, A^n), \psi(\bar{y}, A^m)$$

are true, by Lemma 2A.1.

Notice first that

(1) $$\bar{x} \leqslant^* \bar{y} \Leftrightarrow \varphi(\bar{x}, \{\bar{x}' : |\bar{x}'|_\varphi < |\bar{y}|_\psi\});$$

this is immediate from the definitions.

A less obvious equivalence which perhaps requires checking is

$$\text{(2)} \qquad |\bar{y}|_\psi \leqslant |\bar{x}|_\varphi \Leftrightarrow \psi(\bar{y}, \{\bar{y}' : \neg(\bar{x} \leqslant^* \bar{y}')\}).$$

To see this, take cases on whether $\bar{x} \in I_\varphi$ or not. If $\bar{x} \notin I_\varphi$, then the left-hand side is automatically true and $\{\bar{y}' : \neg(\bar{x} \leqslant^* \bar{y}')\} = A^m$, so that the right-hand side is also true by the assumption $\psi(\bar{y}, A^m)$. If $\bar{x} \in I_\varphi$ and $|\bar{x}|_\varphi = \xi$, then clearly

$$\begin{aligned} |\bar{y}|_\psi \leqslant \xi &\Leftrightarrow \psi(\bar{y}, \{\bar{y}' : |\bar{y}'|_\psi < \xi\}) \\ &\Leftrightarrow \psi(\bar{y}, \{\bar{y}' : \neg(\xi \leqslant |\bar{y}'|_\psi)\}) \\ &\Leftrightarrow \psi(\bar{y}, \{\bar{y}' : \neg(\bar{x} \leqslant^* \bar{y}')\}). \end{aligned}$$

We now use (1) and (2) to get a formula χ whose fixed point will be $\leqslant^*$:

$$\begin{aligned} \bar{x} \leqslant^* \bar{y} &\Leftrightarrow \varphi(x, \{\bar{x}' : |\bar{x}'|_\varphi < |\bar{y}|_\psi\}) && \text{(by (1))} \\ &\Leftrightarrow \varphi(\bar{x}, \{\bar{x}' : \neg(|\bar{y}|_\psi \leqslant |\bar{x}'|_\varphi)\}) \\ &\Leftrightarrow \varphi(\bar{x}, \{\bar{x}' : \neg\psi(\bar{y}, \{\bar{y}' : \neg(\bar{x}' \leqslant^* \bar{y}')\})\}). && \text{(by (2)).} \end{aligned}$$

It is easy to see that there is a formula

$$\chi \equiv \chi(\bar{x}, \bar{y}, U)$$

in which U occurs positively such that

$$\chi(\bar{x}, \bar{y}, U) \Leftrightarrow \varphi(\bar{x}, \{\bar{x}' : \neg\psi(\bar{y}, \{\bar{y}' : \neg U(\bar{x}', \bar{y}')\})\});$$

just push the negation sign in $\neg\psi(\bar{y}, \{\bar{y}' : \neg U(\bar{x}', \bar{y}')\})$ through all the quantifiers and connectives (using lower predicate calculus rules) until it applies only to prime formulas and then replace each $\neg\neg U(\bar{x}', \bar{y}')$ by $U(\bar{x}', \bar{y}')$. Then the computation above proves that

$$\bar{x} \leqslant^* \bar{y} \Leftrightarrow \chi(\bar{x}, \bar{y}, \leqslant^*),$$

so that $\leqslant^*$ is a fixed point of χ and hence

$$I_\chi \subseteq \leqslant^*,$$

i.e.

$$(\bar{x}, \bar{y}) \in I_\chi \Rightarrow \bar{x} \leqslant^* \bar{y}.$$

We now complete the proof by showing

$$\text{(3)} \qquad \bar{x} \leqslant^* y \Rightarrow (\bar{x}, \bar{y}) \in I_\chi.$$

Proof of (3) is by transfinite induction on $|\bar{x}|_\varphi$. Assume $\bar{x} \leqslant^* \bar{y}$ and $\neg(\bar{x}, \bar{y}) \in I_\chi$, or equivalently

$$\neg\chi(\bar{x}, \bar{y}, I_\chi),$$

i.e.

(4) $$\neg\varphi(\bar{x}, \{\bar{x}': \neg\psi(\bar{y}, \{\bar{y}': \neg(\bar{x}', \bar{y}') \in I_\chi\})\}).$$

Since $\bar{x} \leqslant^* \bar{y}$, we have in particular $\bar{x} \in I_\varphi$, i.e. with $\xi = |\bar{x}|_\varphi$

(5) $$\varphi(\bar{x}, I_\varphi^{<\xi}).$$

Since $\varphi(\bar{x}, S)$ is monotone in S, (4) and (5) imply that there must be some $\bar{x}' \in I_\varphi^{<\xi}$ such that

(6) $$\psi(\bar{y}, \{\bar{y}': \neg(\bar{x}', \bar{y}') \in I_\chi\});$$

otherwise

$$I_\varphi^{<\xi} \subseteq \{\bar{x}': \neg\psi(\bar{y}, \{\bar{y}': \neg(\bar{x}', \bar{y}') \in I_\chi\})\},$$

and then (5) implies the negation of (4). For this fixed $\bar{x}'$ we have

$$|\bar{x}'|_\varphi < |\bar{x}|_\varphi,$$

so by induction hypothesis for all $\bar{y}'$,

$$\bar{x}' \leqslant^* \bar{y}' \Rightarrow (\bar{x}', \bar{y}') \in I_\chi,$$

i.e.

$$\neg(\bar{x}', \bar{y}') \in I_\chi \Rightarrow \neg(\bar{x}' \leqslant^* \bar{y}').$$

Hence the monotonicity of $\psi(y, T)$ in T and (6) imply

$$\psi(\bar{y}, \{\bar{y}': \neg(\bar{x}' \leqslant^* \bar{y}')\}),$$

which by (2) yields

$$|\bar{y}|_\psi \leqslant |\bar{x}'|_\varphi < |\bar{x}|_\varphi,$$

contradicting the hypothesis $\bar{x} \leqslant^* \bar{y}$.

The construction of some χ' such that

$$\bar{x} <^*_{\varphi,\psi} \bar{y} \Leftrightarrow (\bar{x}, \bar{y}) \in I_{\chi'}$$

is similar and we omit it. ⊣

2B. Closure ordinals and the Closure Theorem

If $\varphi \equiv \varphi(\bar{x}, S)$ is S-positive in the language with relation and individual constants from some infinite set A, we let $\|\varphi\|$ be the *closure ordinal* of the monotone operator defined by φ, i.e.

$$\begin{aligned}\|\varphi\| &= \textit{least } \xi \textit{ such that } I_\varphi^\xi = I_\varphi^{<\xi}\\ &= \textit{supremum } \{|\bar{x}|_\varphi + 1 : \bar{x} \in I_\varphi\}.\end{aligned}$$

For given relations $Q_1, \ldots, Q_m$ on A, we let

$$\kappa(A, Q_1, \ldots, Q_m) = \textit{supremum}\ \{\|\varphi\|: \textit{the only relation constants in } \varphi \textit{ are } =, Q_1, \ldots, Q_m, \textit{ and } Q_1, \ldots, Q_m \textit{ occur positively}\},$$

and finally for a structure $\mathfrak{A} = \langle A, R_1, \ldots, R_l\rangle$ we define *the (closure) ordinal of* $\mathfrak{A}$ by

$$\begin{aligned}\kappa^{\mathfrak{A}} &= \kappa(A, R_1, \neg R_1, \ldots, R_l, \neg R_l)\\ &= \textit{supremum}\ \{\|\varphi\|: \varphi \textit{ is S-positive in the language of } \mathfrak{A}\}.\end{aligned}$$

The chief result of this section is that a fixed point I_φ is hyperelementary on $\mathfrak{A}$ if and only if $\|\varphi\| < \kappa^{\mathfrak{A}}$, i.e. if and only if the induction determined by φ closes before the ordinal of the structure.

Let us first put down an immediate corollary of the Stage Comparison Theorem which is very basic.

2B.1. THEOREM. *Let $\varphi(\bar{x}, S)$ be S-positive in the language of an infinite structure $\mathfrak{A}$. For each $\lambda < \kappa^{\mathfrak{A}}$, the set I_φ^λ is hyperelementary on $\mathfrak{A}$. In particular, if $\|\varphi\| < \kappa^{\mathfrak{A}}$, then I_φ is hyperelementary on $\mathfrak{A}$.*

PROOF. If $\lambda < \kappa^{\mathfrak{A}}$, then by definition there is some $\psi \equiv \psi(\bar{y}, T)$ in the language of $\mathfrak{A}$ and a fixed $\bar{y}^*$ such that

$$\bar{y}^* \in I_\psi^\lambda - I_\psi^{<\lambda}.$$

Applying the Stage Comparison Theorem to φ and ψ we have

$$\bar{x} \in I_\varphi^\lambda \Leftrightarrow \bar{x} \leqslant^*_{\varphi,\psi} \bar{y}^*,$$

and since $\leqslant^*_{\varphi,\psi}$ is inductive so is I_φ^λ. Also

$$\bar{x} \in I_\varphi^\lambda \Leftrightarrow \neg(\bar{y}^* <^*_{\psi,\varphi} \bar{x})$$

and since $<^*_{\varphi,\psi}$ is inductive, I_φ^λ is coinductive.

The second statement follows by noticing that for each φ,

$$I_\varphi = I_\varphi^{\|\varphi\|}. \qquad \dashv$$

To obtain stronger corollaries of the Stage Comparison Theorem we first read an estimate on the closure ordinals of formulas off the Combination Lemma 1C.2.

2B.2. THEOREM. *Let A be an infinite set, $Q, Q_1, \ldots, Q_m$ relations on A, suppose Q is inductive in $Q_1, \ldots, Q_m$ and $Q, Q_1, \ldots, Q_m, T$ all occur positively in the formula*

$$\varphi(\bar{x}, Q, T) \equiv \varphi(\bar{x}, =, Q, Q_1, \ldots, Q_m, T).$$

Then

$$\|\varphi\| \leqslant \kappa(A, Q_1, \ldots, Q_m),$$

i.e. the closure ordinal of φ *is majorized by the closure ordinal of some formula in which only* $=$, $Q_1, \ldots, Q_m$ *occur, and all except* $=$ *occur positively.*

PROOF. Since Q is inductive in $Q_1, \ldots, Q_m$, there is some

$$\psi \equiv \psi(\bar{u}, \bar{y}, S)$$

with only positive occurrences of $Q_1, \ldots, Q_m, S$ and constants $\bar{a}$ such that

$$Q(\bar{y}) \Leftrightarrow (\bar{a}, \bar{y}) \in I_\psi.$$

Take $\chi \equiv \chi(t, \bar{u}, \bar{y}, \bar{x}, U)$ as in the Combination Lemma 1C.2. It is obvious that $Q_1, \ldots, Q_m$ occur positively in χ, so it will be enough to prove that

$$\|\varphi\| \leqslant \|\chi\|.$$

For this again it is enough to verify that

$$\bar{x} \in I_\varphi \Rightarrow (\exists \eta < \|\chi\|)[\bar{x} \in I_\varphi^\eta].$$

Now, if $\bar{x} \in I_\varphi$, then by (3) of 1C.2 we have $(c_1, \bar{u}^*, \bar{y}^*, \bar{x}) \in I_\chi$, so for some $\eta < \|\chi\|$, $(c_1, \bar{u}^*, \bar{y}^*, \bar{x}) \in I_\chi^\eta$, hence by (2) of 1C.2, $\bar{x} \in I_\varphi^\eta$. ⊣

From this it follows that the ordinal of a structure does not increase if we add hyperelementary relations to the structure.

2B.3. COROLLARY. *Let* $\mathfrak{A}$ *be an infinite structure,* $Q_1, \ldots, Q_m$ *inductive relations on* $\mathfrak{A}$. *If* $\varphi(\bar{x}, Q_1, \ldots, Q_m, S)$ *is S-positive in the language of* $(\mathfrak{A}, Q_1, \ldots, Q_m)$ *and* $Q_1, \ldots, Q_m$ *occur positively in* φ, *then*

$$\|\varphi\| \leqslant \kappa^{\mathfrak{A}}.$$

If $Q_1, \ldots, Q_m$ *are hyperelementary on* $\mathfrak{A}$, *then*

$$\kappa^{(\mathfrak{A}, Q_1, \ldots, Q_m)} = \kappa^{\mathfrak{A}}.$$

PROOF. Letting $\mathfrak{A} = \langle A, R_1, \ldots, R_l \rangle$ and taking $m = 2$ for simplicity, we have directly by Theorem 2B.2,

$$\begin{aligned}\|\varphi\| &\leqslant \kappa(A, R_1, \neg R_1, \ldots, R_l, \neg R_l, Q_1, Q_2)\\ &\leqslant \kappa(A, R_1, \neg R_1, \ldots, R_l, \neg R_l, Q_1)\\ &\leqslant \kappa(A, R_1, \neg R_1, \ldots, R_l, \neg R_l) = \kappa^{\mathfrak{A}}.\end{aligned}$$

The second assertion follows immediately. ⊣

We now prove the main result of this section.

2B.4. CLOSURE THEOREM. *Let $\varphi(\bar{x}, S)$ be an S-positive formula in the language of an infinite structure $\mathfrak{A}$. The fixed point I_φ is hyperelementary on $\mathfrak{A}$ if and only if the closure ordinal of φ is smaller than the ordinal of $\mathfrak{A}$, $\|\varphi\| < \kappa^{\mathfrak{A}}$.*

PROOF. That I_φ is hyperelementary if $\|\varphi\| < \kappa = \kappa^{\mathfrak{A}}$ was proved in Theorem 2B.1. Towards proving the converse, assume that I_φ is hyperelementary, let $c_0 \neq c_1$ be distinct elements of A and put

$$\psi(t, \bar{x}, T) \equiv [t = c_0 \;\&\; \varphi(\bar{x}, \{\bar{x}' : T(c_0, \bar{x}')\})] \\ \vee\; [t = c_1 \;\&\; (\forall \bar{x}')[\bar{x}' \in I_\varphi \Rightarrow T(c_0, \bar{x}')]].$$

The formula ψ is in the language of $(\mathfrak{A}, \neg I_\varphi)$ and $\neg I_\varphi$ occurs positively in ψ and $\neg I_\varphi$ is inductive on $\mathfrak{A}$, so by Corollary 2B.3,

$$\|\psi\| \leqslant \kappa$$

and it will be enough to prove that $\|\varphi\|+1 \leqslant \|\psi\|$.

A trivial induction on ξ shows that

$$\bar{x} \in I_\varphi^\xi \Leftrightarrow (c_0, \bar{x}) \in I_\psi^\xi,$$

so taking $\lambda = \|\varphi\|$, we have

$$(c_0, \bar{x}) \in I_\psi^\lambda \Leftrightarrow \bar{x} \in I_\varphi.$$

On the other hand, for each $\xi < \lambda$ there is some $\bar{x} \in I_\varphi$ so that $\bar{x} \notin I_\varphi^{<\xi}$, i.e. $(c_0, \bar{x}) \notin I_\psi^{<\xi}$. The definition of ψ then implies that for any $\bar{x}$,

$$(c_1, \bar{x}) \in I_\psi^\lambda - I_\psi^{<\lambda},$$

so that $\|\psi\| \geqslant \lambda+1$; in fact $\|\psi\| = \lambda+1$. ⊣

We end this section by computing several alternative characterizations of the ordinal $\kappa^{\mathfrak{A}}$ which suggest that it is indeed an ordinal naturally associated with the structure $\mathfrak{A}$.

Recall that if R is a binary relation, i.e. a set of pairs, then the field of R is defined by

$$\mathit{Field}(R) = \{\bar{x} : (\exists \bar{y})(\bar{x}, \bar{y}) \in R \vee (\exists \bar{y})(\bar{y}, \bar{x}) \in R\}.$$

We call R *wellfounded* if each nonempty subset of the field of R has an R-minimal element

$$S \subseteq \mathit{Field}(R) \;\&\; S \neq \emptyset \Rightarrow (\exists \bar{x} \in S)(\forall \bar{y} \in S) \neg (\bar{y}, \bar{x}) \in R.$$

Thus for wellfounded R, $\neg (x, x) \in R$, i.e. R is *strict*. In the context of studying wellfounded relations we will use symbols like "$<$" "$\prec$" to name binary relations and write

$$\bar{x} \prec \bar{y} \textit{ for } (\bar{x}, \bar{y}) \in \prec,$$

but there is no implication that $\prec$ must be transitive or have any of the other properties of strict linear orderings.

If $\prec$ is wellfounded, then there is a unique *rank function*

$$\rho^{\prec}: \mathit{Field}(\prec) \twoheadrightarrow \lambda$$

mapping *Field*($\prec$) onto an ordinal λ and satisfying

$$\rho^{\prec}(\bar{x}) = \mathit{supremum}\{\rho^{\prec}(\bar{y})+1 : \bar{y} \prec \bar{x}\},$$

where as usual

$$\mathit{supremum}(\emptyset) = 0.$$

The range λ of $\rho^{\prec}$ is *the rank of* $\prec$,

$$\mathit{rank}(\prec) = \mathit{supremum}\{\rho^{\prec}(\bar{x})+1 : \bar{x} \in \mathit{Field}(\prec)\}.$$

A very special class of wellfounded relations consists of those which are completely determined by their rank function. We call $\prec$ a *prewellordering* if it is wellfounded and if

$$\bar{x} \prec \bar{y} \Leftrightarrow x, \bar{y} \in \mathit{Field}(\prec) \ \&\ \rho^{\prec}(\bar{x}) < \rho^{\prec}(\bar{y}).$$

Thus a prewellordering with field some set P is determined by starting with a function

$$\sigma: P \twoheadrightarrow \lambda$$

mapping P onto some ordinal λ and putting

$$\bar{x} \prec \bar{y} \Leftrightarrow \bar{x}, \bar{y} \in P \ \&\ \sigma(\bar{x}) < \sigma(\bar{y});$$

for this $\prec$ we then have

$$\sigma = \rho^{\prec}.$$

If $\prec$ is a binary relation on the set A^n of all n-tuples from the space of some structure $\mathfrak{A} = \langle A, R_1, \ldots, R_l \rangle$, we call $\prec$ *inductive*, *coinductive* or *hyperelementary* according as the $2n$-ary relation $\bar{x} \prec \bar{y}$ on A is inductive, coinductive or hyperelementary.

2B.5. THEOREM. *Let $\mathfrak{A}$ be an infinite structure with ordinal $\kappa = \kappa^{\mathfrak{A}}$. Then*

$$\begin{aligned} \kappa &= \mathit{supremum}\{\mathit{rank}(\prec) : \prec \text{ is a hyperelementary prewellordering on some } A^n\} \\ &= \mathit{supremum}\{\mathit{rank}(\prec) : \prec \text{ is a hyperelementary wellfounded relation on some } A^n\} \\ &= \mathit{supremum}\{\mathit{rank}(\prec) : \prec \text{ is a coinductive wellfounded relation on some } A^n\}. \end{aligned}$$

Moreover, none of these three suprema is attained.

PROOF. It is immediate from the definitions that it will be sufficient to prove the following two assertions:

(i) If $\varphi(\bar{x}, S)$ is S-positive in the language of $\mathfrak{A}$ and $\bar{x}^* \in I_\varphi$, then there is a hyperelementary prewellordering of rank $|\bar{x}^*|_\varphi + 1$.

(ii) If $\prec$ is a coinductive wellfounded relation on some A^n, then $rank(\prec) \leqslant \kappa^{\mathfrak{A}}$.

Proof of (i). For fixed $\bar{x}^* \in I_\varphi$, put

$$\bar{x}_1 \prec \bar{x}_2 \Leftrightarrow \bar{x}_1, \bar{x}_2 \in I_\varphi \;\&\; |\bar{x}_1|_\varphi < |\bar{x}_2|_\varphi \;\&\; |\bar{x}_2|_\varphi \leqslant |\bar{x}^*|_\varphi.$$

Clearly $\prec$ is a prewellordering of rank $|\bar{x}^*|_\varphi + 1$. That it is hyperelementary follows by an application of the Stage Comparison Theorem just like that in the proof of Theorem 2B.1.

Proof of (ii). As in the example of Section 1B, put

$$\varphi(\bar{x}, S) \equiv (\forall \bar{y})[\bar{y} \prec \bar{x} \Rightarrow \bar{y} \in S].$$

This formula is S-positive in the language of $(\mathfrak{A}, \prec)$ and $\neg\prec \; = A^{2n} - \prec$ occurs positively in it, so by Corollary 2B.3,

$$\|\varphi\| \leqslant \kappa^{\mathfrak{A}}.$$

On the other hand a trivial induction on ξ shows that

$$\bar{x} \in I_\varphi^\xi \Leftrightarrow \bar{x} \notin Field(\prec) \vee \rho^\prec(\bar{x}) \leqslant \xi,$$

i.e.

$$I_\varphi = A^n,$$

$$\bar{x} \in Field(\prec) \Rightarrow |\bar{x}|_\varphi = \rho^\prec(\bar{x}).$$

Hence

$$\begin{aligned} \|\varphi\| &= supremum\{|\bar{x}|_\varphi + 1 : \bar{x} \in I_\varphi\} \\ &\geqq supremum\{\rho^\prec(\bar{x}) + 1 : \bar{x} \in Field(\prec)\} \\ &= rank(\prec). \end{aligned}$$

To prove that these suprema are not attained, for each wellfounded relation $\prec$ on A^n let c_0, c_1 be distinct objects of A and define the relation $\prec'$ on A^{n+1} by

$$\begin{aligned} (t, \bar{x}) \prec' (s, \bar{y}) \Leftrightarrow & [t = c_0 \;\&\; s = c_0 \;\&\; \bar{x} \prec \bar{y}] \\ & \vee [t = c_0 \;\&\; s = c_1 \;\&\; \bar{x} \in Field(\prec)]. \end{aligned}$$

It is immediate that $\prec'$ is also wellfounded, that $rank(\prec') = rank(\prec) + 1$, that $\prec'$ is a prewellordering if $\prec$ is, and that $\prec'$ is hyperelementary, inductive or coinductive accordingly as $\prec$ is hyperelementary, inductive or coinductive. Thus for each candidate for the supremum in each of the three cases in the theorem there is another candidate with greater rank, so the suprema are not attained. ⊣

A trivial corollary of Theorem 2B.5 is that $\kappa^{\mathfrak{A}}$ is always a limit ordinal. We will see later that for most interesting structures (those that are "acceptable"), $\kappa^{\mathfrak{A}} = \|\varphi\|$ for some S-positive formula in the language of $\mathfrak{A}$.

Exercises for Chapter 2

2.1. Prove that if λ is an infinite ordinal such that

$$\sum_{\nu<\lambda} (\nu+\nu+1) = \lambda,$$

then the structure $\boldsymbol{\lambda} = \langle \lambda, \leqslant \rangle$ admits a hyperelementary pair.

HINT: Consider the Gödel wellordering on pairs,

$$\begin{aligned}(\xi, \eta) < (\xi', \eta') \Leftrightarrow & \; maximum\{\xi, \eta\} < maximum\{\xi', \eta'\} \\ & \vee [maximum\{\xi, \eta\} = maximum\{\xi', \eta'\} \;\&\; \xi < \xi'] \\ & \vee [maximum\{\xi, \eta\} = maximum\{\xi', \eta'\} \;\&\; \xi = \xi' \;\&\; \eta < \eta'].\end{aligned}$$

⊣

2.2. Prove that for every infinite ordinal λ the structure $\boldsymbol{\lambda}$ admits a hyperelementary pair; use Exercise 1.7 to infer that $\boldsymbol{\lambda}$ admits a hyperelementary coding scheme. ⊣

2.3. Suppose $\prec_1, \prec_2$ are hyperelementary (on $\mathfrak{A}$) wellorderings with fields $B_1 \subseteq A^n$, $B_2 \subseteq A^m$ such that there is an order preserving map of $\prec_1$ into $\prec_2$. Prove that there is a hyperelementary order preserving map of $\prec_1$ into $\prec_2$. ⊣

2.4. Let $P(s, t, \bar{x})$ be hyperelementary on the infinite structure $\mathfrak{A}$. Prove that the relation

$$R(\bar{x}) \Leftrightarrow \{(s, t) : P(s, t, \bar{x})\} \text{ is wellfounded}$$

is inductive on $\mathfrak{A}$. ⊣

2.5. Prove that if $\mathfrak{A}$ is infinite and $\lambda < \kappa^{\mathfrak{A}}$, then for each n, $\lambda^n < \kappa^{\mathfrak{A}}$ (ordinal exponentiation). ⊣

2.6. For infinite $\mathfrak{A}$, prove that there exist inductive nonhyperelementary relations on $\mathfrak{A}$ if and only if there is some S-positive formula $\varphi(\bar{x}, S)$ such that $\|\varphi\| = \kappa^{\mathfrak{A}}$. ⊣

2.7. Prove that $\mathfrak{A}$ admits a hyperelementary coding scheme if and only if $\kappa^{\mathfrak{A}} > \omega$ and there exists a hyperelementary one-to-one function $f: A \times A \to A$. ⊣

For each infinite ordinal λ, let

$$\lambda^{(\prime)} = \kappa^{\langle \lambda, \leqslant \rangle}$$

be the ordinal of the structure $\boldsymbol{\lambda} = \langle \lambda, \leqslant \rangle$. The traditional notation for the case $\lambda = \omega$ is

$$\omega_1 = \omega^{(\prime)}.$$

2.8. Prove that if $\mathfrak{A}$ is infinite and $\lambda < \kappa^{\mathfrak{A}}$, then $\lambda^{(\prime)} \leqslant \kappa^{\mathfrak{A}}$. ⊣

CHAPTER 3

STRUCTURE THEORY FOR INDUCTIVE RELATIONS

The Stage Comparison Theorem is stated so that it is directly applicable to the study of fixed points and stages. In order to derive from it consequences about arbitrary inductive relations which need not be fixed points, it is convenient to reformulate it somewhat. We derive in Section 3A the Prewellordering Theorem and then we use it in the remainder of the chapter to develop a fairly rich structure theory for the class of inductive relations on an infinite structure.

The astute reader will notice that in a couple of spots the proofs can be shortened by direct appeals to the Stage Comparison Theorem. We have attempted to use only the Prewellordering Theorem and the closure properties of inductive relations, whenever possible, partly for reasons of elegance but more significantly because these methods generalize directly to the study of other classes of relations. The extra effort will pay off in Chapter 9.

3A. Inductive norms and the Prewellordering Theorem

A *norm* on a set P is a function

$$\sigma : P \twoheadrightarrow \lambda$$

which maps P onto some ordinal λ, the *length* or *rank* of σ. According to the discussion in Section 2B, an ordinal valued function σ with domain P is a norm if and only if it is the rank function of some prewellordering on P.

Each norm $\sigma : P \twoheadrightarrow \lambda$ naturally resolves its domain P into a λ-sequence of sets,

$$P = \bigcup_{\xi < \lambda} P_\sigma^\xi,$$

where for each $\xi < \lambda$,

$$\begin{aligned} P_\sigma^\xi &= \{\bar{x} \in P : \sigma(\bar{x}) \leqslant \xi\} \\ &= \textit{the } \xi^{th} \textit{ resolvent of } P \textit{ relative to } \sigma. \end{aligned}$$

If P is a relation on some infinite structure $\mathfrak{A}$, it is natural and useful to study those norms for which this resolution is "uniformly hyperelementary" on $\mathfrak{A}$. The following precise notion turns out to be the most interesting.

A norm $\sigma: P \twoheadrightarrow \lambda$ is *inductive* (on $\mathfrak{A}$) if there exist relations $J_\sigma(\bar{x}, \bar{y})$, $\breve{J}_\sigma(\bar{x}, \bar{y})$ such that:

(1) *J_σ is inductive and $\breve{J}_\sigma$ is coinductive.*

(2) *If $\bar{y} \in P$, then $(\forall \bar{x})\{[\bar{x} \in P \;\&\; \sigma(\bar{x}) \leqslant \sigma(\bar{y})] \Leftrightarrow J_\sigma(\bar{x}, \bar{y}) \Leftrightarrow \breve{J}_\sigma(\bar{x}, \bar{y})\}$.*

The resolvents of an inductive norm are hyperelementary, since for each $\xi < \lambda$, picking some $\bar{y}_0$ with $\sigma(\bar{y}_0) = \xi$ we have

$$\begin{aligned} \bar{x} \in P_\sigma^\xi &\Leftrightarrow J_\sigma(\bar{x}, \bar{y}_0) \\ &\Leftrightarrow \breve{J}_\sigma(\bar{x}, \bar{y}_0). \end{aligned}$$

But the more important and useful part of the definition is the *uniformity* with which we can write down hyperelementary definitions for the resolvents.

There is an alternative characterization of inductive norms which is also very useful. For any $\sigma: P \twoheadrightarrow \lambda$, put

(3) $$\bar{x} \leqslant_\sigma^* \bar{y} \Leftrightarrow \bar{x} \in P \;\&\; [\bar{y} \notin P \vee \sigma(\bar{x}) \leqslant \sigma(\bar{y})],$$

(4) $$\bar{x} <_\sigma^* \bar{y} \Leftrightarrow \bar{x} \in P \;\&\; [\bar{y} \notin P \vee \sigma(\bar{x}) < \sigma(\bar{y})].$$

3A.1. THEOREM. *Let P be an inductive relation on an infinite structure $\mathfrak{A}$. A norm $\sigma: P \twoheadrightarrow \lambda$ is inductive if and only if both $\leqslant_\sigma^*$, $<_\sigma^*$ are inductive.*

PROOF. If J_σ, $\breve{J}_\sigma$ satisfy (1) and (2), we have

$$\bar{x} \leqslant_\sigma^* \bar{y} \Leftrightarrow \bar{x} \in P \;\&\; [J_\sigma(\bar{x}, \bar{y}) \vee \neg \breve{J}_\sigma(\bar{y}, \bar{x})],$$
$$\bar{x} <_\sigma^* \bar{y} \Leftrightarrow \bar{x} \in P \;\&\; \neg \breve{J}_\sigma(\bar{y}, \bar{x}).$$

On the other hand, if $\leqslant_\sigma^*$, $<_\sigma^*$ are inductive, we can take

(5) $$J_\sigma(\bar{x}, \bar{y}) \Leftrightarrow \bar{x} \leqslant_\sigma^* \bar{y},$$

(6) $$\breve{J}_\sigma(\bar{x}, \bar{y}) \Leftrightarrow \neg(\bar{y} <_\sigma^* \bar{x}).$$ ⊣

The last two equivalences of the proof are one reason why the characterization of inductive norms via $\leqslant_\sigma^*$, $<_\sigma^*$ is more useful—it is easier to define suitable J_σ, $\breve{J}_\sigma$ from $\leqslant_\sigma^*$, $<_\sigma^*$ than vice versa. We will tend to work with these relations in computations. They are also particularly suited for seeing immediately an important corollary of the Stage Comparison Theorem 2A.1.

3A.2. THEOREM. *Let $\mathfrak{A}$ be an infinite structure, $\varphi \equiv \varphi(\bar{x}, S)$ an S-positive formula in the language of $\mathfrak{A}$. The stage-assigning function*

$$|\bar{x}|_\varphi = \textit{least } \xi \textit{ such that } \bar{x} \in I_\varphi \quad (\bar{x} \in I_\varphi)$$

is an inductive norm on I_φ.

PROOF. Taking $\varphi \equiv \psi$ in the Stage Comparison Theorem 2A.1 and letting

$$\sigma(\bar{x}) = |\bar{x}|_\varphi \qquad (\bar{x} \in I_\varphi),$$

it is immediate from the definitions that

$$\bar{x} \leqslant^*_\sigma \bar{y} \Leftrightarrow \bar{x} \leqslant^*_{\varphi,\varphi} \bar{y},$$

$$\bar{x} <^*_\sigma \bar{y} \Leftrightarrow \bar{x} <^*_{\varphi,\varphi} \bar{y}.$$

The result follows by the Stage Comparison Theorem 2A.1, where we proved that both $\leqslant^*_{\varphi,\varphi}$ and $<^*_{\varphi,\varphi}$ are inductive. ⊣

3A.3. PREWELLORDERING THEOREM. *Every inductive relation on an infinite structure admits an inductive norm.*

PROOF. Suppose

$$\bar{x} \in P \Leftrightarrow (\bar{a}, \bar{x}) \in I_\varphi,$$

where $\varphi \equiv \varphi(\bar{x}, S)$ is S-positive in the language of the structure and $\bar{a} = a_1, \ldots, a_k$ is a sequence of constants. We know that $|\ |_\varphi$ is an inductive norm on I_φ, so we would like to put

$$\sigma(\bar{x}) = |\bar{a}, \bar{x}|_\varphi \qquad (\bar{x} \in P).$$

The trouble is that if we define σ in this way, it need not be *onto* an ordinal, so we must "collapse" first the values of $|\ |_\varphi$ on $\{(\bar{a}, \bar{x}) : \bar{x} \in P\}$.

Suppose $|\ |_\varphi$ maps I_φ onto λ. Put

$$\mathscr{C} = \{|\bar{a}, \bar{x}|_\varphi : \bar{x} \in P\};$$

now $\mathscr{C} \subseteq \lambda$, so there is a unique order preserving map

$$\rho : \mathscr{C} \twoheadrightarrow \lambda'$$

mapping $\mathscr{C}$ onto an ordinal $\lambda' \leqslant \lambda$. The function σ given by

$$\sigma(\bar{x}) = \rho(|\bar{a}, \bar{x}|_\varphi)$$

is clearly a norm on P of length λ'. It is inductive, since obviously

$$\bar{x} \leqslant^*_\sigma \bar{y} \Leftrightarrow (\bar{a}, \bar{x}) \leqslant^*_{\varphi,\varphi} (\bar{a}, \bar{y}),$$

$$\bar{x} <^*_\sigma \bar{y} \Leftrightarrow (\bar{a}, \bar{x}) <^*_{\varphi,\varphi} (\bar{a}, \bar{y}).$$ ⊣

There will be many consequences of the Prewellordering Theorem in the remainder of this chapter, but here are two of the simplest ones.

3A.4. REDUCTION THEOREM. *Let P, Q be inductive n-ary relations on an infinite structure $\mathfrak{A}$. There exist inductive n-ary relations P_1, Q_1 such that*

$$P_1 \subseteq P, \qquad Q_1 \subseteq Q,$$
$$P \cup Q = P_1 \cup Q_1,$$
$$P_1 \cap Q_1 = \emptyset.$$

(See Fig. 3.1.)

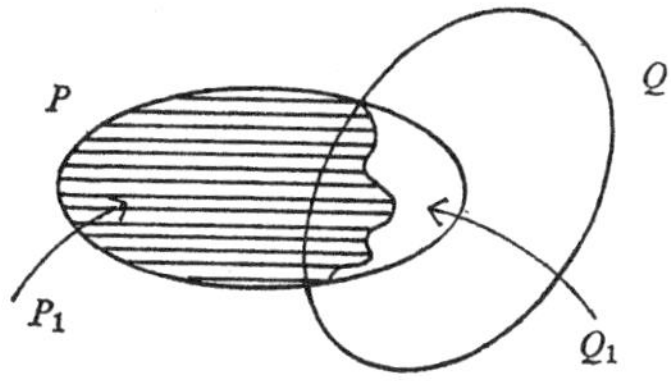

Fig. 3.1. Reduction.

PROOF. Let c_0, c_1 be distinct elements of A and put

$$R(y, \bar{x}) \Leftrightarrow [y = c_0 \;\&\; P(\bar{x})] \vee [y = c_1 \;\&\; Q(\bar{x})].$$

Clearly R is inductive, so let σ be an inductive norm on R.
Put

$$\bar{x} \in P_1 \Leftrightarrow (c_0, \bar{x}) \leqslant^*_\sigma (c_1, \bar{x}),$$
$$\bar{x} \in Q_1 \Leftrightarrow (c_1, \bar{x}) <^*_\sigma (c_0, \bar{x});$$

it is trivial to check that P_1, Q_1 have the required properties. ⊣

3A.5. SEPARATION THEOREM. *Let P, Q be disjoint coinductive n-ary relations on an infinite structure $\mathfrak{A}$. There exists a hyperelementary relation R which separates P from Q, i.e.*

$$P \subseteq R, \qquad Q \cap R = \emptyset.$$

(See Fig. 3.2).

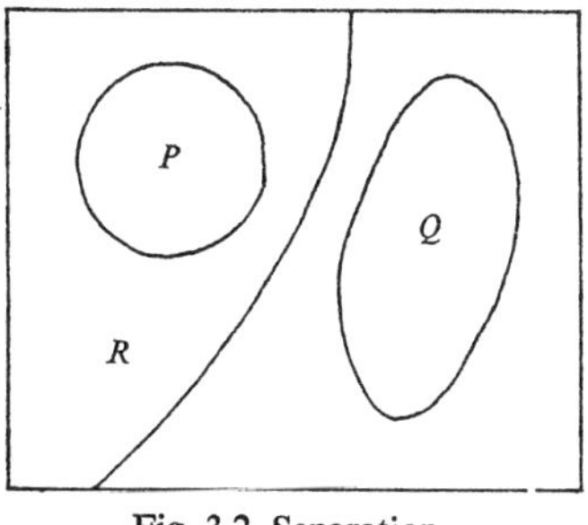

Fig. 3.2. Separation.

PROOF. Let $A^n - P_1$, $A^n - Q_1$ reduce $A^n - P$, $A^n - Q$ by 3A.4, i.e. $A^n - P_1$, $A^n - Q_1$ are inductive and

$$A^n - P_1 \subseteq A^n - P, \qquad A^n - Q_1 \subseteq A^n - Q,$$
$$(A^n - P) \cup (A^n - Q) = (A^n - P_1) \cup (A^n - Q_1)$$
$$(A^n - P_1) \cap (A^n - Q_1) = \emptyset.$$

De Morgan's laws and the hypothesis $P \cap Q = \emptyset$ turn these relationships to

$$P \subseteq P_1, \qquad Q \subseteq Q_1,$$
$$P \cap Q = \emptyset = P_1 \cap Q_1,$$
$$P_1 \cup Q_1 = A^n,$$

so we can take $R = P_1$. ⊣

3B. Making hyperelementary selections

The results in this section show that in certain cases we can find a nonempty hyperelementary subset of some inductive set, and we can do so "uniformly" in the parameters present. Such results are often called *uniformization* or *selection* theorems.

3B.1. HYPERELEMENTARY SELECTION THEOREM. *Suppose $P(\bar{x}, \bar{y})$ is an inductive relation on an infinite structure. There are inductive relations $P^*(\bar{x}, \bar{y})$, $P^{**}(\bar{x}, \bar{y})$ such that*

(1) $$P^* \subseteq P,$$

(2) $$(\exists \bar{y})P(\bar{x}, \bar{y}) \Rightarrow (\exists \bar{y})P^*(\bar{x}, \bar{y}),$$

(3) $$(\exists \bar{y})P(\bar{x}, \bar{y}) \Rightarrow (\forall \bar{y})[P^*(\bar{x}, \bar{y}) \Leftrightarrow \neg P^{**}(\bar{x}, \bar{y})].$$

(See Fig. 3.3.)

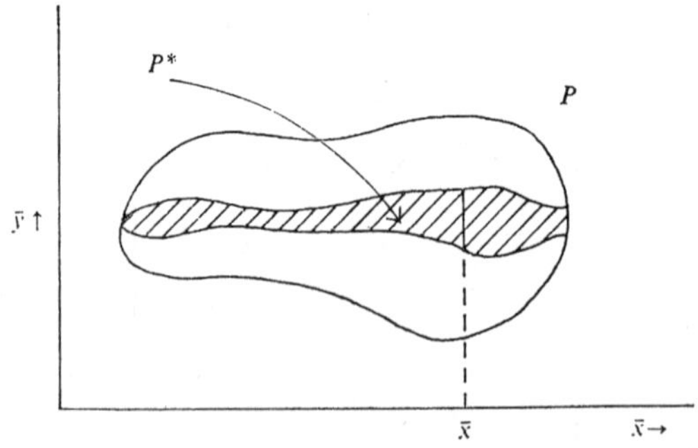

Fig. 3.3. Uniformization by a set with hyperelementary sections.

PROOF. Let σ be an inductive norm on P and put

$$P^*(\bar{x}, \bar{y}) \Leftrightarrow P(\bar{x}, \bar{y}) \;\&\; (\forall \bar{y}')[P(\bar{x}, \bar{y}') \Rightarrow \sigma(\bar{x}, \bar{y}) \leqslant \sigma(\bar{x}, \bar{y}')].$$

Now (1) and (2) are obvious and that P^* is inductive follows from

$$P^*(\bar{x}, \bar{y}) \Leftrightarrow (\forall \bar{y}')[(\bar{x}, \bar{y}) \leqslant^*_\sigma (\bar{x}, \bar{y}')].$$

We can prove (3) easily with

$$P^{**}(\bar{x}, \bar{y}) \Leftrightarrow (\exists \bar{y}')[(\bar{x}, \bar{y}') <^*_\sigma (\bar{x}, \bar{y})].$$ ⊣

To see what this means, consider the *projection* of P,

$$proj(P) = \{\bar{x} \colon (\exists \bar{y})P(\bar{x}, \bar{y})\}.$$

The theorem asserts that if $P(\bar{x}, \bar{y})$ is inductive, then we can find an inductive $P^* \subseteq P$ with the same projection and such that for each $\bar{x}$ in the common projection, the section $\{\bar{y} \colon P^*(\bar{x}, \bar{y})\}$ is hyperelementary—and uniformly in $\bar{x}$. We state separately the immediate corollaries that look more like selection principles.

3B.2. COROLLARY. *Let $P(\bar{x}, \bar{y})$ be inductive on the infinite structure $\mathfrak{A} = \langle A, R_1, \ldots, R_l \rangle$, let $B \subseteq A^n$ be hyperelementary. If $(\forall \bar{x} \in B)(\exists \bar{y})P(\bar{x}, \bar{y})$, then there exists a hyperelementary $P^* \subseteq P$ such that $(\forall \bar{x} \in B)(\exists \bar{y})P^*(\bar{x}, \bar{y})$.*

PROOF. Apply the theorem to

$$P'(\bar{x}, \bar{y}) \Leftrightarrow \bar{x} \in B \;\&\; P(\bar{x}, \bar{y}).$$ ⊣

3B.3. COROLLARY. *Let $P(\bar{x}, \bar{y})$ be inductive on an infinite structure $\mathfrak{A}$, suppose that $B \subseteq A^m$ is hyperelementary and admits a hyperelementary wellordering. Then*

$$(\forall \bar{x})(\exists \bar{y} \in B)P(\bar{x}, \bar{y}) \Rightarrow (\exists f)[f \colon A^n \to B, f \textit{ is hyperelementary and } (\forall \bar{x})P(\bar{x}, f(\bar{x}))].$$

(See Fig. 3.4.)

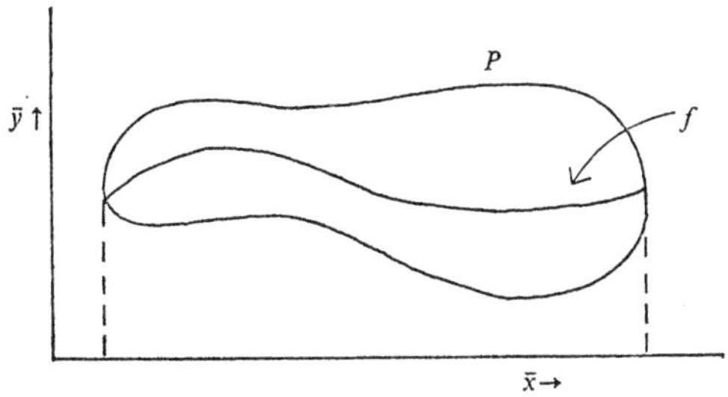

Fig. 3.4. Uniformization by a function.

PROOF. Choose a hyperelementary P^* by Corollary 3B.2 such that $P^* \subseteq P$, $(\forall \bar{x})(\exists \bar{y})[\bar{y} \in B \;\&\; P^*(\bar{x}, \bar{y})]$ and then put

$$(\bar{x}, \bar{y}) \in f \Leftrightarrow P^*(\bar{x}, \bar{y}) \;\&\; (\forall \bar{y}')[\bar{y}' \prec \bar{y} \Rightarrow \neg P^*(\bar{x}, \bar{y}')],$$

where $\prec$ is the assumed hyperelementary wellordering of B. ⊣

We now apply these selection principles to the comparison of wellfounded relations. The main theorem is a bit technical in the most general version that we prove, but this is the statement with widest applicability.

3B.4. RANK COMPARISON THEOREM. *Let $\mathfrak{A}$ be an infinite structure, $\prec_1$ a coinductive wellfounded relation on A^n, $\prec_2$ an inductive wellfounded relation on A^m. There exists an inductive $P \subseteq A^{n+m}$ such that*

$$\bar{x} \in Field(\prec_1) \Rightarrow (\forall \bar{y})\{P(\bar{x}, \bar{y}) \Leftrightarrow [\bar{y} \in Field(\prec_2) \;\&\; \rho^{\prec_1}(\bar{x}) \leqslant \rho^{\prec_2}(\bar{y})]\}.$$

PROOF. Put

$$\varphi(\bar{x}, \bar{y}, S) \Leftrightarrow \bar{y} \in Field(\prec_2) \;\&\; (\forall \bar{x}')\{\bar{x}' \prec_1 \bar{x} \Rightarrow (\exists \bar{y}')[\bar{y}' \prec_2 \bar{y} \;\&\; S(\bar{x}', \bar{y}')]\}.$$

Clearly φ is S-positive and we will prove that for any $\bar{x}$ in $Field(\prec_1)$ and all $\bar{y}$,

$$(*) \qquad (\bar{x}, \bar{y}) \in I_\varphi \Leftrightarrow \bar{y} \in Field(\prec_2) \;\&\; \rho^{\prec_1}(\bar{x}) \leqslant \rho^{\prec_2}(\bar{y}).$$

This will complete the proof since the relations $A^n - \prec_1, \prec_2$ occur positively in φ, so I_φ is inductive in $A^n - \prec_1, \prec_2$ and hence inductive on $\mathfrak{A}$ by the Transitivity Theorem 1C.2.

Proof of direction ($\Rightarrow$) *of* (*) is by induction on $\rho^{\prec_1}(\bar{x})$. Assume for some $\bar{y}$ that $(\bar{x}, \bar{y}) \in I_\varphi$. Then clearly $\bar{y} \in Field(\prec_2)$. By the definition of I_φ and the induction hypothesis,

$$(\forall \bar{x}')\{\bar{x}' \prec_1 \bar{x} \Rightarrow (\exists \bar{y}')[\bar{y}' \prec_2 \bar{y} \;\&\; \rho^{\prec_1}(\bar{x}') \leqslant \rho^{\prec_2}(\bar{y}')]\};$$

thus

$$\begin{aligned} \rho^{\prec_1}(\bar{x}) &= supremum\{\rho^{\prec_1}(\bar{x}')+1 : \bar{x}' \prec_1 \bar{x}\} \\ &\leqslant supremum\{\rho^{\prec_2}(\bar{y}')+1 : \bar{y}' \prec_2 \bar{y}\} = \rho^{\prec_2}(\bar{y}). \end{aligned}$$

Proof of direction ($\Leftarrow$) *of* (*) is again by induction on $\rho^{\prec_1}(\bar{x})$. Assuming the right-hand side of (*), we know that $\rho^{\prec_1}(\bar{x}) \leqslant \rho^{\prec_2}(\bar{y})$, so that by the definition of rank,

$$(\forall \bar{x}')\{\bar{x}' \prec_1 \bar{x} \Rightarrow (\exists \bar{y}')[\bar{y}' \prec_2 \bar{y} \;\&\; \rho^{\prec_1}(\bar{x}') \leqslant \rho^{\prec_2}(\bar{y}')]\}$$

which by induction hypothesis gives

$$(\forall \bar{x}')\{\bar{x}' \prec_1 \bar{x} \Rightarrow (\exists \bar{y}')[\bar{y}' \prec_2 \bar{y} \;\&\; (\bar{x}', \bar{y}') \in I_\varphi]\}$$

which implies $(\bar{x}, \bar{y}) \in I_\varphi$. ⊣

3C. The Boundedness and Covering Theorems

The first of these results is the correct extension of the Closure Theorem 2B.4 to all inductive relations (not just the fixed points). The second has both a clear geometrical meaning and wide applicability.

3C.1. BOUNDEDNESS THEOREM. *Let P be an inductive relation on some infinite structure $\mathfrak{A}$ with ordinal $\kappa = \kappa^{\mathfrak{A}}$, let $\sigma : P \twoheadrightarrow \lambda$ be an inductive norm on P. Then*:

(1) $\lambda \leqslant \kappa$.

(2) $\lambda < \kappa \Leftrightarrow P$ *is hyperelementary on* $\mathfrak{A}$.

PROOF. To prove (1), for each $\bar{y} \in P$ consider the prewellordering $\prec_{\bar{y}}$ defined by

$$\bar{x}_1 \prec_{\bar{y}} \bar{x}_1 \Leftrightarrow \bar{x}_1, \bar{x}_2 \in P \;\&\; \sigma(\bar{x}_1) < \sigma(\bar{x}_2) \;\&\; \sigma(\bar{x}_2) < \sigma(\bar{y}).$$

Clearly $rank(\prec_{\bar{y}}) = \sigma(\bar{y})$, except for the trivial case $\sigma(\bar{y}) = 1$ which is irrelevant to the argument since $\kappa > 1$. Also $\prec_{\bar{y}}$ is hyperelementary, since easily

$$\begin{aligned} \bar{x}_1 \prec_{\bar{y}} \bar{x}_2 &\Leftrightarrow \bar{x}_1 <^*_\sigma \bar{x}_2 \;\&\; \bar{x}_2 <^*_\sigma \bar{y} \\ &\Leftrightarrow \neg(\bar{y} \leqslant^*_\sigma \bar{x}_2) \;\&\; \neg(\bar{x}_2 \leqslant^*_\sigma \bar{x}_1). \end{aligned}$$

Thus by Theorem 2B.5 for each $\bar{y}$, $\sigma(\bar{y}) < \kappa$, hence $\lambda \leqslant \kappa$.

Proof of direction ($\Leftarrow$) *of* (2). Put

$$\bar{x}_1 \prec \bar{x}_2 \Leftrightarrow \bar{x}_1, \bar{x}_2 \in P \;\&\; \sigma(\bar{x}_1) < \sigma(\bar{x}_2).$$

As before, $\prec$ is a hyperelementary prewellordering with $rank(\prec) = \lambda$, so by 2B.5, $\lambda < \kappa$.

Proof of direction ($\Rightarrow$) *of* (2). Choose a hyperelementary wellfounded relation $\prec$ on A^m with $rank(\prec) > \lambda$, pick a fixed $\bar{y}_0 \in Field(\prec)$ such that $\rho^{\prec}(\bar{y}_0) = \lambda$ and put

$$R_1(\bar{y}, \bar{x}) \Leftrightarrow [\bar{y} \prec \bar{y}_0 \;\&\; \bar{x} \in P \;\&\; \rho^{\prec}(\bar{y}) \leqslant \rho(\bar{x})].$$

Now

$$\sigma(\bar{x}) = \rho^{\prec_2}(\bar{x})$$

if we take for $\prec_2$ the inductive prewellordering

$$\begin{aligned} \bar{x}_1 \prec_2 \bar{x}_2 &\Leftrightarrow \bar{x}_1, \bar{x}_2 \in P \;\&\; \sigma(\bar{x}_1) < \sigma(\bar{x}_2) \\ &\Leftrightarrow \bar{x}_2 \in P \;\&\; \bar{x}_1 <^*_\sigma \bar{x}_2, \end{aligned}$$

hence 3B.4 applies and R_1 agrees with some inductive $R(\bar{y}, \bar{x})$ whenever $\bar{y} \prec \bar{y}_0$. By hypothesis,

$$(\forall \bar{y} \prec \bar{y}_0)(\exists \bar{x})R(\bar{y}, \bar{x}),$$

so by Corollary 3B.2 there is a hyperelementary $R^* \subseteq R$ such that

$$(\forall \bar{y} \prec \bar{y}_0)(\exists \bar{x})R^*(\bar{y}, \bar{x}).$$

It is now immediate that

$$\begin{aligned} P(\bar{x}) &\Leftrightarrow (\exists \bar{y})(\exists \bar{x}')[\bar{y} \prec \bar{y}_0 \;\&\; R^*(\bar{y}, \bar{x}') \;\&\; \sigma(\bar{x}) \leqslant \sigma(\bar{x}')] \\ &\Leftrightarrow (\exists \bar{y})(\exists \bar{x}')[\bar{y} \prec \bar{y}_0 \;\&\; R^*(\bar{y}, \bar{x}') \;\&\; \neg(\bar{x}' <_\sigma^* \bar{x})] \end{aligned}$$

which implies that P is coinductive, hence hyperelementary. ⊣

The Boundedness Theorem is a direct generalization of the Closure Theorem 2B.4, since by Theorem 3A.2 the stage-assigning function $|\;|_\varphi$ is an inductive norm on the fixed point I_φ.

3C.2. COVERING THEOREM. *Let P be an n-ary inductive relation on some infinite structure with ordinal $\kappa = \kappa^{\mathfrak{A}}$, let $\sigma: P \twoheadrightarrow \lambda$ be an inductive norm on P, let Q be a coinductive m-ary relation and assume that*

$$f: A^m \to A^n$$

is a hyperelementary function such that $f[Q] \subseteq P$. Then there exists some $\xi < k$ such that

$$\bar{y} \in Q \Rightarrow \sigma(f(\bar{y})) \leqslant \xi.$$

PROOF. If P is hyperelementary, then $\lambda < \kappa$ by 3C.1 and we can take $\xi = \lambda$. If P is not coinductive and all the hypotheses held but the conclusion failed, we would have

$$\begin{aligned} \bar{x} \in P &\Leftrightarrow (\exists \bar{y})[\bar{y} \in Q \;\&\; \sigma(\bar{x}) \leqslant \sigma(f(\bar{y}))] \\ &\Leftrightarrow (\exists \bar{y})[\bar{y} \in Q \;\&\; \neg(f(\bar{y}) <_\sigma^* \bar{x})] \end{aligned}$$

which implies that P is coinductive, contrary to hypothesis. (See Fig. 3.5.) ⊣

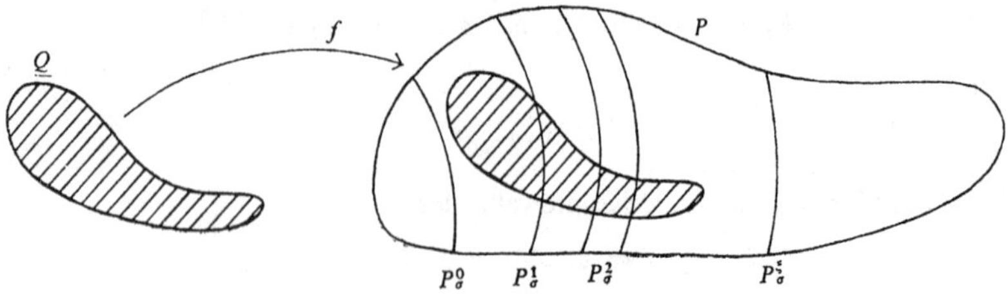

Fig. 3.5.

Using the notation of resolvents relative to σ that we introduced in the beginning of Section 3A, the Covering Theorem asserts that if a hyperelementary function f maps a coinductive set Q into P, then $f[Q]$ is wholly contained in one of the resolvents. We have stated this in the most general case suitable for applications, but it is worth putting down explicitly the special case when f is the identity.

3C.3. THEOREM. *Let P be an inductive relation on some infinite structure $\mathfrak{A}$ with ordinal $\kappa = \kappa^{\mathfrak{A}}$, let $\sigma: P \twoheadrightarrow \lambda$ be an inductive norm on P with resolvents*

$$P_\sigma^\xi = \{\bar{x}: \bar{x} \in P \;\&\; \sigma(\bar{x}) \leqslant \xi\}, \qquad \xi < \kappa.$$

Then every coinductive subset of P is wholly contained in some resolvent P_σ^ξ. ⊣

3D. Expanding a structure by an inductive relation

If P is hyperelementary on $\mathfrak{A}$, then the expanded structure $(\mathfrak{A}, P)$ has the same inductive relations and the same closure ordinal as $\mathfrak{A}$. What happens if we expand $\mathfrak{A}$ by an inductive relation which is not hyperelementary? Spector [1955] proved two very pretty and useful theorems about this situation. His proofs were about $\mathbb{N}$, but they adapt easily to the abstract case.

3D.1. THEOREM. *If P, Q are inductive, nonhyperelementary relations on the infinite structure $\mathfrak{A}$, then the expanded structures $(\mathfrak{A}, P)$, $(\mathfrak{A}, Q)$ have the same inductive relations; in particular, P is hyperelementary on $(\mathfrak{A}, Q)$ and Q is hyperelementary on $(\mathfrak{A}, P)$.*

PROOF. It will be enough to prove the last assertion. From it we get that $(\mathfrak{A}, Q)$ has the same inductive relations as $(\mathfrak{A}, Q, P)$ by Theorem 1D.2 and symmetrically that $(\mathfrak{A}, P)$ has the same inductive relations as $(\mathfrak{A}, P, Q)$, which proves the first assertion.

Suppose then that

$$P(\bar{x}) \Leftrightarrow (\bar{a}, \bar{x}) \in I_\varphi,$$
$$Q(\bar{y}) \Leftrightarrow (\bar{b}, \bar{y}) \in I_\psi$$

and that Q is not hyperelementary. The Boundedness Theorem 3C.1 implies immediately that

$$\textit{supremum}\{|\bar{b}, \bar{y}|_\psi : Q(\bar{y})\} = \kappa^{\mathfrak{A}},$$

since from the contrary hypothesis we can easily construct an inductive norm on Q of rank less than $\kappa^{\mathfrak{A}}$. Hence

$$P(\bar{x}) \Leftrightarrow (\exists \bar{y})[Q(\bar{y}) \;\&\; |\bar{a}, \bar{x}|_\varphi \leqslant |\bar{b}, \bar{y}|_\psi]$$
$$\Leftrightarrow (\exists \bar{y})[Q(\bar{y}) \;\&\; \neg(\bar{b}, \bar{y}) <^*_{\psi,\varphi} (\bar{a}, \bar{x})]$$

which proves P coinductive on $(\mathfrak{A}, Q)$ by the Stage Comparison Theorem 2A.2. Since P is inductive on $\mathfrak{A}$ and hence inductive on $(\mathfrak{A}, Q)$, this completes the proof. ⊣

The second result says something about the ordinal of an expansion by an arbitrary relation.

3D.2. THEOREM. *Let Q be inductive, nonhyperelementary on the infinite structure $\mathfrak{A} = \langle A, R_1, \ldots, R_l \rangle$, let P be an arbitrary relation on A. Then*

$$Q \text{ is hyperelementary on } (\mathfrak{A}, P) \Leftrightarrow \kappa^{\mathfrak{A}} < \kappa^{(\mathfrak{A},P)}.$$

PROOF. Assume first that $\kappa^{\mathfrak{A}} < \kappa^{(\mathfrak{A},P)}$. If $\varphi(\bar{x}, S)$ is S-positive in the language of $\mathfrak{A}$, then $\|\varphi\| \leqslant \kappa^{\mathfrak{A}} < \kappa^{(\mathfrak{A},P)}$. Since φ is also in the language of $(\mathfrak{A}, P)$, the Closure Theorem 2B.4 implies that I_φ is hyperelementary on $(\mathfrak{A}, P)$. Thus every fixed point of $\mathfrak{A}$ is hyperelementary on $(\mathfrak{A}, P)$, hence every inductive relation on $\mathfrak{A}$ is hyperelementary on $(\mathfrak{A}, P)$.

Conversely, if Q is hyperelementary on $(\mathfrak{A}, P)$, then every inductive relation on $\mathfrak{A}$ is hyperelementary on $(\mathfrak{A}, P)$ by Theorem 3D.1 and the Transitivity Theorem 1C.3, and in particular every fixed point I_φ of $\mathfrak{A}$ is hyperelementary on $(\mathfrak{A}, P)$. By the Closure Theorem again, this means that for each S-positive $\varphi(x, S)$ in the language of $\mathfrak{A}$ we have $\|\varphi\| < \kappa^{(\mathfrak{A},P)}$. Now there must be some φ such that $\|\varphi\| = \kappa^{\mathfrak{A}}$, or else every fixed point and hence every inductive relation on $\mathfrak{A}$ is hyperelementary, contradicting the fact that Q is not; choosing such a φ, we have $\kappa^{\mathfrak{A}} = \|\varphi\| < \kappa^{(\mathfrak{A},P)}$. ⊣

3E. Generalization of the theory to richer languages

The theory of the first three chapters can be generalized directly to languages richer than the first order predicate calculus we have been studying. Consider the following two important examples, where in each case we allow the equality symbol $=$ and arbitrary constants from a fixed infinite set A.

(1) The language $\mathscr{L}^A_{\omega_1,\omega}$ which admits countable conjunctions and disjunctions, with constants from A and individual and relation variables ranging over A. More generally, we can take the richer classical languages $\mathscr{L}^A_{\kappa,\lambda}$ interpreted over A or the finer Barwise languages $\mathscr{L}^A_M$, one for each admissible set M. (For the definitions of these notions see Keisler [1971].)

(2) The language $\mathscr{L}^A(\mathrm{Q})$ which extends $\mathscr{L}^A$ by a *unary quantifier* Q and its dual $\mathrm{Q}^\cup$, so that if φ is a formula, so are

$$(\mathrm{Q}x)\varphi, \qquad (\mathrm{Q}^\cup x)\varphi.$$

We interpret Q by any *nontrivial, monotone* collection of subsets of A, i.e.

$$\emptyset \subsetneqq \mathrm{Q} \subsetneqq \mathit{Power}(A),$$

$$S \in \mathrm{Q}\ \&\ S \subseteq T \Rightarrow T \in \mathrm{Q}.$$

This extends the definition of truth for this language by the clauses

$(\mathrm{Q}x)\varphi$ *is true* if $\{x: \varphi(x)\} \in \mathrm{Q}$,

$(\mathrm{Q}^{\cup}x)$ *is true* $\Leftrightarrow \neg(\mathrm{Q}x)\neg\varphi$ *is true*

$\Leftrightarrow A - \{x: \neg\varphi\} \notin \mathrm{Q}.$

More generally, we may adjoin a monotone quantifier Q to any of the languages described in (1) or (2) and take, e.g., $\mathscr{L}^A_{\omega_1,\omega}(\mathrm{Q})$ or $\mathscr{L}^A_M(\mathrm{Q})$. Particular Q's which have been studied include

$\mathrm{Q} = \{S \subseteq A: S \textit{ is uncountable}\}$,

$\mathrm{Q} = \{S \subseteq A: |S| = |A|\}$.

For more information about these languages see Keisler [1970].

There is a natural interpretation of the formulas in all these languages over the set A. The languages have symbols for $\neg$, &, $\vee$, $\exists$, $\forall$ and extend $\mathscr{L}^A$. For each of them we can give a precise definition of the notion

the relation symbol S occurs positively in $\varphi(S)$,

so that if this holds, then

$$\varphi(S)\ \&\ S \subseteq S' \Rightarrow \varphi(S').$$

Thus if $\varphi \equiv \varphi(\bar{x}, S)$ is a formula with $x_1, \ldots, x_n$ the only free variables and S the only (n-ary) relation variable and if S occurs positively in φ, we can define the stages I^{ξ}_{φ} and the set I_{φ} built up by φ exactly as we did in Section 1B.

Let $\mathscr{L}$ be any of these languages and let $R, Q_1, \ldots, Q_m$ be relations on A. We call R *positive* $\mathscr{L}$*-inductive in* $Q_1, \ldots, Q_m$ if there is an S-positive $\varphi(\bar{u}, \bar{x}, S)$ in $\mathscr{L}$ and constants $\bar{a}$ such that

$$R(\bar{x}) \Leftrightarrow (\bar{a}, \bar{x}) \in I_{\varphi}.$$

For a structure $\mathfrak{A} = \langle A, R_1, \ldots, R_l\rangle$ with domain A we define the *positive* $\mathscr{L}$*-inductive* and the *hyper-*$\mathscr{L}$*-definable* relations from this relative notion exactly as we did in Section 1D.

With these definitions and the obvious generalizations of the other concepts we have defined, *all the results of the first three chapters extend with the same proofs*. (In Theorem 1D.3 we must define $\mathscr{L}$-Π^1_1 relations using $\mathscr{L}$-definable relations in the matrix.)

All one needs do is to substitute "positive $\mathscr{L}$-inductive" for "inductive" in all the definitions and proofs.

A further generalization can be obtained if we notice that *the only property of positive formulas we have used up till now is monotonicity.* Letting again $\mathscr{L}$ be $\mathscr{L}^A$ or any of the languages in (1), (2), and for a fixed structure $\mathfrak{A} = \langle A, R_1, \ldots, R_l \rangle$, let us say that *the relation variable S occurs monotonically in $\varphi(S)$ relative to* $\mathfrak{A}$ if in the natural interpretation on $\mathfrak{A}$,

$$\varphi(S) \;\&\; S \subseteq S' \Rightarrow \varphi(S').$$

We can use formulas $\varphi(\bar{x}, S)$ with monotone occurrences of S to build sets I_φ and then we can define the notion

R is monotone-$\mathscr{L}$-inductive on $\mathfrak{A}$

in the obvious way. Again with this definition, *all the results of the first three chapters extend with the same proofs.*

The reason we have restricted ourselves to positive formulas in this book will become obvious in the next chapter and still more obvious in Chapter 5. The relation between the syntactical notion of *positivity* and the semantical notion of *monotonicity* is interesting but very little is known about it. (See Exercises 1.14 and 8.8.)

It should be pointed out that the generalization of the results we have proved so far to monotone inductive definability and to positive inductive definability relative to infinitary languages depends on the Aczel–Kunen version of the proof of the key Stage Comparison Theorem 2A.2. My own version of this proof depended on the results about positive formulas in the next chapter and yielded direct extensions of the theory only for the (syntactically) finitary languages in (2) above.

There are many other natural notions of inductive definability for which the theory of the first three chapters does not extend directly. For example, take all S-positive Σ^1_1 formulas $\varphi(\bar{x}, S)$ in the language of a structure $\mathfrak{A}$ and use these to define fixed points and *positive Σ^1_1-inductive* relations. Our proof of the Stage Comparison Theorem 2A.2 does not extend to this case, because it defines $\leqslant^*_{\varphi,\psi}$ by an induction that has more quantifier alterations than either φ or ψ. The notions of inductive definability that usually arise in recursion theory are often *restricted* in a like manner. Still other notions involve *nonmonotone inductions* which we have not even defined here—in any case our proofs in both the Transitivity Theorem 1C.2 and the Stage Comparison Theorem 2A.2 depend heavily on monotonicity. Recently there have been some very exciting results about nonmonotone inductions, e.g. see Richter [1971], Aczel and Richter [1972], [1973], Aanderaa [1973].

It is often the case that when the proofs we gave here fail to extend to some more general notion of inductive definability, the results we have proved are still true. One would guess that almost any reasonable notion of inductive definability will satisfy appropriate versions of the very elementary structure properties we have been studying in these first three chapters.

Exercises for Chapter 3

3.1. Let $R(\bar{x})$ be a Π_1^1 relation on $\mathbb{N}$. Prove that there is an elementary $P(u, v, \bar{x})$ such that:

(a) *For each* $\bar{x}$, $\{(u, v): P(u, v, \bar{x})\}$ *is a linear ordering.*

(b) $R(\bar{x}) \Leftrightarrow \{(u, v): P(u, v, \bar{x})\}$ *is a wellordering.*

HINT: Use the representation of (1.12) and in the notation of that problem put

$$\begin{aligned} P(u, v, \bar{x}) \Leftrightarrow \textit{for suitable } & u_1, \ldots, u_m, v_1, \ldots, v_k, \\ & u = \langle u_1, \ldots, u_m\rangle \;\&\; v = \langle v_1, \ldots, v_k\rangle \\ & \&\; \neg\psi(u, \bar{x}) \;\&\; \neg\psi(v, \bar{x}) \\ & \&\; \{[u_1 < v_1] \\ & \vee\; [u_1 = v_1 \;\&\; u_2 < v_2] \\ & \vee\; \ldots \\ & \vee\; [u_1 = v_1 \;\&\; u_2 = v_2 \;\&\ldots\&\; u_k = v_k \;\&\; m > k]\}. \end{aligned}$$ ⊣

3.2. Using the representation in Exercise 3.1 for a Π_1^1 relation on $\mathbb{N}$, put

$$\sigma(\bar{x}) = \textit{order type of } \{(u, v): P(u, v, \bar{x})\} \quad \textit{if } \bar{x} \in R.$$

Prove that σ is an inductive norm on R.

HINT. Use Exercise 3.1 and the obvious way of comparing wellorderings. ⊣

This is the classical easy proof for the Prewellordering Theorem on $\mathbb{N}$. We will see in Chapter 8 that it cannot be extended to many other structures.

3.3. Suppose $\prec_1, \prec_2$ are hyperelementary wellfounded relations on the infinite structure $\mathfrak{A}$ such that $\textit{rank}(\prec_1) \leqslant \textit{rank}(\prec_2)$. Prove that the relation

$$P(\bar{x}, \bar{y}) \Leftrightarrow \bar{x} \in \textit{Field}(\prec_1) \;\&\; \bar{y} \in \textit{Field}(\prec_2) \;\&\; \rho^{\prec_1}(\bar{x}) = \rho^{\prec_2}(\bar{y})$$

is hyperelementary. ⊣

We will use the following convenient notation in the next two problems: if $R \subseteq A^n \times A^n$, put

$$\textit{rank}(R) = \begin{cases} \textit{rank}(R) & \textit{if } R \textit{ is wellfounded,} \\ |A|^+ & \textit{if } R \textit{ is not wellfounded.} \end{cases}$$

For infinite A, clearly

$$R \textit{ is wellfounded} \Leftrightarrow \textit{rank}(R) < |A|^+.$$

3.4. Let $P_1(\bar{u}_1, \bar{v}_1, \bar{x}_1)$, $P_2(\bar{u}_2, \bar{v}_2, \bar{x}_2)$ be hyperelementary relations on the infinite structure $\mathfrak{A}$. Prove that the relations

$$\begin{aligned} Q(\bar{x}_1, \bar{x}_2) \Leftrightarrow & \{(\bar{u}_1, \bar{v}_1) : P_1(\bar{u}_1, \bar{v}_1, \bar{x}_1)\} \textit{ is wellfounded} \\ & \,\&\, \textit{rank}(\{(\bar{u}_1, \bar{v}_1) : P(\bar{u}_1, \bar{v}_1, \bar{x}_1)\}) \\ & \quad \leqslant \textit{rank}(\{(\bar{u}_2, \bar{v}_2) : P_2(\bar{u}_2, \bar{v}_2, \bar{x}_2)\}), \\ R(\bar{x}_1, \bar{x}_2) \Leftrightarrow & \{(\bar{u}_1, \bar{v}_1) : P_1(\bar{u}_1, \bar{v}_1, \bar{x}_1)\} \textit{ is wellfounded} \\ & \,\&\, \textit{rank}(\{(\bar{u}_1, \bar{v}_1) : P_1(\bar{u}_1, \bar{v}_1, \bar{x}_1)\}) \\ & \quad < \textit{rank}(\{(\bar{u}_2, \bar{v}_2) : P_2(\bar{u}_2, \bar{v}_2, \bar{x}_2)\}) \end{aligned}$$

are inductive. ⊣

3.5. Prove that if $P(\bar{u}, \bar{v}, \bar{x})$ is hyperelementary on the infinite structure $\mathfrak{A}$ and $\xi < \kappa^{\mathfrak{A}}$, then the relation

$$Q(\bar{x}) \Leftrightarrow \textit{rank}(\{(\bar{u}, \bar{v}) : P(\bar{u}, \bar{v}, \bar{x})\}) \leqslant \xi$$

is hyperelementary. ⊣

3.6. Let $\varphi(\bar{x}, S)$ be an S-positive formula in the language over an infinite set A. Prove that the set I_φ built up by φ is *the unique* relation P which admits a norm $\sigma : P \twoheadrightarrow \lambda$ such that for every $\bar{x}$,

$$\bar{x} \in P \Leftrightarrow \varphi(\bar{x}, \{\bar{y} : \bar{y} <_{\sigma}^{*} \bar{x}\}).$$ ⊣

CHAPTER 4

GAMES AND GAME QUANTIFIERS

We establish here a basic connection between positive inductive definitions and open, infinite games. This is useful conceptually for the new insight that it brings into inductive definability, but it will also prove a powerful technical tool.

4A. Interpreting quantifier strings via games

Suppose R is an n-ary relation on some set A and

$$\bar{\mathrm{Q}} = \mathrm{Q}_1, \mathrm{Q}_2, \ldots, \mathrm{Q}_n$$

is a string of n quantifiers, i.e. each Q_i is $\exists$ or $\forall$. With $\bar{\mathrm{Q}}$ and R we associate the *two person perfect information game* $G(\bar{\mathrm{Q}}, R)$, described as follows. There are two players, call them $(\exists)$ and $(\forall)$ and a *run* of the game consists of their choosing a sequence $x_1, \ldots, x_n$ of elements of A. For each i, $1 \leqslant i \leqslant n$, x_i is chosen by $(\exists)$ if $\mathrm{Q}_i = \exists$ and x_i is chosen by $(\forall)$ if $\mathrm{Q}_i = \forall$. The game is one *of perfect information* in that the player who chooses x_i is allowed to see $x_1, \ldots, x_{i-1}$ before he makes his move. At the end of the run, $(\exists)$ wins if $R(x_1, \ldots, x_n)$ and $(\forall)$ wins if $\neg R(x_1, \ldots, x_n)$.

A *winning strategy* for one of the players is a systematic way of playing which will produce a win for that player in every run of the game. More precisely, a *strategy* for player (Q) ($\mathrm{Q} = \exists$ or $\mathrm{Q} = \forall$) is a set

$$\mathscr{S} = \{f_i \colon \mathrm{Q}_i = \mathrm{Q}\}$$

of functions, one for each i at which it is (Q)'s turn to play and such that f_i has $i-1$ arguments. (A function of 0 arguments is simply an element of A.) The player (Q) *follows* strategy $\mathscr{S}$ if he plays

$$x_i = f_i(x_1, \ldots, x_{i-1})$$

for each i such that $\mathrm{Q}_i = \mathrm{Q}$. We call $\mathscr{S}$ a *winning strategy* for (Q) if (Q) wins every run in which he follows $\mathscr{S}$.

For example, if

$$\bar{\mathrm{Q}} = \exists, \forall, \forall, \exists,$$

then a winning strategy for ∃ is any pair $f_1 \in A$, $f_4: A \times A \times A \to A$ such that

$$(\forall x_2)(\forall x_3)R(f_1, x_2, x_3, f_4(f_1, x_2, x_3)),$$

i.e. any pair of *Skolem functions* for the assertion

$$(\bar{\mathrm{Q}}\bar{x})R(\bar{x}) \Leftrightarrow (\exists x_1)(\forall x_2)(\forall x_3)(\exists x_4)R(x_1, x_2, x_3, x_4).$$

This is true for an arbitrary string $\bar{\mathrm{Q}}$ and an arbitrary relation R, so that in fact

$$\begin{aligned}(\bar{\mathrm{Q}}\bar{x})R(\bar{x}) &\Leftrightarrow (\mathrm{Q}_1 x_1) \ldots (\mathrm{Q}_n x_n)R(x_1, \ldots, x_n)\\ &\Leftrightarrow (\exists)\ \textit{has a winning strategy in } G(\bar{\mathrm{Q}}, R)\\ &\Leftrightarrow (\exists)\ \textit{wins } G(\bar{\mathrm{Q}}, R).\end{aligned}$$

Notice the distinction between *winning a run of* $G(\bar{\mathrm{Q}}, R)$ and *winning* $G(\bar{\mathrm{Q}}, R)$, i.e. having a winning strategy in $G(\bar{\mathrm{Q}}, R)$.

One nice thing about this interpretation of truth for assertions in prenex form is that it works equally well for infinite strings of quantifiers. If

$$\bar{\mathrm{Q}} = \mathrm{Q}_0, \mathrm{Q}_1, \ldots, \mathrm{Q}_i, \ldots$$

is any infinite string and $R \subseteq {}^{\omega}A$ is any relation of infinitely many arguments on A, we may define the game $G(\bar{\mathrm{Q}}, R)$ exactly as before, only now it is an infinite game: a run of it produces an infinite sequence $f = (x_0, x_1, \ldots, x_i, \ldots)$ and (∃) wins the run precisely if $R(f)$. Strategies are defined exactly as in the finite case, except of course that $\{i: \mathrm{Q}_i = \mathrm{Q}\}$ may be infinite. And as before,

$$\begin{aligned}(\bar{\mathrm{Q}}f)R(f) &\Leftrightarrow \{(\mathrm{Q}_0 x_0)(\mathrm{Q}_1 x_1) \ldots (\mathrm{Q}_i x_i) \ldots\}R(x_0, x_1, \ldots, x_i, \ldots)\\ &\Leftrightarrow (\exists)\ \textit{wins } G(\bar{\mathrm{Q}}, R);\end{aligned}$$

this can be taken as the definition of truth for assertions in prenex form with an infinite prefix, or it can be easily verified if another definition is given, e.g. via Skolem functions.

Infinite strings of quantifiers and their game-theoretic interpretation have been studied in Henkin [1961] and Keisler [1965]. The advantage of considering such games is that they often provide a direct and intuitive understanding of complicated arguments involving strings of quantifiers, whether infinite or finite. As a simple example, consider the equivalence

$$\begin{aligned}(\mathrm{Q}_0 x_0) \ldots (\mathrm{Q}_{n-1} x_{n-1})\{(\mathrm{Q}_n x_n)(\mathrm{Q}_{n+1} x_{n+1}) \ldots\}R(x_0, x_1, \ldots, x_n, x_{n+1}, \ldots)\\ \Leftrightarrow \{(\mathrm{Q}_0 x_0)(\mathrm{Q}_1 x_1) \ldots (\mathrm{Q}_n x_n) \ldots\}\quad R(x_0, x_1, \ldots, x_n, x_{n+1}, \ldots),\end{aligned}$$

which asserts that our interpretation of quantification by infinite strings allows for *absorption* of finite strings at the beginning. The left-hand side means that (∃) wins the game

$$G(Q_0, \ldots, Q_{n-1}, P)$$

determined by the string $Q_0, Q_1, \ldots, Q_{n-1}$ and the relation

$$P(x_0, x_1, \ldots, x_{n-1}) \Leftrightarrow \{(Q_n x_n)(Q_{n+1} x_{n+1}) \ldots\} R(x_0, x_1, \ldots, x_n, x_{n+1}, \ldots),$$

while the right-hand side means that (∃) wins the game

$$G(\bar{Q}, R)$$

determined by the string $\bar{Q} = Q_0, Q_1, \ldots, Q_n, Q_{n+1}, \ldots$ and the relation R. Assume the left-hand side; now (∃) can win $G(\bar{Q}, R)$ by playing first to win $G(Q_0, \ldots, Q_{n-1}, P)$ and once $x_0, x_1, \ldots, x_{n-1}$ have been played, then playing to win the game that insures $P(x_0, x_1, \ldots, x_{n-1})$ is true. Similarly, if the right-hand side is true, then (∃) can win $G(Q_0, \ldots, Q_{n-1}, P)$ simply by playing to win $G(\bar{Q}, R)$ for the first n moves; because once $x_0, x_1, \ldots, x_{n-1}$ have been determined in this way, then (∃) can go on to play and insure that $R(x_0, x_1, \ldots, x_n, x_{n+1}, \ldots)$ holds, which means precisely that $P(x_0, x_1, \ldots, x_{n-1})$ is true.

A similar but much simpler argument shows that if $P(y_1, \ldots, y_m)$, $R(y_1, \ldots, y_m, x_0, x_1, x_2, \ldots)$ are relations on A, the first m-ary, the second of infinitely many variables, then

$$P(y_1, \ldots, y_m) \mathbin{\&} \{(Q_0 x_0)(Q_1 x_1) \ldots\} R(y_1, \ldots, y_m, x_0, x_1, \ldots)$$
$$\Leftrightarrow \{(Q_0 x_0)(Q_1 x_1) \ldots\}[P(y_1, \ldots, y_m) \mathbin{\&} R(y_1, \ldots, y_m, x_0, x_1, \ldots)]$$

and similarly for disjunction,

$$P(y_1, \ldots, y_m) \vee \{(Q_0 x_0)(Q_1 x_1) \ldots\} R(y_1, \ldots, y_m, x_0, x_1, \ldots)$$
$$\Leftrightarrow \{(Q_0 x_0)(Q_1 x_1) \ldots\}[P(y_1, \ldots, y_m) \vee R(y_1, \ldots, y_m, x_0, x_1, \ldots)].$$

These simple facts suggest that we can treat infinite strings formally much as we manipulate finite strings. We will often use these properties tacitly, often with a vague reference to "ordinary logic".

We must be careful though, because not all formal rules that are obeyed by finite strings hold also for the infinite ones. The main exception is the transformation which allows us to push the negation sign through a string,

$$\neg(\forall x)(\exists y)R(x, y) \Leftrightarrow (\exists x)(\forall y)\neg R(x, y).$$

We now look into this a bit more carefully.

The *dual* of a finite or infinite string

$$\bar{Q} = Q_0, Q_1, \ldots, Q_i, \ldots$$

is defined by

$$\bar{Q}^{\cup} = Q_0^{\cup}, Q_1^{\cup}, \ldots, Q_i^{\cup}, \ldots,$$

where

$$\exists^{\cup} = \forall, \qquad \forall^{\cup} = \exists.$$

From the interpretation we have immediately

$$(\forall) \textit{ wins } G(\bar{Q}, R) \Leftrightarrow (\exists) \textit{ wins } G(\bar{Q}^{\cup}, \neg R),$$

$$(\exists) \textit{ wins } G(\bar{Q}, R) \Leftrightarrow (\forall) \textit{ wins } G(\bar{Q}^{\cup}, \neg R).$$

Hence *if either* $(\forall)$ *or* $(\exists)$ *wins* $G(\bar{Q}, R)$, we have

$$\begin{aligned}\neg(\bar{Q}f)R(f) &\Leftrightarrow (\exists) \textit{ does not win } G(\bar{Q}, R)\\ &\Leftrightarrow (\forall) \textit{ wins } G(\bar{Q}, R)\\ &\Leftrightarrow (\exists) \textit{ wins } G(\bar{Q}^{\cup}, \neg R)\\ &\Leftrightarrow (\bar{Q}^{\cup}f)\neg R(f),\end{aligned}$$

i.e. we can push the negation through a string of quantifiers by changing the string to its dual.

If $\bar{Q}$ is a finite string, then surely either $(\bar{Q}\bar{x})R(\bar{x})$ or $(\bar{Q}^{\cup}\bar{x})\neg R(\bar{x})$ by ordinary logic, so that either $(\exists)$ or $(\forall)$ wins $G(\bar{Q}, R)$. This is not true for all infinite strings and all R as is shown in Gale and Stewart [1953], but it does hold in the important special cases of open or closed R.

An infinitary relation $R \subseteq^{\omega} A$ is *open* if for suitably chosen $R_0, R_1, \ldots, R_i, \ldots$,

$$\begin{aligned}R(x_0, x_1, \ldots, x_i, \ldots) &\Leftrightarrow R_0(x_0) \vee R_1(x_0, x_1) \vee \ldots \vee R_i(x_0, x_1, \ldots, x_i) \vee \ldots\\ &\Leftrightarrow \textstyle\bigvee_{i\in\omega} R_i(x_0, \ldots, x_i),\end{aligned}$$

i.e. R can be written as an infinite disjunction of finitary relations. Similarly, $R \subseteq^{\omega} A$ is *closed* if it can be written as an infinite conjunction of finitary relations,

$$\begin{aligned}R(x_0, x_1, \ldots, x_i, \ldots) &\Leftrightarrow R_0(x_0) \,\&\, R_1(x_0, x_1) \,\&\ldots\&\, R_i(x_0, x_1, \ldots, x_1) \,\&\ldots\\ &\Leftrightarrow \textstyle\bigwedge_{i\in\omega} R_i(x_0, \ldots, x_i).\end{aligned}$$

4A.1. GALE–STEWART THEOREM. *If* $R \subseteq {}^{\omega}A$ *is either open or closed and* $\bar{Q} = Q_0, Q_1, \ldots, Q_i, \ldots$ *is an infinite string of quantifiers, then either* $(\exists)$ *or* $(\forall)$ *wins* $G(\bar{Q}, R)$ *and hence*

$$\neg(\bar{Q}f)R(f) \Leftrightarrow (\bar{Q}^{\cup}f)\neg R(f).$$

PROOF. It is enough to consider the case of open R; because if R is closed, then $\neg R$ is open and hence either $(\exists)$ or $(\forall)$ wins $G(\bar{Q}^{\cup}, \neg R)$, i.e. either $(\forall)$ or $(\exists)$ wins $G(\bar{Q}, R)$. Assume then that

$$R(x_0, x_1, \ldots) \Leftrightarrow R_0(x_0) \vee R_1(x_0, x_1) \vee \ldots$$

and

$$(1) \qquad \neg\{(\mathrm{Q}_0x_0)(\mathrm{Q}_1x_1)\ldots\}[R_0(x_0) \vee R_1(x_0, x_1) \vee \ldots],$$

i.e. (∃) has no winning strategy in $G(\bar{\mathrm{Q}}, R)$. We will describe informally a strategy for (∀) so that at each step i of the game, when $x_0, x_1, \ldots, x_i$ have been determined,

$$(2)_i \qquad \neg R_0(x_0), \neg R_1(x_0, x_1), \ldots, \neg R_i(x_0, x_1, \ldots, x_i),$$

$$(3)_i \qquad \neg\{(\mathrm{Q}_{i+1}x_{i+1})(\mathrm{Q}_{i+2}x_{i+2})\ldots\}[R_{i+1}(x_0, \ldots, x_{i+1}) \vee R_{i+2}(x_0, \ldots, x_{i+2}) \vee \ldots].$$

To start the game, we can pull out the first quantifier from (1),

$$\neg(\mathrm{Q}_0x_0)\{(\mathrm{Q}_1x_1)(\mathrm{Q}_2x_2)\ldots\}[R_0(x_0) \vee R_1(x_0, x_1) \vee R_2(x_0, x_1, x_2) \vee \ldots],$$

whence we get by ordinary logic

$$(\mathrm{Q}_0^{\smile}x_0)\neg\{(\mathrm{Q}_1x_1)(\mathrm{Q}_2x_2)\ldots\}[R_0(x_0) \vee R_1(x_0, x_1) \vee R_2(x_0, x_1, x_2) \vee \ldots].$$

If $\mathrm{Q}_0 = \exists$, then $\mathrm{Q}_0^{\smile} = \forall$, so no matter which x_0 (∃) picks we have

$$(4) \quad \neg\{(\mathrm{Q}_1x_1)(\mathrm{Q}_2x_2)\ldots\}[R_0(x_0) \vee R_1(x_0, x_1) \vee R_2(x_0, x_1, x_2) \vee \ldots].$$

If $\mathrm{Q}_0 = \forall$, then $\mathrm{Q}_0^{\smile} = \exists$, so (∀) can play some x_0 so that (4) holds. Hence at step 0 of the game we have (4), which by ordinary logic implies

$$\neg R_0(x_0),$$

$$\neg\{(\mathrm{Q}_1x_1)(\mathrm{Q}_2x_2)\ldots\}[R_1(x_0, x_1) \vee R_2(x_0, x_1, x_2) \vee \ldots],$$

i.e. precisely $(2)_0$, $(3)_0$.

The argument is similar for arbitrary i. Assuming $(2)_i$, $(3)_i$, either $\mathrm{Q}_{i+1} = \exists$ and then no matter which x_{i+1} is picked $(2)_{i+1}$, $(3)_{i+1}$ hold, or $\mathrm{Q}_{i+1} = \forall$ and then (∀) can pick some x_{i+1} so $(2)_{i+1}$, $(3)_{i+1}$ hold. At the end of the run all $(2)_i$ hold, which implies that (∀) has won. ⊣

4B. A canonical form for positive formulas

The key to the connection between positive inductive definitions and open games is the following simple result of predicate logic.

4B.1. Canonical form for positive formulas. *Let $\varphi(S)$ be an S-positive formula in the language $\mathscr{L}^A$ over a set A, where S is n-ary. Then there is a quantifier free formula*

$$\theta(\bar{z}, \bar{u}) \equiv \theta(z_1, \ldots, z_m, u_1, \ldots, u_n)$$

with free variables $\bar{z}, \bar{u}$ *and the free variables of* φ, *and a string*

$$\bar{\mathrm{Q}} = \mathrm{Q}_1, \ldots, \mathrm{Q}_m$$

of quantifiers, such that whenever $S \subsetneqq A^n$,

$$\varphi(S) \Leftrightarrow (\mathrm{Q}_1 z_1) \ldots (\mathrm{Q}_m z_m)(\forall u_1) \ldots (\forall u_n)[\theta(z_1, \ldots, z_m, u_1, \ldots, u_n) \vee S(u_1, \ldots, u_n)];$$

in abbreviated form, for $S \subsetneqq A^n$,

$$(*) \qquad \varphi(S) \Leftrightarrow (\bar{\mathrm{Q}}\bar{z})(\forall \bar{u})[\theta(\bar{z}, \bar{u}) \vee S(\bar{u})].$$

PROOF is by induction on the definition of formulas in which S occurs positively.

Case 1: S does not occur in φ. Then for $S \subsetneqq A^n$,

$$\varphi \Leftrightarrow [\varphi \vee (\forall \bar{u})S(\bar{u})]$$

from which we get (*) by putting φ in prenex normal form and bringing all the quantifiers in its prefix and then $(\forall \bar{u})$ to the front.

Case 2: $\varphi(S) \equiv S(\bar{t})$. Then

$$\varphi(S) \Leftrightarrow (\forall \bar{u})[\bar{u} \neq \bar{t} \vee S(\bar{u})].$$

Case 3: $\varphi(S) \equiv (\mathrm{Q}x)\psi(x, S)$, where $\mathrm{Q} = \exists$ or $\mathrm{Q} = \forall$. By induction hypothesis there is some θ and some $\bar{\mathrm{Q}}$ so that if $S \subsetneqq A^n$, then

$$\psi(x, S) \Leftrightarrow (\bar{\mathrm{Q}}\bar{z})(\forall \bar{u})[\theta(x, \bar{z}, \bar{u}) \vee S(\bar{u})],$$

from which

$$\varphi(S) \Leftrightarrow (\mathrm{Q}x)(\bar{\mathrm{Q}}\bar{z})(\forall \bar{u})[\theta(x, \bar{z}, \bar{u}) \vee S(\bar{u})]$$

follows immediately.

Case 4: $\varphi(S) \equiv \varphi_1(S) \;\&\; \varphi_2(S)$ or $\varphi(S) \equiv \varphi_1(S) \vee \varphi_2(S)$. By induction hypothesis there are $\theta_1, \theta_2, \bar{\mathrm{Q}}_1, \bar{\mathrm{Q}}_2$ such that

$$\varphi_1(S) \Leftrightarrow (\bar{\mathrm{Q}}_1\bar{z}_1)(\forall \bar{u}_1)[\theta_1(\bar{z}_1, \bar{u}_1) \vee S(\bar{u}_1)],$$

$$\varphi_2(S) \Leftrightarrow (\bar{\mathrm{Q}}_2\bar{z}_2)(\forall \bar{u}_2)[\theta_2(\bar{z}_2, \bar{u}_2) \vee S(\bar{u}_2)],$$

for $S \subsetneqq A^n$, where we may assume that all the variables in the list $\bar{z}_1, \bar{z}_2, \bar{u}_1, \bar{u}_2$ are distinct.

The case for conjunction is trivial, since for $S \subsetneqq A^n$ we clearly have

$$\varphi_1(S) \;\&\; \varphi_2(S) \Leftrightarrow (\bar{\mathrm{Q}}_1\bar{z}_1)(\bar{\mathrm{Q}}_2\bar{z}_2)(\forall \bar{u})\{[\theta_1(\bar{z}_1, \bar{u}) \;\&\; \theta_2(\bar{z}_2, \bar{u})] \vee S(\bar{u})\}.$$

For the case of disjunction, first verify by direct inspection that if $S \subsetneq A^n$, then

$$S(\bar{u}_1) \vee S(\bar{u}_2) \Leftrightarrow (\exists\bar{z})(\forall\bar{u})\{[(\bar{z} = \bar{u}_1 \vee \bar{z} = \bar{u}_2) \mathbin{\&} \bar{z} \neq \bar{u}] \vee S(\bar{u})\};$$

here $\bar{z}$, $\bar{u}$ are fresh lists of variables of length n. Now take

$$\theta(\bar{z},\bar{z}_1,\bar{u}_1,\bar{z}_2,\bar{u}_2,\bar{u}) \Leftrightarrow \theta_1(\bar{z}_1,\bar{u}_1) \vee \theta_2(\bar{z}_2,\bar{u}_2) \vee [(\bar{z} = \bar{u}_1 \vee \bar{z} = \bar{u}_2) \mathbin{\&} \bar{z} \neq \bar{u}]$$

and verify directly that for $S \subsetneq A^n$,

$$\varphi_1(S) \vee \varphi_2(S) \Leftrightarrow (\bar{Q}_1\bar{z}_1)(\bar{Q}_2\bar{z}_2)(\forall\bar{u}_1)(\forall\bar{u}_2)(\exists\bar{z})(\forall\bar{u})[\theta(\bar{z},\bar{z}_1,\bar{u}_1,\bar{z}_2,\bar{u}_2,\bar{u}) \vee S(\bar{u})]. \qquad \dashv$$

The restriction in equivalence (*) to $S \subsetneq A^n$ is essential, since the right-hand side is automatically true if $S = A^n$ while the left-hand side may be false, e.g. if $\varphi(S)$ is a false formula in which S does not occur. However, Lemma 2A.1 implies that for the purpose of studying the induction determined by a formula $\varphi(\bar{x}, S)$ we may as well assume that $\varphi(\bar{x}, A^n)$ is true, in which case $\varphi(\bar{x}, S)$ is equivalent to a formula in canonical positive form for all S.

4C. Explicit formulas for inductive relations

4C.1. THEOREM. *Let A be an infinite set and*

$$\varphi(\bar{x}, S) \equiv (\bar{Q}\bar{z})(\forall\bar{u})[\theta(\bar{x},\bar{z},\bar{u}) \vee S(\bar{u})]$$

a formula in the language $\mathcal{L}^A$ over A, where $\bar{x} = x_1, \ldots, x_n$, $\bar{z} = z_1, \ldots, z_m$, $\bar{u} = \bar{u}_1, \ldots, u_n$, S is an n-ary relation symbol and $\bar{Q} = Q_1, \ldots, Q_m$ is a string of m quantifiers, let I_φ be the relation built up by φ. Then

$$\text{(1)} \quad \bar{x} \in I_\varphi \Leftrightarrow \{(\bar{Q}\bar{z}_1)(\forall\bar{u}_1)(\bar{Q}\bar{z}_2)(\forall\bar{u}_2)\ldots\}[\theta(\bar{x},\bar{z}_1,\bar{u}_1) \vee \theta(\bar{u}_1,\bar{z}_2,\bar{u}_2) \vee \theta(\bar{u}_2,\bar{z}_3,\bar{u}_3) \vee \ldots].$$

PROOF. We first prove direction ($\Leftarrow$) of (1) by showing its contrapositive,

$$\bar{x} \notin I_\varphi \Rightarrow \neg\{(\bar{Q}\bar{z}_1)(\forall\bar{u}_1)(\bar{Q}\bar{z}_2)(\forall\bar{u}_2)\ldots\}[\theta(\bar{x},\bar{z}_1,\bar{u}_1) \vee \theta(\bar{u}_1,\bar{z}_2,\bar{u}_2) \vee \ldots].$$

For this it will be sufficient to assume $\bar{x} \notin I_\varphi$ and then describe a winning strategy for ($\forall$) in the game determined by the right-hand side of (1), call it G.

Since I_φ is a fixed point of the operator defined by $\varphi(\bar{x}, S)$, we have

$$\bar{x} \in I_\varphi \Leftrightarrow \varphi(\bar{x}, I_\varphi),$$

so taking negations and using the canonical form for $\varphi(\bar{x}, S)$,

$$\text{(2)} \quad \bar{x} \notin I_\varphi \Leftrightarrow (\bar{Q}^\cup\bar{z})(\exists\bar{u})[\neg\theta(\bar{x},\bar{z},\bar{u}) \mathbin{\&} \bar{u} \notin I_\varphi].$$

Now by assumption the right-hand side of (2) holds, so (∃) has a winning strategy for the finite game determined by it. *Let* (∀) *play by this strategy in G until* $\bar{z}_1, \bar{u}_1$ *are determined.* We then have

$$\neg\theta(\bar{x}, \bar{z}_1, \bar{u}_1) \;\&\; \bar{u}_1 \notin I_\varphi,$$

so the right-hand side of (2) holds with $\bar{x} = \bar{u}_1$ and (∀) can play again following (∃)'s winning strategy in this game until $\bar{z}_2, \bar{u}_2$ are determined so that

$$\neg\theta(\bar{u}_1, \bar{z}_2, \bar{u}_2) \;\&\; \bar{u}_2 \notin I_\varphi.$$

It is clear that if (∀) continues to play in this manner by the strategies guaranteed to him by (2), he will insure that

$$\neg\theta(\bar{x}, \bar{z}_1, \bar{u}_1) \;\&\; \neg\theta(\bar{u}_1, \bar{z}_2, \bar{u}_2) \;\&\; \neg\theta(\bar{u}_2, \bar{z}_3, \bar{u}_3) \;\&\; \ldots$$

and hence win G.

Putting

$$R(\bar{x}) \Leftrightarrow \{(\bar{Q}\bar{z}_1)(\forall\bar{u}_1)(\bar{Q}\bar{z}_2)(\forall\bar{u}_2)\ldots\}[\theta(\bar{x}, \bar{z}_1, \bar{u}_1) \vee \theta(\bar{u}_1, \bar{z}_2, \bar{u}_2) \vee \ldots],$$

we have now shown that

$$R \subseteq I_\varphi. \tag{3}$$

To complete the proof it will be enough to verify that R is a fixed point of $\varphi(\bar{x}, S)$, since I_φ is the least fixed point of $\varphi(\bar{x}, S)$, so that we will then also have $I_\varphi \subseteq R$. We compute:

$$\begin{aligned} \varphi(\bar{x}, R) &\Leftrightarrow (\bar{Q}\bar{z})(\forall\bar{u})[\theta(\bar{x}, \bar{z}, \bar{u}) \vee R(\bar{u})] \\ &\Leftrightarrow (\bar{Q}\bar{z})(\forall\bar{u})[\theta(\bar{x}, \bar{z}, \bar{u}) \vee \{(\bar{Q}\bar{z}_1)(\forall\bar{u}_1)\ldots\}[\theta(\bar{u}, \bar{z}_1, \bar{u}_1) \\ &\qquad\qquad \vee \theta(\bar{u}_1, \bar{z}_2, \bar{u}_2) \vee \ldots]] \\ &\Leftrightarrow (\bar{Q}\bar{z})(\forall\bar{u})[\{(\bar{Q}\bar{z}_1)(\forall\bar{u}_1)\ldots\}[\theta(\bar{x}, \bar{z}, \bar{u}) \vee \theta(\bar{u}, \bar{z}_1, \bar{u}_1) \vee \ldots]] \\ &\Leftrightarrow \{(\bar{Q}\bar{z})(\forall\bar{u})(\bar{Q}\bar{z}_1)(\forall\bar{u}_1)\ldots\}[\theta(\bar{x}, \bar{z}, \bar{u}) \vee \theta(\bar{u}, \bar{z}_1, \bar{u}_1) \vee \ldots] \\ &\Leftrightarrow R(\bar{x}), \end{aligned}$$

where the last two steps are by the trivial logical properties of infinite quantifier strings which we discussed in Section 4A. ⊣

As an immediate corollary of this theorem we get an elegant characterization of inductive relations on a structure.

Recall that P is a *fixed point on* $\mathfrak{A} = \langle A, R_1, \ldots, R_l\rangle$ if there is some S-positive formula $\varphi(\bar{x}, S)$ in the language of $\mathfrak{A}$ such that $P = I_\varphi$.

4C.2. Fixed Point Normal Form Theorem. *A relation P on an infinite structure $\mathfrak{A}$ is a fixed point if and only if there is a formula*

$$\theta(\bar{x}, \bar{z}, \bar{u}) = \theta(x_1, \ldots, x_n, z_1, \ldots, z_m, u_1, \ldots, u_n)$$

in the language of $\mathfrak{A}$ *and a string of quantifiers*

$$\bar{Q} = Q_1, \ldots, Q_m$$

such that

$$(*)\quad P(\bar{x}) \Leftrightarrow \{(\bar{Q}\bar{z}_1)(\forall \bar{u}_1)(\bar{Q}\bar{z}_2)(\forall \bar{u}_2)\ldots\}[\theta(\bar{x}, \bar{z}_1, \bar{u}_1) \vee \theta(\bar{u}_1, \bar{z}_2, \bar{u}_2) \vee \ldots]$$

$$\Leftrightarrow \{(\bar{Q}\bar{z}_1)(\forall \bar{u}_1)(\bar{Q}\bar{z}_2)(\forall \bar{u}_2)\ldots\}[\theta(\bar{x}, \bar{z}_1, \bar{u}_1) \vee \bigvee_{i \geqslant 1} \theta(\bar{u}_i, \bar{z}_{i+1}, \bar{u}_{i+1})].$$

In fact, every fixed point satisfies (*) *with a quantifier free* θ.

PROOF. If P is a fixed point, then $P = I_\varphi$ and we can assume by 4B.1 and the Stage Comparison Theorem 2A.1 that

$$(4)\qquad \varphi(\bar{x}, S) \Leftrightarrow (\bar{Q}\bar{z})(\forall \bar{u})[\theta(\bar{x}, \bar{z}, \bar{u}) \vee S(\bar{u})]$$

for some quantifier free θ, from which (*) follows by 4C.1. On the other hand, if (*) holds with any θ, then we can take $\varphi(\bar{x}, S)$ to be defined by (4) and (*) follows again by Theorem 4C.1. ⊣

The result of course gives immediately a normal form for inductive relations which is a bit more complicated than that for fixed points.

These normal forms are in terms of infinite formulas, but we should emphasize the regularity of the infinite strings of quantifiers and the infinite disjunctions involved. Looking at (*), the infinite string is *repeating*, i.e. of the form

$$\bar{Q}, \forall^n, \bar{Q}, \forall^n, \bar{Q}, \forall^n, \ldots,$$

where $\bar{Q}$ is a string of length m and $\forall^n = \forall\forall \ldots \forall$ (n times). The matrix is an infinite disjunction of substitution instances of the same elementary formula, the substitutions themselves being of an obvious regular pattern. It is this canonical form of the infinitary expression in (*) which allows us to prove that any P thus defined is a fixed point.

Exercises for Chapter 4

4.1. Prove that every Π^1_1 relation $P(\bar{x})$ on the structure $\mathbb{N}$ satisfies

$$P(\bar{x}) \Leftrightarrow \{(\bar{Q}\bar{z}_1)(\forall \bar{y}_1)(\bar{Q}\bar{z}_2)(\forall \bar{y}_2)\ldots\}[\theta(\bar{a}, \bar{x}, \bar{z}_1, \bar{y}_1) \vee \theta(\bar{y}_1, \bar{z}_2, \bar{y}_2) \vee \ldots]$$

with a string $\bar{Q} = Q_1, \ldots, Q_m$ of quantifiers, a sequence $\bar{a} = a_1, \ldots, a_k$ of integers and a quantifier free formula

$$\theta(\bar{u}, \bar{z}, \bar{y}) \Leftrightarrow \theta(u_1, \ldots, u_l, z_1, \ldots, z_m, y_1, \ldots, y_l)$$

built up only from the symbols $=$, $\leqslant$ and variables. ⊣

4.2. Let λ be an infinite cardinal, let

$$F: \lambda \twoheadrightarrow L_\lambda$$

be some canonical mapping of λ onto the sets constructible before λ, e.g. the function F in Gödel [1940]. Prove that the relation

$$P(\eta, \xi) \Leftrightarrow F(\eta) \in F(\xi)$$

is inductive on $\boldsymbol{\lambda} = \langle \lambda, \leqslant \rangle$. (You will need a bit of set theory to do this.) $\dashv$

4.3. Prove that if x is a set of ordinals, then

x *is constructible* $\Leftrightarrow (\exists\lambda)$ [x *is hyperelementary on* $\langle \lambda, \leqslant \rangle$]. $\dashv$

If $f: A \to A$ and $a \in A$, put

$$Orbit(f, a) = \{a, f(a), f(f(a)), f(f(f(a))), \ldots\}.$$

4.4. Prove that if $f: A \to A$ is hyperelementary on $\mathfrak{A} = \langle A, R_1, \ldots, R_k \rangle$, then for each $a \in A$ the set $Orbit(f, a)$ is inductive. $\dashv$

We outline the proofs of some simple model theoretic facts which will be useful for the construction of examples and counterexamples.

A *type* on x over a structure $\mathfrak{A}$ is a collection Φ of formulas in the language of $\mathfrak{A}$ such that every formula $\varphi \equiv \varphi(x)$ in Φ has no free variables, except perhaps x. We say that Φ is *finitely satisfiable* in $\mathfrak{A}$ if for every finite set $\varphi_1, \ldots, \varphi_n$ of formulas in Φ there is some c in A so that

$$\varphi_1(c) \;\&\; \varphi_2(c) \;\&\; \ldots \;\&\; \varphi_n(c)$$

is true. We say that Φ is *realized* in $\mathfrak{A}$ if there is a fixed c in A such that $\varphi(c)$ is true for every $\varphi(x)$ in Φ.

If $\mathscr{T}$ is a collection of types over $\mathfrak{A}$, we call $\mathfrak{A}$ *$\mathscr{T}$-saturated* if every type in $\mathscr{T}$ which is finitely satisfiable is realized. This is the interesting and useful notion. For example, $\mathfrak{A}$ is *$\aleph_0$-saturated* if $\mathfrak{A}$ is $\mathscr{T}$-saturated with

$\Phi \in \mathscr{T} \Leftrightarrow$ *there are fixed constants* $a_1, \ldots, a_k$ *such that the constants in every* $\varphi \in \Phi$ *are among* $a_1, \ldots, a_k$.

There are many theorems on the existence of saturated models, but for our purpose two simple ones will suffice.

4.5. Let $f: A \rightarrowtail\!\!\!\twoheadrightarrow A$ be a *permutation* on A (one-to-one, onto function) such that for each $a \in A$ both $Orbit(f, a)$ and $Orbit(f^{-1}, a)$ are infinite and there are infinitely many distinct orbits. Prove that the structure $\mathfrak{A} = \langle A, G_f \rangle$ is $\aleph_0$-saturated, where

$$G_f(x, y) \Leftrightarrow y = f(x).$$

HINT: Use the Compactness Theorem to prove that if $a_1, \ldots, a_k$ are the only constants in $\varphi(x, a_1, \ldots, a_k)$ and $\{x: \varphi(x, a_1, \ldots, a_k)\}$ contains infinitely many members of $Orbit(f, a_i) \cup Orbit(f^{-1}, a_i)$, then for every b not in $Orbit(f, a_j) \cup Orbit(f^{-1}, a_j)$, $j = 1, \ldots, k$, we have $Orbit(f, b) \cup Orbit(f^{-1}, b) \subseteq \{x: \varphi(x, a_1, \ldots, a_k)\}$. ⊣

A collection of formulas Ψ is a *pretype on* $x, v_1, \ldots, v_k$ *over* $\mathfrak{A}$ if every formula φ in Ψ has its free variables among $x, v_1, \ldots, v_k$. For each $a_1, \ldots, a_k$ in A, Ψ then defines the type

$$\Psi^{a_1, \ldots, a_k} = \{\varphi(x, a_1, \ldots, a_k): \varphi(x, v_1, \ldots, v_k) \in \Psi\}.$$

If $\mathscr{K}$ is a collection of pretypes in the language of $\mathfrak{A}$, then the *collection of types generated by* $\mathscr{K}$ is defined by

$$\mathscr{K}(\mathfrak{A}) = \{\Psi^{a_1, \ldots, a_k}: \Psi \textit{ is a pretype in } \mathscr{K} \textit{ on } x, v_1, \ldots, v_k \textit{ and } a_1, \ldots, a_k \in A\}.$$

4.6. Let $\mathfrak{A}$ be a countable structure and $\mathscr{K}$ a countable collection of pretypes on $\mathfrak{A}$. Prove that there is a countable elementary extension $\mathfrak{B}$ of $\mathfrak{A}$ which is $\mathscr{K}(\mathfrak{B})$-saturated.

HINT: Use the Henkin-type argument of adding witnesses for the proof of the Completeness Theorem to obtain a complete theory which extends the theory of $\mathfrak{A}$ and defines the needed structure. ⊣

We now come to the result which ties up these ideas with inductive definitions.

4.7. Prove that there is a fixed countable collection $\mathscr{K}$ of pretypes in the language with relation constants $R_1, \ldots, R_l$ and no individual constants such that whenever $\mathfrak{A} = \langle A, R_1, \ldots, R_l\rangle$ is $\mathscr{K}(\mathfrak{A})$-saturated, then for every S-positive $\varphi(\bar{x}, S)$ in the language of $\mathfrak{A}$,

$$I_\varphi = \bigcup_{n \in \omega} I_\varphi^n.$$

If we put φ in canonical form,

$$\varphi(\bar{x}, S) \Leftrightarrow (\bar{Q}\bar{z})(\forall\bar{u})[\theta(\bar{x}, \bar{z}, \bar{u}) \vee u \in S],$$

the above equation becomes

$$(*) \qquad x \in I_\varphi \Leftrightarrow (\bar{Q}\bar{z}_1)(\forall\bar{u}_1)\theta(\bar{x}, \bar{z}_1, \bar{u}_1) \vee (\bar{Q}\bar{z}_1)(\forall\bar{u}_1)(\bar{Q}\bar{z}_2)(\forall\bar{u}_2)[\theta(\bar{x}, \bar{z}_1, \bar{u}_1) \vee \theta(\bar{u}_1, \bar{z}_2, \bar{u}_2)] \vee \ldots.$$

(Keisler [1965].)

HINT: Prove first that (*) holds if $\mathfrak{A}$ is $\aleph_0$-saturated and then check your proof to see how many types you actually needed realized. ⊣

We will call $\mathfrak{A}$ *induction saturated* if it is $\mathscr{K}(\mathfrak{A})$-saturated with some $\mathscr{K}$ satisfying problem 4.6. Clearly $\aleph_0$-saturated structures are induction saturated.

4.8. Prove that there is a countable structure $\mathfrak{A}$ such that every hyperelementary relation on $\mathfrak{A}$ is elementary but there are nonelementary inductive relations on $\mathfrak{A}$.
HINT: Use an induction saturated structure. ⊣

4.9. Let G be a group, $a_1, \ldots, a_k \in G$ and let $[a_1, \ldots, a_k]$ be the subgroup generated by $a_1, \ldots, a_k$. Prove that $[a_1, \ldots, a_k]$ is inductive in the structure $\langle G, P \rangle$, where

$$P(x, y, z) \Leftrightarrow z = x \cdot y.$$

Give examples of infinite groups where $[a_1, \ldots, a_k]$ is always hyperelementary and where $[a_1, \ldots, a_k]$ need not be hyperelementary. ⊣

4.10. Let F be an algebraically closed field, $a_1, \ldots, a_k \in F$ and let $[a_1, \ldots, a_k]$ be the smallest algebraically closed subfield of F containing $a_1, \ldots, a_k$. Prove that $[a_1, \ldots, a_k]$ is inductive in the obvious field structure of F. Give examples where $[a_1, \ldots, a_k]$ is always hyperelementary and where $[a_1, \ldots, a_k]$ need not be hyperelementary. ⊣

In these two examples from algebra the inductions in question close at ω. The next problem gives a natural and useful example of an inductive definition in group theory which in general will not close at ω. It is due to K. J. Barwise.

4.11. A p-group is an abelian group G in which every element has order some power p^n of the prime number p. We call G *divisible* if for every x there is some y such that $py = x$. Every p-group has a largest divisible subgroup H; show that this H is coinductive in the natural group structure $\langle G, P \rangle$, but not in general hyperelementary. ⊣

The natural coinductive definition of the largest divisible subgroup plays a central role in the study of p-groups—e.g. see Kaplansky [1954].

CHAPTER 5

ACCEPTABLE STRUCTURES

The first four chapters just about cover all that is known now about positive elementary inductive definability on completely arbitrary structures. In order to prove some of the deeper and more interesting results of the theory, we must restrict ourselves to structures that satisfy certain definability conditions. For example, it is possible that every inductive set on $\mathfrak{A}$ is elementary (see Exercise 1.15) and hence we must place restrictions on $\mathfrak{A}$ to ensure the existence of nontrivial inductive sets.

The useful hypothesis seems to be that we can code finite sequences from A by single elements of A so that both the coding and decoding functions are elementary. We introduce here *acceptable* structures where this can be done and prove that in such structures there exist nontrivial hyperelementary sets as well as "universal" inductive relations.

Some of the ideas of this chapter have been introduced in the exercises at the end of Chapter 1.

5A. Coding schemes

A *coding scheme* for a set A is a triple

$$\mathscr{C} = \langle N^{\mathscr{C}}, \leqslant^{\mathscr{C}}, \langle\,\rangle^{\mathscr{C}}\rangle$$

such that:

(1) $N^{\mathscr{C}} \subseteq A$, $\leqslant^{\mathscr{C}}$ is an ordering on $N^{\mathscr{C}}$ and the structure $\langle N^{\mathscr{C}}, \leqslant^{\mathscr{C}}\rangle$ is isomorphic to the integers $\{0, 1, 2, \ldots\}$ with their usual ordering,

(2) $\langle\,\rangle^{\mathscr{C}}$ is a one-to-one function mapping the set $\bigcup_{n\geqslant 0} A^n$ of all finite sequences from A into A.

In the second condition we include the empty sequence in the domain of $\langle\,\rangle^{\mathscr{C}}$, as the only sequence of length 0, i.e. $A^0 = \{\emptyset\}$.

With each coding scheme there are naturally associated the following *decoding* relations and functions:

(3) $Seq^{\mathscr{C}}(x) \Leftrightarrow$ *for some* $x_1, \ldots, x_n$, $x = \langle x_1, \ldots, x_n\rangle^{\mathscr{C}}$ (where the case $x = \langle\emptyset\rangle^{\mathscr{C}}$ of the code of the empty sequence is covered by the convention that $x_1, \ldots, x_n = \emptyset$ if $n = 0$).

$$(4) \qquad lh^{\mathscr{C}}(x) = \begin{cases} 0 & \textit{if } \neg Seq^{\mathscr{C}}(x), \\ n & \textit{if } Seq^{\mathscr{C}}(x) \,\&\, x = \langle x_1, \ldots, x_n \rangle^{\mathscr{C}}. \end{cases}$$

Here $lh^{\mathscr{C}}$ maps A into $N^{\mathscr{C}}$, so that $0, 1, \ldots$ are the elements of $N^{\mathscr{C}}$ which correspond to the integers $0, 1, \ldots$ under the unique isomorphism of $\langle N^{\mathscr{C}}, \leqslant^{\mathscr{C}} \rangle$ with the integers.

$$(5) \qquad q^{\mathscr{C}}(x, i) = (x)_i^{\mathscr{C}} = \begin{cases} x_i & \textit{if for some } x_1, \ldots, x_n, \\ & x = \langle x_1, \ldots, x_n \rangle^{\mathscr{C}} \textit{ and } 1 \leqslant i \leqslant n, \\ 0 & \textit{otherwise.} \end{cases}$$

We call the coding scheme $\mathscr{C}$ *elementary* (or *hyperelementary*) on a structure $\mathfrak{A} = \langle A, R_1, \ldots, R_l \rangle$ if the relations and functions $N^{\mathscr{C}}, \leqslant^{\mathscr{C}}, Seq^{\mathscr{C}}, lh^{\mathscr{C}}, q^{\mathscr{C}}$ are all elementary (or hyperelementary). It follows then that each of the functions

$$p_n^{\mathscr{C}}(x_1, \ldots, x_n) = \langle x_1, \ldots, x_n \rangle$$

is elementary or hyperelementary accordingly, since

$$p_n^{\mathscr{C}}(x_1, \ldots, x_n) = u \Leftrightarrow Seq^{\mathscr{C}}(u) \,\&\, lh^{\mathscr{C}}(u) = n \\ \&\, q^{\mathscr{C}}(u, 1) = x_1 \,\&\, \ldots \,\&\, q^{\mathscr{C}}(u, n) = x_n.$$

We call $\mathfrak{A}$ *acceptable* if it admits an elementary coding scheme.

In practice we will be working with a fixed coding scheme in a given situation and we will not bother to put in the superscripts in the objects N, $\leqslant$, $\langle\,\rangle$, *Seq*, *lh*, *q*. It is, however, a sticky technical point that some of the definitions and results we will state for acceptable structures make explicit reference to a particular coding scheme, while others are *coding free*, i.e. they only depend (for their proof) on the availability of *some* elementary coding scheme. It is sometimes important to keep this clearly in mind.

Relative to a fixed coding scheme we have the obvious *successor function* on N,

$$s(n) = n+1 = m \Leftrightarrow [n < m \,\&\, (\forall i < m)(i \leqslant n)],$$

which is of course elementary if $\mathscr{C}$ is elementary. One of the nice things about acceptable structures is that functions defined by recursion on N are elementary.

5A.1. THEOREM. *Let $\mathfrak{A}$ be an acceptable structure, let $\mathscr{C}$ be a fixed elementary coding scheme on $\mathfrak{A}$, let $g: A^n \to A$, $h: A^{n+2} \to A$ be elementary functions and define $f: A^{n+1} \to A$ by the following recursion on N:*

$$f(t, \bar{x}) = 0 \quad \textit{if } t \notin N,$$
$$f(0, \bar{x}) = g(\bar{x}),$$
$$f(k+1, \bar{x}) = h(f(k, \bar{x}), k, \bar{x}).$$

Then f is elementary.

PROOF. Check that

$$\begin{aligned} f(k, \bar{x}) = z \Leftrightarrow & [k \notin N \mathbin{\&} z = 0] \\ & \vee (\exists w)\{(w)_1 = g(\bar{x}) \mathbin{\&} (\forall i < k)[(w)_{i+2} = h((w)_{i+1}, i, \bar{x})] \\ & \qquad \mathbin{\&} (w)_{k+1} = z\}, \end{aligned}$$

where the abbreviations are obvious. This implies trivially that f is elementary. ⊣

It follows easily from this that all recursive functions on N to N are elementary, and then that all arithmetical relations and functions on N are elementary, where these notions are relative to a fixed elementary coding scheme on an acceptable structure. We will not need these facts in their full generality, but we will use them in simple specific cases without apology, e.g. to assert that addition or multiplication is elementary.

We have already mentioned in the exercises of Chapter 1 that the structures $\mathbb{N}$ and $\mathbb{R}$ of arithmetic and analysis are acceptable. There are many other interesting structures, however, which are not acceptable, e.g.

$$\aleph_1 = \langle \aleph_1, \leqslant \rangle.$$

But all of these are *almost acceptable*, in the sense that there is a finite list $R'_1, \ldots, R'_m$ of hyperelementary relations such that $\mathfrak{A}' = (\mathfrak{A}, R'_1, \ldots, R'_m)$ is acceptable. We know then that the expanded structure $\mathfrak{A}'$ has the same inductive and hyperelementary relations as $\mathfrak{A}$, it has the same closure ordinal, and for all practical purposes we can substitute $\mathfrak{A}'$ for $\mathfrak{A}$ in studying inductive definability on $\mathfrak{A}$. We leave these results for the exercises since they are not hard, but some of them are very important for the applications of the theory.

5B. Satisfaction is hyperelementary

We code here the formulas of the language of an acceptable structure, using a fixed coding scheme, and we show that the satisfaction relation is hyperelementary but not elementary.

Suppose then that $\mathscr{C}$ is an elementary coding scheme on $\mathfrak{A} = \langle A, R_1, \ldots, R_l \rangle$, where for $1 \leqslant i \leqslant l$, R_i is n_i-ary. Recall that the language of $\mathfrak{A}$ has a constant $\mathbf{c}$ for each $c \in A$. Let us assume for definiteness and simplicity that the variables of the languages are $v_1, v_2, \ldots$ and that the only logical symbols are $\neg$, &, $\exists$, the others being abbreviations.

To each variable v_i we assign $\langle 0, i \rangle$ as code and to each constant $\mathbf{c}$ we assign $\langle 1, c \rangle$ as code. In symbols,

$$\ulcorner v_i \urcorner = \langle 0, i \rangle,$$

$$\ulcorner \mathbf{c} \urcorner = \langle 1, c \rangle.$$

For the prime formulas, put

$$\ulcorner R_i(t_1, \ldots, t_{n_i})\urcorner = \langle 2, i, \ulcorner t_1\urcorner, \ldots, \ulcorner t_{n_i}\urcorner\rangle \qquad (1 \leqslant i \leqslant l)$$
$$\ulcorner t = s\urcorner = \langle 3, \ulcorner t\urcorner, \ulcorner s\urcorner\rangle.$$

Finally, for more complicated formulas we proceed by induction,

$$\ulcorner \neg\varphi\urcorner = \langle 4, \ulcorner\varphi\urcorner\rangle,$$
$$\ulcorner \varphi \mathbin{\&} \psi\urcorner = \langle 5, \ulcorner\varphi,\urcorner \ulcorner\psi\urcorner\rangle,$$
$$\ulcorner (\exists v_i)\varphi\urcorner = \langle 6, i, \ulcorner\varphi\urcorner\rangle.$$

Of course these definitions are relative to the fixed scheme $\mathscr{C}$, but we will indicate this (by a subscript) only when we need to emphasize it.

5B.1. LEMMA. *Let $\mathscr{C}$ be a fixed elementary coding scheme on the acceptable structure $\mathfrak{A}$, put*

$$Fml^{\mathscr{C}}(a) \Leftrightarrow a \text{ is the code of some formula.}$$

Then $Fml^{\mathscr{C}}$ is elementary.

PROOF. It is easy to verify that

$$PrFml(a) \Leftrightarrow a \text{ is the code of some prime formula}$$

is elementary. We then have by the usual analysis of induction,

$$\begin{aligned} Fml(a) \Leftrightarrow (\exists w)(\exists k)\{(w)_{k+1} &= a \\ \mathbin{\&} (\forall j \leqslant k)[PrFml((w)_{j+1}) & \\ \vee (\exists m < j)[(w)_{j+1} &= \langle 4, (w)_{m+1}\rangle] \\ \vee (\exists m < j)(\exists n < j)[(w)_{j+1} &= \langle 5, (w)_{m+1}, (w)_{n+1}\rangle] \\ \vee (\exists i)(\exists m < j)[(w)_{j+1} &= \langle 6, i+1, (w)_{m+1}\rangle]]\}. \quad \dashv \end{aligned}$$

Recall that

$$x_1, x_2, x_3, \ldots \models \varphi$$

means that φ is true if we interpret v_1 by x_1, v_2 by x_2, etc.

5B.2. THEOREM. *Let $\mathscr{C}$ be a fixed elementary coding scheme on the acceptable structure $\mathfrak{A}$, put*

$$Sat^{\mathscr{C}}(a, x) \Leftrightarrow a \text{ is the code of some formula } \varphi \mathbin{\&} Seq(x) \text{ and } (x)_1, (x)_2, \ldots, (x)_{lh(x)}, 0, 0, \ldots \models \varphi.$$

Then $Sat^{\mathscr{C}}$ is hyperelementary but not elementary.

PROOF. Put

$$Val(a, x, t) \Leftrightarrow a \text{ is the code of some formula } \varphi \text{ and } Seq(x) \text{ and}$$
$$\{[t = 0 \;\&\; (x)'_1, (x)'_2, \ldots \models \varphi] \vee [t = 1 \;\&\; (x)'_1, (x)'_2, \ldots \models \neg\varphi]\},$$

where we have used the convenient notation

$$(x)'_i = \begin{cases} (x)_i & \text{if } i \leqslant lh(x), \\ 0 & \text{if } i > lh(x). \end{cases}$$

We will prove that *Val* is inductive. From this the theorem follows immediately, since

$$Sat(a, x) \Leftrightarrow Val(a, x, 0)$$
$$\Leftrightarrow Fml(a) \;\&\; \neg Val(a, x, 1).$$

It is easy (though tedious) to verify that the relation

$$PrVal(a, x, t) \Leftrightarrow a \text{ is the code of some prime formula } \varphi \text{ and}$$
$$\{[((x)'_1, (x)'_2, \ldots \models \varphi) \;\&\; t = 0]$$
$$\vee\ [((x)'_1, (x)'_2, \ldots \models \neg\varphi) \;\&\; t = 1]\}$$

is elementary. We will omit this computation.

Put

$$\chi(a, x, t, S) \Leftrightarrow Fml(a) \;\&\; \{PrVal(a, x, t)$$
$$\vee\ [a = \langle 4, (a)_2\rangle \;\&\; S((a)_2, x, 1) \;\&\; t = 0]$$
$$\vee\ [a = \langle 4, (a)_2\rangle \;\&\; S((a)_2, x, 0) \;\&\; t = 1]$$
$$\vee\ [a = \langle 5, (a)_2, (a)_3\rangle \;\&\; S((a)_2, x, 0) \;\&\; S((a)_3, x, 0) \;\&\; t = 0]$$
$$\vee\ [a = \langle 5, (a)_2, (a)_3\rangle \;\&\; [S((a)_2, x, 1) \vee S((a)_3, x, 1)] \;\&\; t = 1]$$
$$\vee\ [a = \langle 6, (a)_2, (a)_3\rangle$$
$$\&\; (\exists y)[(\forall i \neq (a)_2)[(y)'_i = (x)'_i] \;\&\; S((a)_3, y, 0)] \;\&\; t = 0]$$
$$\vee\ [a = \langle 6, (a)_2, (a)_3\rangle$$
$$\&\; (\forall y)[(\forall i \neq (a)_2)[(y)'_i = (x)'_i] \Rightarrow S((a)_3, y, 1)] \;\&\; t = 1]\}.$$

Clearly χ is S-positive in the language of $\mathfrak{A}$. It is easy to check

$$(a, x, t) \in I^{\xi}_{\chi} \Rightarrow Val(a, x, t)$$

by induction on ξ, and

$$Val(a, x, t) \Rightarrow (a, x, t) \in I_{\chi}$$

by induction on the complexity of the formula that a codes.

To see that *Sat* is not elementary, consider the relation

$$R(x) \Leftrightarrow \neg Sat(x, \langle x \rangle);$$

if *Sat* were elementary, *R* would be elementary, so letting *a* be the code of some formula φ which defines *R* with the free variable v_1, we would have for all *x*

$$R(x) \Leftrightarrow Sat(a, \langle x \rangle)$$

which is absurd for $x = a$. ⊣

From this we get immediately:

5B.3. COROLLARY. *If* $\mathfrak{A}$ *is an acceptable structure, then there exists a relation on* $\mathfrak{A}$ *which is hyperelementary but not elementary.* ⊣

This is a *coding-free* result, i.e. its statement does not refer to any coding scheme for the structure.

5C. The quantifier G

Let $\mathscr{C}$ be a fixed coding scheme on the acceptable structure $\mathfrak{A} = \langle A, R_1, \ldots, R_l \rangle$ and for each $n+1$-ary relation $R(z, \bar{x})$ on A put

$$(*) \quad (\mathsf{G}^{\mathscr{C}} z)R(z, \bar{x}) \Leftrightarrow \{(\forall s_1)(\exists t_1)(\forall s_2)(\exists t_2) \ldots\}[R(\langle \emptyset \rangle, \bar{x}) \vee R(\langle s_1, t_1 \rangle, \bar{x}) \vee R(\langle s_1, t_1, s_2, t_2 \rangle, \bar{x}) \vee \ldots]$$
$$\Leftrightarrow \{(\forall s_1)(\exists t_1)(\forall s_2)(\exists t_2) \ldots\} \bigvee_{m \in \omega} R(\langle s_1, t_1, \ldots, s_m, t_m \rangle, \bar{x}).$$

This defines a quantifier $\mathsf{G} = \mathsf{G}^{\mathscr{C}}$ which is associated with the structure $\mathfrak{A}$ and the coding scheme $\mathscr{C}$. We proceed to show that if $\mathscr{C}$ is elementary on $\mathfrak{A}$, then the inductive relations are precisely those of the form $(\mathsf{G}z)R(z, \bar{x})$ with *R* elementary.

5C.1. THEOREM. *Let* $\mathscr{C}$ *be an elementary coding scheme on* $\mathfrak{A} = \langle A, R_1, \ldots, R_l \rangle$, *let* $R \subseteq A^{n+1}$ *be elementary, put*

$$P(\bar{x}) \Leftrightarrow (\mathsf{G}z)R(z, \bar{x})$$
$$\Leftrightarrow \{(\forall s_1)(\exists t_1)(\forall s_2)(\exists t_2) \ldots\} \bigvee_{m \in \omega} R(\langle s_1, t_1, \ldots, s_m, t_m \rangle, \bar{x}).$$

Then P is inductive.

PROOF. Choose φ so that

$$\varphi(w, \bar{x}, S) \Leftrightarrow Seq(w) \;\&\; [lh(w) \text{ is even}] \;\&\; \{R(w, \bar{x}) \vee (\forall s)(\exists t)S(w^\frown\langle s, t\rangle, \bar{x})\},$$

where $w^\frown u$ is the obvious concatenation function,

$$z = w^\frown u \Leftrightarrow Seq(w) \;\&\; Seq(u) \;\&\; Seq(z) \;\&\; lh(z) = lh(w)+lh(u)$$
$$\&\; (\forall i < lh(w))[(z)_{i+1} = (w)_{i+1}]$$
$$\&\; (\forall i < lh(w)+lh(u))[lh(w) \leqslant i \Rightarrow (z)_{i+1} = (u)_{i-lh(w)+1}].$$

We will prove

(1) $$(\mathrm{G}z)R(z, \bar{x}) \Leftrightarrow (\langle\emptyset\rangle, \bar{x}) \in I_\varphi$$

in two steps.

Step 1: $(\langle s_1, t_1, \ldots, s_k, t_k\rangle, \bar{x}) \in I_\varphi^\xi$

$$\Rightarrow \{(\forall s_{k+1})(\exists t_{k+1}) \ldots\} \bigvee_{m\in\omega} R(\langle s_1, t_1, \ldots, s_{k+m}, t_{k+m}\rangle, \bar{x}).$$

Proof is by induction on ξ. The hypothesis gives

(2) $$R(\langle s_1, t_1, \ldots, s_k, t_k\rangle, \bar{x}) \vee (\forall s_{k+1})(\exists t_{k+1})(\langle s_1, t_1, \ldots, s_{k+1}, t_{k+1}\rangle, \bar{x}) \in \bigcup_{\eta<\xi} I_\varphi$$

which by the induction hypothesis and ordinary logic yields

$$(\forall s_{k+1})(\exists t_{k+1})\{(\forall s_{k+2})(\exists t_{k+2}) \ldots\}[R(\langle s_1, t_1, \ldots, s_k, t_k\rangle, \bar{x}) \vee \bigvee_{m\in\omega} R(\langle s_1, t_1, \ldots, s_{k+1+m}, t_{k+1+m}\rangle, \bar{x})]$$

which in turn gives the right-hand side of Step 1 by absorbing $(\forall s_{k+1})(\exists t_{k+1})$ into the infinite string.

Now taking $k = 0$ in Step 1 we get

(3) $$(\langle\emptyset\rangle, \bar{x}) \in I_\varphi \Rightarrow (\mathrm{G}z)R(z, \bar{x})$$

which is half of the equivalence (1).

Step 2: $(\langle\emptyset\rangle, \bar{x}) \notin I_\varphi \Rightarrow \neg(\mathrm{G}z)R(z, \bar{x})$.

Proof. We show that if $(\langle\emptyset\rangle, \bar{x}) \notin I_\varphi$, then $(\forall)$ wins the game determined by $(\mathrm{G}z)R(z, \bar{x})$, i.e. $(\exists)$ wins the game determined by

$$\{(\exists s_1)(\forall t_1)(\exists s_2)(\forall t_2) \ldots\} \bigwedge_m \neg R(\langle s_1, t_1, \ldots, s_m, t_m\rangle, \bar{x}).$$

Since $(\langle\emptyset\rangle, \bar{x}) \notin I_\varphi$, we have $\neg\varphi(\langle\emptyset\rangle, \bar{x}, I_\varphi)$, which immediately implies

$$\neg R(\langle\emptyset\rangle, \bar{x}) \;\&\; (\exists s_1)(\forall t_1)(\langle s_1, t_1\rangle, \bar{x}) \notin I_\varphi.$$

Let $(\exists)$ play some s_1 such that

$$(\forall t_1)(\langle s_1, t_1\rangle, \bar{x}) \notin I_\varphi.$$

For any t_1 that (∀) plays, we have $(\langle s_1, t_1\rangle, \bar{x}) \notin I_\varphi$, so again $\neg\varphi(\langle s_1, t_1\rangle, \bar{x}, I_\varphi)$, hence

$$\neg R(\langle s_1, t_1\rangle, \bar{x}) \;\&\; (\exists s_2)(\forall t_2)(\langle s_1, t_1, s_2, t_2\rangle, \bar{x}) \notin I_\varphi.$$

Now let (∃) play some s_2 such that

$$(\forall t_2)(\langle s_1, t_1, s_2, t_2\rangle, \bar{x}) \notin I_\varphi,$$

etc. Clearly (∃) will win by this strategy. ⊣

5C.2. THEOREM. *Let $\mathscr{C}$ be an elementary coding scheme on $\mathfrak{A}$, let $P \subseteq A^n$ be inductive. Then there exists an elementary $R \subseteq A^{n+1}$ such that*

$$P(\bar{x}) \Leftrightarrow (\mathrm{G}z)R(z, \bar{x}).$$

PROOF. Assume first that P is a fixed point. By 4C.2 we know that there is an elementary $\theta(\bar{x}, \bar{z}, \bar{y})$ and a finite string $\bar{\mathrm{Q}}$ so that

$$P(\bar{x}) \Leftrightarrow \{(\bar{\mathrm{Q}}\bar{z}_1)(\forall\bar{y}_1)(\bar{\mathrm{Q}}\bar{z}_2)(\forall\bar{y}_2)\ldots\}[\theta(\bar{x}, \bar{z}_1, \bar{y}_1) \vee \bigvee_{m=1}^{\infty} \theta(\bar{y}_m, \bar{z}_{m+1}, \bar{y}_{m+1})].$$

By adding extraneous quantifiers we can assume that $\bar{\mathrm{Q}}$ is alternating of even length starting with ∀, i.e.

$$P(\bar{x}) \Leftrightarrow \{(\forall u_1)(\exists v_1)\ldots(\forall u_t)(\exists v_t)(\forall y_1)\ldots(\forall y_n)(\forall u_{t+1})(\exists v_{t+1})\ldots$$
$$(\forall u_{2t})(\exists v_{2t})\ldots\}[\theta(\bar{x}, u_1, v_1, \ldots, u_t, v_t, y_1, \ldots, y_n) \vee \bigvee_{m=1}^{\infty}\ldots].$$

Now the obvious idea is to introduce n vacuous quantifiers in each of the blocks of the form $(\forall\bar{y}_m)$ so that the string becomes alternating, like that in the definition of G, and then check what happens to the matrix. What does happen is that we can take

$$R(z, \bar{x}) \Leftrightarrow Seq(z) \;\&\; (\exists i)\{lh(z) = (i+1)(2t+2n) \;\&\; [R_1(i, z, \bar{x}) \vee R_2(i, z, \bar{x})]\},$$

where

$$R_1(i, z, \bar{x}) \Leftrightarrow i = 0 \;\&\; \theta(\bar{x}, (z)_1, (z)_2, \ldots, (z)_{2t}, (z)_{2t+1}, (z)_{2t+3}, \ldots, (z)_{2t+2n-1}),$$
$$R_2(i, z, \bar{x}) \Leftrightarrow i > 0 \;\&\; \theta((z)_{2it+2(i-1)n+1}, (z)_{2it+2(i-1)n+3}, \ldots, (z)_{2it+2in-1},$$
$$(z)_{2it+2in+1}, (z)_{2it+2in+2}, \ldots, (z)_{2it+2in+2t},$$
$$(z)_{2it+2in+2t+1}, (z)_{2it+2in+2t+3}, \ldots, (z)_{2it+2in+2t+2n-1}),$$

and then verify immediately as soon as we manage to read this formula that

$$P(\bar{x}) \Leftrightarrow (\mathrm{G}z)R(z, \bar{x})$$

with this R. We leave the verification as an exercise.

If P is not a fixed point, then

$$P(\bar{x}) \Leftrightarrow Q(\bar{a}, \bar{x})$$

with Q a fixed point and fixed constants $\bar{a}$, hence

$$P(\bar{x}) \Leftrightarrow (\mathsf{G}z)R(z, \bar{a}, \bar{x})$$

with R elementary, so that if

$$R^*(z, \bar{x}) \Leftrightarrow R(z, \bar{a}, \bar{x}),$$

then

$$P(\bar{x}) \Leftrightarrow (\mathsf{G}z)R^*(z, \bar{x}). \qquad \dashv$$

A relation P is called G_1 if it satisfies

$$P(\bar{x}) \Leftrightarrow (\mathsf{G}z)R(z, \bar{x})$$

with an elementary R and $\mathsf{G} = \mathsf{G}^{\mathscr{C}}$ for *some* elementary coding scheme $\mathscr{C}$. Thus the two theorems of this section assert that *on an acceptable structure the* G_1 *relations are precisely the inductive relations.* It will turn out that this version of 5C.2 is not only elegant but also very useful.

The *quantifier* $\mathsf{G}^{\smile}$ *dual to* G is naturally defined by

$$\begin{aligned}(\mathsf{G}^{\smile}z)R(z, \bar{x}) &\Leftrightarrow \{(\exists s_1)(\forall t_1)(\exists s_2)(\forall t_2)\ldots\} \textstyle\bigwedge_{m\in\omega} R(\langle s_1, t_1, \ldots, s_m, t_m\rangle, \bar{x})\\ &\Leftrightarrow \neg(\mathsf{G}z)\neg R(z, \bar{x}).\end{aligned}$$

Then the coinductive relations on an acceptable structure are precisely the $\mathsf{G}_1^{\smile}$ relations, where $\mathsf{G}_1^{\smile}$ is defined relative to any elementary coding scheme. The hyperelementary relations are those which are both G_1 and $\mathsf{G}_1^{\smile}$, i.e. $\mathsf{G}_1 \cap \mathsf{G}_1^{\smile}$.

5D. Parametrizations and universal sets

If $G \subseteq \mathscr{X} \times \mathscr{Y}$, then for each $x \in \mathscr{X}$ we define the x-section of G by

$$G_x = \{y \in \mathscr{Y} : (x, y) \in \mathscr{X}\}.$$

(See Fig. 5.1.)

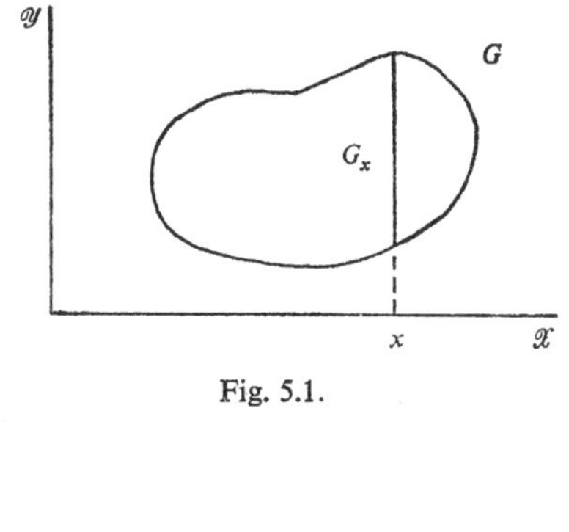

Fig. 5.1.

We say that G *parametrizes* a collection $\mathscr{F}$ of subsets of $\mathscr{Y}$ if

$$\mathscr{F} = \{G_x : x \in \mathscr{X}\};$$

in this case $\mathscr{X}$ is often called the *domain* of G and we call G an *$\mathscr{X}$-parametrization* of $\mathscr{F}$.

It is a very common problem in many areas of mathematics to search for "nice" parametrizations of interesting classes of sets. We solve this problem here for the elementary, hyperelementary and inductive relations on an acceptable structure.

5D.1. THEOREM. *Let $\mathfrak{A}$ be an acceptable structure. For each $n \geqslant 1$, there is a hyperelementary $E^n \subseteq A^{n+1}$ which parametrizes the elementary n-ary relations on $\mathfrak{A}$; i.e., if $R \subseteq A^n$, then*

$$R \text{ is elementary} \Leftrightarrow \text{for some } a \in A,\ R = E^n_a = \{\bar{x} \in A^n : (a, \bar{x}) \in E^n\}.$$

PROOF. Let $\mathscr{C}$ be a fixed elementary coding scheme for $\mathfrak{A}$ and put

$$E^n(a, \bar{x}) \Leftrightarrow Sat^{\mathscr{C}}(a, \langle\bar{x}\rangle).$$

Now for each fixed a, the *section* E^n_a is elementary, since either a does not code a formula and $E^n_a = \emptyset$ or a codes some formula $\varphi(v_1, \ldots, v_m)$ whose free variables are among $v_1, \ldots, v_m$ and

$$\bar{x} \in E^n_a \Leftrightarrow (a, \bar{x}) \in E^n \Leftrightarrow Sat^{\mathscr{C}}(a, \langle\bar{x}\rangle)$$
$$\Leftrightarrow \varphi((\langle\bar{x}\rangle)'_1, (\langle\bar{x}\rangle)'_2, \ldots, (\langle\bar{x}\rangle)'_m).$$

On the other hand, if $R(\bar{x})$ is elementary, we can take a to be the code of some $\varphi(v_1, \ldots, v_n)$ which defines R with free variables $v_1, \ldots, v_n$ and we have

$$R(\bar{x}) \Leftrightarrow Sat(a, \langle\bar{x}\rangle) \Leftrightarrow E^n(a, \bar{x}) \Leftrightarrow E^n_a(\bar{x}). \qquad \dashv$$

We can consider this result as a coding-free version of the fact that $Sat^{\mathscr{C}}$ is hyperelementary, no matter which elementary coding scheme $\mathscr{C}$ we choose.

It is easier to parametrize the inductive sets before we go to the hyperelementary case.

5D.2. PARAMETRIZATION THEOREM (for inductive relations). *Let $\mathfrak{A}$ be an acceptable structure. For each $n \geqslant 1$ there is an inductive $U^n \subseteq A^{n+1}$ which parametrizes the inductive n-ary relations on $\mathfrak{A}$.*

PROOF. Choose hyperelementary $E^n \subseteq A^{n+1}$ which parametrize the elementary relations by 5D.1 and set

$$U^n(a, \bar{x}) \Leftrightarrow (\mathrm{G}z)E^{n+1}(a, z, \bar{x}),$$

where of course G is defined relative to some elementary coding scheme $\mathscr{C}$.

Now the coding scheme $\mathscr{C}$ is elementary on the structure $(\mathfrak{A}, E^{n+1})$, hence U^n is inductive on $\mathfrak{A}$ by Theorem 1D.2. On the other hand, if P is inductive on $\mathfrak{A}$, then

$$P(\bar{x}) \Leftrightarrow (\mathrm{G}z)R(z, \bar{x})$$

for some elementary R by 4C.2, hence for some fixed $a \in A$,

$$P(\bar{x}) \Leftrightarrow (\mathrm{G}z)E^{n+1}(a, z, \bar{x}) \Leftrightarrow U^n(a, \bar{x}). \qquad \dashv$$

The relations U^n that satisfy this theorem are called *universal inductive sets.* The idea is that U^n is inductive and every n-ary inductive relation is "reducible" to it.

5D.3. Corollary. *Let $\mathfrak{A}$ be acceptable, let $U^n \subseteq A^{n+1}$ be a universal inductive set; then U^n is not hyperelementary, so in particular not every inductive relation on $\mathfrak{A}$ is hyperelementary.*

Proof. Put

$$P(x_1, \ldots, x_n) \Leftrightarrow \neg U^n(x_1, x_1, x_2, \ldots, x_n).$$

If U^n were hyperelementary, then P would be inductive, so that for some a

$$P(x_1, \ldots, x_n) \Leftrightarrow U^n(a, x_1, \ldots, x_n)$$

which is absurd if $x_1 = a$. $\dashv$

The construction of parametrizations for the hyperelementary relations is a bit more complicated, partly because we want a very strong result. The key tool is the Covering Theorem 3C.2.

5D.4. Parametrization Theorem (for hyperelementary relations). *Let $\mathfrak{A}$ be acceptable. For each $n \geqslant 1$ there is an inductive $H^n \subseteq A^{n+1}$ which parametrizes the n-ary hyperelementary relations on $\mathfrak{A}$. Moreover, there is an inductive, non-hyperelementary set $I^n \subseteq A$ and a coinductive $\breve{H}^n \subseteq A^{n+1}$, such that:*

(i) *If $R \subseteq A^n$ is hyperelementary, then $R = H^n_a$ for some $a \in I^n$.*

(ii) *If $a \in I^n$, then $H^n_a = \breve{H}^n_a$.*

Proof. For simplicity we assume $n = 1$, the general case being only a notational variant. Let $U^1 \subseteq A^2$ be a universal inductive set, let $\sigma: U^1 \twoheadrightarrow \kappa$ be an inductive norm on U^1 by 4A.3, and let $\leqslant^*_\sigma$, $<^*_\sigma$ be the inductive relations associated with σ in Section 4A. Put

$$I^1(a) \Leftrightarrow U^1((a)_1, (a)_2),$$

$$H^1(a, x) \Leftrightarrow I^1(a) \;\&\; ((a)_3, x) \leqslant^*_\sigma ((a)_1, (a)_2),$$

$$\breve{H}^1(a, x) \Leftrightarrow \neg[((a)_1, (a)_2) <^*_\sigma ((a)_3, x)].$$

It is immediate that I^1 and H^1 are inductive, that I^1 is not hyperelementary by 5D.3 and that $\breve{H}^1$ is coinductive.

For every a, either $\neg I^1(a)$ and $H_a^1 = \emptyset$ or $I^1(a)$ and in this case

$$\begin{aligned} H_a^1(x) &\Leftrightarrow \sigma((a)_3, x) \leqslant \sigma((a)_1, (a)_2) \\ &\Leftrightarrow ((a)_3, x) \leqslant_\sigma^* ((a)_1, (a)_2) \\ &\Leftrightarrow \neg[((a)_1, (a)_2) <_\sigma^* ((a)_3, x)] \\ &\Leftrightarrow \breve{H}_a^1(x). \end{aligned}$$

This proves that each H_a^1 is hyperelementary and it also proves (ii).

To complete the proof it will be enough to prove (i), so let $R \subseteq A$ be a hyperelementary set. Since R is in particular inductive and U^1 is universal, there is a fixed b such that for all x,

$$R(x) \Leftrightarrow U^1(b, x).$$

Now the function

$$f(x) = (b, x)$$

is elementary, so the Covering Theorem 3C.2 applies and there must exist some $(c, d) \in U^1$ such that

$$R(x) \Rightarrow \sigma(b, x) \leqslant \sigma(c, d).$$

Put

$$a = \langle c, d, b \rangle;$$

now

$$\begin{aligned} H_a^1(x) &\Leftrightarrow U^1(c, d) \;\&\; (b, x) \leqslant_\sigma^* (c, d) \\ &\Leftrightarrow U^1(b, x) \\ &\Leftrightarrow R(x). \end{aligned}$$ ⊣

The coinductive set $\breve{H}^n$ together with H^n give a hyperelementary definition of H_a^n, "uniformly" for $a \in I^n$.

Exercises for Chapter 5

5.1. Prove that a structure $\mathfrak{A} = \langle A, R_1, \ldots, R_l \rangle$ is acceptable if and only if $\mathfrak{A}$ has an elementary copy of ω, $\mathfrak{A}$ admits an elementary pair and the class of elementary functions on $\mathfrak{A}$ is closed under definition by *primitive recursion* on the copy of ω (see Exercise 1.4). ⊣

5.2. Let $\mathsf{G} = \mathsf{G}^{\mathscr{C}}$ be the game quantifier associated with a coding scheme $\mathscr{C}$ on the structure $\mathfrak{A}$. Prove the following formal properties of G:

$$(\mathsf{G}x)P(x) \ \& \ (\mathsf{G}y)Q(y) \Leftrightarrow (\mathsf{G}x)(\mathsf{G}y)[P(x) \ \& \ Q(y)],$$

$$(\mathsf{G}x)P(x) \vee (\mathsf{G}y)Q(y) \Leftrightarrow (\mathsf{G}x)(\mathsf{G}y)[P(x) \vee Q(y)],$$

$$(\forall x)P(x) \Leftrightarrow (\mathsf{G}x)P((x)_1),$$

$$(\exists x)P(x) \Leftrightarrow (\mathsf{G}x)P((x)_2),$$

$$(\mathsf{G}x)(\mathsf{G}y)P(x, y) \Leftrightarrow (\mathsf{G}x)\{(\exists m \leqslant lh(z))[m \text{ is even} \\ \& \ P(\langle (z)_1, \ldots, (z)_m\rangle, \langle (z)_{m+1}, \ldots, (z)_{lh(z)}\rangle)]\}.$$

Put

$$(\mathsf{S}x)P(x) \Leftrightarrow \{(\forall x_1)(\forall x_2) \ldots\} \bigvee_k P(\langle x_1, x_2, \ldots, x_k\rangle)$$

and prove

$$(\mathsf{S}x)P(x) \Leftrightarrow (\mathsf{G}x)P(\langle (x_1), (x)_3, \ldots, (x)_{lh(x)-1}\rangle). \qquad ⊣$$

5.3. Prove that if $\mathfrak{A} = \langle A, R_1, \ldots, R_l\rangle$ is acceptable, then there exist disjoint inductive subsets P, Q of A which cannot be separated by a hyperelementary R, i.e. there is no hyperelementary R such that $P \subseteq R$, $Q \cap R = \emptyset$. (*Inseparability Theorem.*)

HINT: Take $P = \{a: ((a)_1, a) \in U^1\}$, $Q = \{a: ((a)_2, a) \in U^1\}$ and then reduce P and Q by 3A.4, where U^1 is a universal inductive set. ⊣

If Σ is a set of finite sequences (of any length) from A and $\mathscr{C}$ is a coding scheme on A, put

$$\langle \Sigma\rangle^{\mathscr{C}} = \{\langle x_1, \ldots, x_n\rangle^{\mathscr{C}}: (x_1, \ldots, x_n) \in \Sigma\}$$

5.4. Prove that if $\mathscr{C}_1, \mathscr{C}_2$ are elementary coding schemes on $\mathfrak{A}$ and Σ is a set of finite sequences on A, then $\langle \Sigma\rangle^{\mathscr{C}_1}$ is elementary if and only if $\langle \Sigma\rangle^{\mathscr{C}_2}$ is elementary. ⊣

Given $\mathfrak{A} = \langle A, R_1, \ldots, R_l\rangle$, let us call $\mathfrak{B} = \langle B, P_1, \ldots, P_m\rangle$ an *acceptable extension of* $\mathfrak{A}$ if $\mathfrak{B}$ is acceptable, $A \subseteq B$ and $A, R_1, \ldots, R_l$ are all hyperelementary on $\mathfrak{B}$. It is easy to characterize the relations on A which are inductive on every acceptable extension of $\mathfrak{A}$ in the following manner.

Choose some object $0 \notin A$ and a pairing function so that neither 0 nor any element of A is a pair, i.e. a set $B \supseteq A \cup \{0\}$ and a one-to-one function

$$p: B \times B \to B$$

such that $0 \notin p[B \times B]$, $A \cap p[B \times B] = \emptyset$. Let A^* be *the closure of* $A \cup \{0\}$

under p, i.e. A^* is the smallest set containing $A \cup \{0\}$ and closed under p. We associate an element n^* of A^* with each integer n by the induction

$$0^* = 0,$$
$$(n+1)^* = p(n^*, 0)$$

and we put

$$N^* = \{0^*, 1^*, 2^*, \ldots\}.$$

Let

$$\mathfrak{A}^* = \langle A^*, A, R_1, \ldots, R_l, N^*, G_p \rangle,$$

where G_p is the graph of the pairing function p,

$$(x, y, z) \in G_p \Leftrightarrow p(x, y) = z.$$

5.5. Prove that for each infinite $\mathfrak{A}$, the structure $\mathfrak{A}^*$ admits a hyperelementary coding scheme. Moreover, a relation R on A is inductive on $\mathfrak{A}^*$ if and only if R is inductive on every acceptable extension of $\mathfrak{A}$.

HINT: It will be enough to show that every acceptable extension of $\mathfrak{A}$ contains a hyperelementary copy of $\mathfrak{A}^*$. ⊣

The problem implies in particular that if $\mathfrak{A}$ is acceptable, then a relation R on A is inductive on $\mathfrak{A}$ if and only if it is inductive on $\mathfrak{A}^*$.

In the papers Moschovakis [1969a], [1969b], [1969c] we took the point of view that in constructing inductive definitions on a structure $\mathfrak{A}$ we should be allowed to quantify over finite sequences of A, or (what comes to the same thing) we should be free to use the first order language over $\mathfrak{A}^*$. Thus a relation on A is *semihyperprojective on* $\mathfrak{A}$ in the terminology of [1969b] if and only if it is *inductive on* $\mathfrak{A}^*$ in the present terminology. This was a reasonable approach from the recursion theoretic point of view and it leads to the same class of relations for acceptable structures by Exercise 5.4. Our approach here is more suited to comparing explicit with inductive definability on arbitrary structures and it is, of course, more general.

5.6. Give an example of an infinite structure $\mathfrak{A} = \langle A, R_1, \ldots, R_l \rangle$ and a set $P \subseteq A$ which is hyperelementary on $\mathfrak{A}^*$ but not inductive on $\mathfrak{A}$. ⊣

CHAPTER 6

INDUCTIVE SECOND ORDER RELATIONS

We have been studying ordinary relations of finitely many arguments on some fixed set A, i.e. subsets of A^n, for some n. Even if these are our ultimate objects of interest, their study naturally leads to certain *collections of sets*, or more generally *second order relations* (*with relation arguments*). For example, take a fixed structure $\mathfrak{A}$ and consider the class

$$\mathscr{HE}^1(\mathfrak{A}) = \{S \subseteq A: S \textit{ is hyperelementary on } \mathfrak{A}.\}$$

We will see that with the proper definition, $\mathscr{HE}^1(\mathfrak{A})$ is an inductive class of sets (if $\mathfrak{A}$ is acceptable) and that this fact is useful in understanding the structure of hyperelementary relations on $\mathfrak{A}$. Another interesting *relation on binary relations* is *wellfoundedness*,

$$\mathscr{WF}^1(X) \Leftrightarrow X \textit{ is a wellfounded binary relation on } A;$$

this too will be inductive.

The definition of inductive relations of relations depends on the simple but important method of *relativization*. We will give it in Section 6A, where we will also look at some examples. In Section 6B we prove some basic transitivity and substitutivity results about inductive second order relations and in Section 6C we review the theory we have developed and establish that practically all the results extend directly to this wider class of relations. In Section 6D we study the class $\mathscr{HE}^1(\mathfrak{A})$ for acceptable $\mathfrak{A}$.

6A. Relativization of inductive definitions; examples

We introduce a new crop of variables over finitary relations,

$$X, Y, Z, X_1, Y_1, Z_1, \ldots .$$

In a given context each of these will vary over the n-ary relations over some set A. As before, barred letters name sequences,

$$\bar{Y} = Y_1, \ldots, Y_k,$$

where of course we allow $k = 0$ so that $\bar{Y}$ is the empty sequence.

Suppose $\mathscr{P}(\bar{x}, \bar{Y})$ is a second order relation over some set A. A formula $\varphi \equiv \varphi(\bar{x}, \bar{Y})$ in the language $\mathscr{L}^A$ (see Section 1B) *defines* $\mathscr{P}$ if $\bar{x}, \bar{Y}$ are the only free variables of φ and

(1) $$\mathscr{P}(\bar{x}, \bar{Y}) \Leftrightarrow \varphi(\bar{x}, \bar{Y}).$$

If all the relation constants in φ are among the relations of a structure $\mathfrak{A} = \langle A, R_1, \ldots, R_l \rangle$, we then call $\mathscr{P}$ *elementary* on $\mathfrak{A}$.

For example, the following second order relations are elementary:

$$\mathscr{P}(x, Y) \Leftrightarrow Y(x),$$

$$\mathscr{R}(X, Y) \Leftrightarrow X \subseteq Y \Leftrightarrow (\forall t)[t \in X \Rightarrow t \in Y],$$

$$\mathscr{Q}(X) \Leftrightarrow R_1 \cap X = \emptyset \Leftrightarrow (\forall t)[t \notin R_1 \vee t \notin X],$$

where in the last example R_1 is one of the relations in the structure $\mathfrak{A} = \langle A, R_1, \ldots, R_l \rangle$.

This is the simplest example of *relativization*—the term is used because we obtain a new object, a relation of relations, by considering as variable (relativizing) objects which up until now we have always kept fixed, i.e. some of the relation symbols in φ.

We now define inductive second order relations by carrying this relativization process through the stages of the induction.

Suppose then that

$$\varphi \equiv \varphi(\bar{x}, \bar{Y}, S) \equiv \varphi(x_1, \ldots, x_n, Y_1, \ldots, Y_k, S, =, Q_1, \ldots, Q_m)$$

is a formula in the language $\mathscr{L}^A$ over a set A, where $Q_1, \ldots, Q_m$ are relation constants, $Y_1, \ldots, Y_k$ relation variables, S is n-ary and occurs positively. To each ordinal ξ and each sequence $\bar{Y}$ of relations we assign the set $I^\xi_\varphi(\bar{Y})$ by the induction

$$I^\xi_\varphi(\bar{Y}) = \{\bar{x}: \varphi(\bar{x}, \bar{Y}, \textstyle\bigcup_{\eta<\xi} I^\eta_\varphi(\bar{Y})\},$$

and we put

$$I^{<\xi}_\varphi(\bar{Y}) = \textstyle\bigcup_{\eta<\xi} I^\eta_\varphi(\bar{Y}),$$

$$I_\varphi(\bar{Y}) = \textstyle\bigcup_\xi I^\xi_\varphi(\bar{Y}).$$

Of course each $I^\xi_\varphi(\bar{Y})$ depends on $\xi, \bar{Y}$ and $Q_1, \ldots, Q_m$, but we choose to regard $Q_1, \ldots, Q_m$ as held constant for the discussion and $Y_1, \ldots, Y_k$ as variable.

Each $I^\xi_\varphi(\bar{Y})$ is an n-ary relation on A. We obtain a second order relation on A for each ξ by putting

$$\mathscr{I}^\xi_\varphi = \{(\bar{x}, \bar{Y}): \bar{x} \in I^\xi_\varphi(\bar{Y})\},$$

$$\mathscr{I}^{<\xi}_\varphi = \textstyle\bigcup_{\eta<\xi} \mathscr{I}^\eta_\varphi,$$

$$\mathscr{I}_\varphi = \textstyle\bigcup_\xi \mathscr{I}^\xi_\varphi.$$

We call $\mathscr{P}(\bar{x}, \bar{Y})$ *inductive in* $Q_1, \ldots, Q_m$ if there is a formula

$$\varphi \equiv \varphi(\bar{u}, \bar{x}, \bar{Y}, S, =, Q_1, \ldots, Q_m)$$

in which $S, Q_1, \ldots, Q_m$ occur positively and constants $\bar{a}$ from A such that

$$\mathscr{P}(\bar{x}, \bar{Y}) \Leftrightarrow (\bar{a}, \bar{x}, \bar{Y}) \in \mathscr{I}_\varphi$$
$$\Leftrightarrow (\bar{a}, \bar{x}) \in I_\varphi(\bar{Y}).$$

Suppose $\mathfrak{A} = \langle A, R_1, \ldots, R_l \rangle$ is an infinite structure. A second order relation $\mathscr{P}(\bar{x}, \bar{Y})$ is a *fixed point* of $\mathfrak{A}$ if there is an S-positive formula $\varphi(\bar{x}, \bar{Y}, S)$ in the language of $\mathfrak{A}$ (i.e. with relation constants among $=, R_1, \ldots, R_l$) such that $\mathscr{P} = \mathscr{I}_\varphi$. We call $\mathscr{P}$ *inductive on* $\mathfrak{A}$ if there is a fixed point $\mathscr{I}_\varphi$ and constants $\bar{a}$ such that

$$\mathscr{P}(\bar{x}, \bar{Y}) \Leftrightarrow (\bar{a}, \bar{x}, \bar{Y}) \in \mathscr{I}_\varphi,$$

i.e. if $\mathscr{P}$ is inductive in $R_1, \neg R_1, \ldots, R_l, \neg R_l$. We call $\mathscr{P}$ *coinductive* if $\neg\mathscr{P}$ is inductive and we call $\mathscr{P}$ *hyperelementary* if it is both inductive and coinductive.

A common and succinct way of summarizing the relativization process involved here is to say that $\mathscr{P}(\bar{x}, \bar{Y})$ is inductive if for each $\bar{Y}$ the relation

$$P_{\bar{Y}}(\bar{x}) \Leftrightarrow \mathscr{P}(\bar{x}, \bar{Y})$$

is *inductive and uniformly in* $\bar{Y}$. "Uniformly" simply means that all the relations $P_{\bar{Y}}$ are defined inductively by the same S-positive formula φ, in which $\bar{Y}$ occur as parameters. This picture becomes a bit blurred if the relation $\mathscr{P}$ has only relation arguments, as each $P_{\bar{Y}}$ is then a constant!

Consider first some important and nontrivial examples.

6A.1. THEOREM. *The class* $\mathscr{W}\mathscr{F}^1$ *of wellfounded binary relations on an infinite structure* $\mathfrak{A}$ *is inductive.*

PROOF. Letting Y vary over binary relations on A, choose distinct elements c_0, c_1 of A and choose φ so that

$$\varphi(t, x, Y, S) \Leftrightarrow \{t = c_0 \;\&\; (\forall y)[(y, x) \in Y \Rightarrow (c_0, y) \in S]\}$$
$$\vee \; \{t = c_1 \;\&\; (\forall y)[y \in Field(Y) \Rightarrow (c_0, y) \in S] \;\&\; x = c_1\}.$$

Here of course

$$y \in Field(Y) \Leftrightarrow (\exists z)[(y, z) \in Y \vee (z, y) \in Y].$$

We now claim that

$$(*) \qquad Y \in \mathscr{W}\mathscr{F}^1 \Leftrightarrow (c_1, c_1) \in I_\varphi(Y)$$
$$\Leftrightarrow (c_1, c_1, Y) \in \mathscr{I}_\varphi,$$

so that $\mathscr{W}\mathscr{F}^1$ is inductive.

Proof of direction ($\Rightarrow$) *of* (*). Let $\rho = \rho^Y$ be the rank function of Y. An easy transfinite induction on $\rho(x)$ proves that

$$x \in Field(Y) \Rightarrow (c_0, x) \in I_\varphi(Y),$$

so that eventually for some ordinal λ we have

$$y \in Field(Y) \Rightarrow (c_0, y) \in I_\varphi^\lambda(Y).$$

Now by the definition of φ we have $(c_1, c_1) \in I_\varphi^{\lambda+1}(Y)$.

Proof of direction ($\Leftarrow$) *of* (*). Since $(c_1, c_1) \in I_\varphi(Y)$, by definition we have $\varphi(c_1, c_1, Y, I_\varphi(Y))$, i.e.

$$(\forall y)[y \in Field(Y) \Rightarrow (c_0, y) \in I_\varphi(Y)].$$

This induces a norm on *Field*(Y) by

$$\sigma(y) = \textit{least } \xi \textit{ such that } (c_0, y) \in I_\varphi^\xi(Y) \qquad (y \in Field(Y)).$$

To show that Y is wellfounded, it will be enough to prove

$$(y, x) \in Y \Rightarrow \sigma(y) < \sigma(x);$$

but this is evident, since if $\sigma(x) = \xi$, we have $(c_0, x) \in I_\varphi^\xi(Y)$, hence $\varphi(c_0, x, Y, I_\varphi^{<\xi}(Y))$, hence by the definition of φ,

$$(\forall y')[(y', x) \in Y \Rightarrow (c_0, y') \in I_\varphi^{<\xi}(Y)],$$

which implies $\sigma(y) < \xi = \sigma(x)$. ⊣

This proof is rather typical of many arguments about inductive second order relations. Notice that Y was carried along as a parameter throughout the proof.

Another interesting and typically inductive class of sets is that corresponding to the game quantifier G on an acceptable structure.

6A.2. THEOREM. *Let* $\mathscr{C}$ *be an elementary coding scheme on the structure* $\mathfrak{A}$, *put*

$$\mathsf{G}^{\mathscr{C}} = \{X \subseteq A: \{(\forall s_1)(\exists t_1)\ldots\} \bigvee_{m \in \omega} [\langle s_1, t_1, \ldots, s_m, t_m\rangle \in X]\},$$

where $\langle\,\rangle$ *is the tuple-coding of* $\mathscr{C}$. *Then* $\mathsf{G}^{\mathscr{C}}$ *is an inductive class.*

PROOF. Choose φ so that

$$\varphi(w, X, S) \Leftrightarrow Seq(w) \;\&\; [lh(w) \textit{ is even}]$$
$$\&\; [w \in X \vee (\forall s)(\exists t)[w^\frown\langle s, t\rangle \in S]]$$

as in the proof of Theorem 5C.1. It is now simple to prove by the method of that proof that

$$X \in \mathrm{G}^{\mathscr{C}} \Leftrightarrow \langle\emptyset\rangle \in I_\varphi(X)$$
$$\Leftrightarrow (\langle\emptyset\rangle, X) \in \mathscr{I}_\varphi,$$

so that $\mathrm{G}^{\mathscr{C}}$ is inductive. ⊣

6B. Transitivity, substitutivity and positive induction completeness

We give here some basic results about second order inductive relations whose proofs depend on combining inductions, as in the Combination Lemma 1C.2 and the Transitivity Theorem 1C.3. First comes a direct relativization of that result.

6B.1. TRANSITIVITY THEOREM. *Let A be an infinite set, $Q, Q_1, \ldots, Q_m$ relations on A, and $\mathscr{R}$ a second order relation on A. If $\mathscr{R}$ is inductive in $Q, Q_1, \ldots, Q_m$ and Q is inductive in $Q_1, \ldots, Q_m$, then $\mathscr{R}$ is inductive in $Q_1, \ldots, Q_m$.*

PROOF. By the hypothesis there are formulas

$$\psi \equiv \psi(\bar{u}, \bar{y}, T) \equiv \psi(\bar{u}, \bar{y}, =, Q_1, \ldots, Q_m, T),$$
$$\varphi \equiv \varphi(\bar{v}, \bar{x}, \bar{Y}, Q, S) \equiv \varphi(\bar{v}, \bar{x}, \bar{Y}, =, Q, Q_1, \ldots, Q_m, S)$$

in which $Q, Q_1, \ldots, Q_m, T, S$ occur positively, and constants $\bar{b}, \bar{a}$ such that

$$\bar{y} \in Q \Leftrightarrow (\bar{b}, \bar{y}) \in I_\psi,$$
$$(\bar{x}, \bar{Y}) \in \mathscr{R} \Leftrightarrow (\bar{a}, \bar{x}, \bar{Y}) \in \mathscr{I}_\varphi \Leftrightarrow (\bar{a}, \bar{x}) \in I_\varphi(\bar{Y}).$$

Let χ be the formula that is assigned to φ and ψ by the Combination Lemma 1C.2. There are various constants in χ, but $Q_1, \ldots, Q_m$ occur positively in χ and

$$\chi \equiv \chi(t, \bar{u}, \bar{y}, \bar{v}, \bar{x}, \bar{Y}, U)$$

has the variables $t, \bar{u}, \bar{y}, \bar{v}, \bar{x}, \bar{Y}$. By (5) of 1C.2, for each $\bar{Y}$,

$$(\bar{v}, \bar{x}) \in I_\varphi(\bar{Y}) \Leftrightarrow (c_1, \bar{u}^*, \bar{y}^*, \bar{v}, \bar{x}) \in I_\chi(\bar{Y}).$$

Hence

$$(\bar{x}, \bar{Y}) \in \mathscr{R} \Leftrightarrow (\bar{a}, \bar{x}, \bar{Y}) \in \mathscr{I}_\varphi \Leftrightarrow (c_1, \bar{u}^*, \bar{y}^*, \bar{a}, \bar{x}, \bar{Y}) \in \mathscr{I}_\chi,$$

and $\mathscr{R}$ is inductive in $Q_1, \ldots, Q_m$. ⊣

This is the most obvious way to relativize the proof of 1C.2 and obtain a result about second order relations. But there are other ways of looking at that proof which lead to *substitution theorems*, of particular interest for second order relations.

Suppose $\mathscr{R}(\bar{x}, Z, \bar{Y})$ is inductive in $Q_1, \ldots, Q_m$, where Z varies over j-ary relations on A. We say that *$\mathscr{R}$ depends positively on Z* if there is a formula

(1) $$\varphi(\bar{u}, \bar{x}, Z, \bar{Y}, S) = \varphi(\bar{u}, \bar{x}, Z, \bar{Y}, S, =, Q_1, \ldots, Q_m)$$

in which $Z, S, Q_1, \ldots, Q_m$ all occur positively and constants $\bar{a}$ such that

(2) $$\mathscr{R}(\bar{x}, Z, \bar{Y}) \Leftrightarrow (\bar{a}, \bar{x}, Z, \bar{Y}) \in \mathscr{I}_\varphi.$$

It would be more proper linguistically to say that "$\mathscr{R}$ depends positively on its first (or n^{th}) relation argument," but no confusion will be caused by this slight abuse of syntactical terminology.

The idea is that we can substitute inductive relations for the variables on which $\mathscr{R}$ depends positively and the result is still inductive. In order to get the most mileage out of this idea we prove a simple representation theorem which allows us to deal with "negative" or even arbitrary dependence.

6B.2. THEOREM. *Suppose $\mathscr{R}(\bar{x}, Z, \bar{Y})$ is inductive in fixed relations $Q_1, \ldots, Q_m$ on an infinite set A. There exists a relation $\mathscr{R}^\dagger(\bar{x}, Z_1, Z_2, \bar{Y})$, positively dependent on Z_1, Z_2 and inductive in $Q_1, \ldots, Q_m$, such that*

$$\mathscr{R}(\bar{x}, Z, \bar{Y}) \Leftrightarrow \mathscr{R}^\dagger(\bar{x}, Z, \neg Z, \bar{Y})$$
$$\Leftrightarrow \mathscr{R}^\dagger(\bar{x}, Z, \{\bar{y} : \bar{y} \notin Z\}, \bar{Y}).$$

PROOF. We first assign to each formula $\varphi(Z)$ and variable Z a formula $\varphi^\dagger(Z_1, Z_2)$ in which Z_1, Z_2 both occur positively and such that

$$\varphi(Z) \Leftrightarrow \varphi^\dagger(Z, \neg Z);$$

we do this by pushing the negation sign $\neg$ through all the logical symbols until it applies only to prime formulas and then replacing each $Z(\bar{t})$ by $Z_1(\bar{t})$ and each $\neg Z(\bar{t})$ by $Z_2(\bar{t})$. If

$$\varphi(Z) \equiv \varphi(\bar{x}, Z, S)$$

is S-positive, it follows by a trivial induction on ξ that

$$\bar{x} \in I_\varphi(Z) \Leftrightarrow \bar{x} \in I_{\varphi^\dagger}(Z, \neg Z).$$

Hence if

$$\mathscr{R}(\bar{x}, Z, \bar{Y}) \Leftrightarrow (\bar{a}, \bar{x}, Z, \bar{Y}) \in \mathscr{I}_\varphi$$

is inductive in $Q_1, \ldots, Q_m$, we can take

$$\mathscr{R}^\dagger(\bar{x}, Z_1, Z_2, \bar{Y}) \Leftrightarrow (\bar{a}, \bar{x}, Z_1, Z_2, \bar{Y}) \in \mathscr{I}_{\varphi^\dagger}$$

and we have the desired equivalence. ⊣

6B.3. SUBSTITUTION THEOREM. *Suppose* $\mathscr{R}(\bar{x}, Z, \bar{Y})$, $\mathscr{Q}(\bar{y}, \bar{x}, \bar{Y})$ *are second order relations on the infinite set A, where Z varies over j-ary relations and* $\bar{y}$ *varies over j-tuples, consider the substitution*

$$\mathscr{P}(\bar{x}, \bar{Y}) \Leftarrow \mathscr{R}(\bar{x}, \{\bar{y}\colon \mathscr{Q}(\bar{y}, \bar{x}, \bar{Y})\}, \bar{Y}).$$

If $\mathscr{R}, \mathscr{P}$ *are inductive in* $Q_1, \ldots, Q_m$ *on A and* $\mathscr{R}$ *depends positively on Z, then* $\mathscr{P}$ *is inductive in* $Q_1, \ldots, Q_m$.

If $\mathscr{R}$ *is inductive in* $Q_1, \ldots, Q_m$ *on A and* $\mathscr{Q}$ *is hyperelementary in* $Q_1, \ldots, Q_m$, *then* $\mathscr{R}$ *is inductive in* $Q_1, \ldots, Q_m$.

PROOF. It is convenient to first prove the following slightly different-looking substitution rule.

Lemma. If $\mathscr{Q}(\bar{z}, \bar{y}, \bar{Y})$, $\mathscr{R}(\bar{x}, Z, \bar{Y})$ *are inductive in* $Q_1, \ldots, Q_m$ *and* $\mathscr{R}$ *depends positively on Z, and if* $\mathscr{P}$ *is defined by*

$$\mathscr{P}(\bar{z}, \bar{x}, \bar{Y}) \Leftrightarrow \mathscr{R}(\bar{x}, \{\bar{y}\colon \mathscr{Q}(\bar{z}, \bar{y}, \bar{Y})\}, \bar{Y}),$$

then $\mathscr{P}$ *is inductive in* $Q_1, \ldots, Q_m$.

Proof. By hypothesis there are formulas

$$\psi \equiv \psi(\bar{u}, \bar{z}, \bar{y}, \bar{Y}, S),$$

$$\varphi \equiv \varphi(\bar{v}, \bar{x}, Z, \bar{Y}, T),$$

and constants $\bar{b}$, $\bar{c}$ such that

$$\mathscr{Q}(\bar{z}, \bar{y}, \bar{Y}) \Leftrightarrow (\bar{b}, \bar{z}, \bar{y}) \in I_\psi(\bar{Y}),$$

$$\mathscr{R}(\bar{x}, Z, \bar{Y}) \Leftrightarrow (\bar{c}, \bar{x}) \in I_\varphi(Z, \bar{Y}).$$

The proof hinges on the observation that we can relativize the proof of the Combination Lemma 1C.2 not only by adding relation variables throughout, but also by considering both the relation Q and the constants $\bar{a}$ as variables. Substitute then throughout the proof of 1C.2,

$$Z \textit{ for } Q,$$

$$\bar{b}, \bar{z} \textit{ for } \bar{a}.$$

We get a formula

$$\chi \equiv \chi(t, \bar{u}, \bar{z}, \bar{y}, \bar{v}, \bar{x}, \bar{Y}, U)$$

with several additional constants in which $Q_1, \ldots, Q_m$ occur positively and which has the following property: for each fixed $\bar{z}, \bar{Y}$, if

$$Q_{\bar{z},\bar{Y}}(\bar{y}) \Leftrightarrow (\bar{b}, \bar{z}, \bar{y}) \in I_\psi(\bar{Y}),$$

then

$$(\bar{v}, \bar{x}) \in I_\varphi(Q_{\bar{z},\bar{Y}}, \bar{Y}) \Leftrightarrow (c_1, \bar{u}^*, \bar{z}^*, \bar{y}^*, \bar{v}, \bar{x}) \in I_\chi(\bar{Y}).$$

Taking $\bar{v} = \bar{c}$, we get

$$\mathscr{R}(\bar{x}, Q_{\bar{z},\bar{Y}}, \bar{Y}) \Leftrightarrow (c_1, \bar{u}^*, \bar{z}^*, \bar{y}^*, \bar{c}, \bar{x}, \bar{Y}) \in \mathscr{I}_\chi,$$

which completes the proof of the lemma, since

$$\begin{aligned}\mathscr{P}(\bar{z}, \bar{x}, \bar{Y}) &\Leftrightarrow \mathscr{R}(\bar{x}, \{\bar{y}\colon Q(\bar{z}, \bar{y}, \bar{Y})\}, \bar{Y})\\ &\Leftrightarrow \mathscr{R}(\bar{x}, \{\bar{y}\colon (\bar{b}, \bar{z}, \bar{y}) \in I_\psi(\bar{Y})\}, \bar{Y})\\ &\Leftrightarrow \mathscr{R}(\bar{x}, Q_{\bar{z},\bar{Y}}, \bar{Y}).\end{aligned}$$

The first assertion of the theorem follows easily from the lemma by verifying directly from the definition that if $\mathscr{Q}(\bar{y}, \bar{x}, \bar{Y})$ is inductive in $Q_1, \ldots, Q_m$, then so is $\mathscr{Q}'(\bar{x}, \bar{y}, \bar{Y})$ defined by

$$\mathscr{Q}'(\bar{x}, \bar{y}, \bar{Y}) \Leftrightarrow \mathscr{Q}(\bar{y}, \bar{x}, \bar{Y})$$

and that if $\mathscr{P}(\bar{z}, \bar{x}, \bar{Y})$ is inductive, then so is $\mathscr{P}'(\bar{x}, \bar{Y})$ defined by

$$\mathscr{P}'(\bar{x}, \bar{Y}) \Leftrightarrow \mathscr{P}(\bar{x}, \bar{x}, \bar{Y}).$$

Both of these are trivial.

The second assertion follows from the first by an application of Theorem 6B.2. We have

$$\mathscr{R}(\bar{x}, Z, \bar{Y}) \Leftrightarrow \mathscr{R}^\dagger(\bar{x}, Z, \neg Z, \bar{Y}).$$

Now by the first assertion,

$$\mathscr{P}_1(\bar{x}, Z_2, \bar{Y}) \Leftrightarrow \mathscr{R}^\dagger(\bar{x}, \{\bar{y}\colon \mathscr{Q}(\bar{y}, \bar{x}, \bar{Y})\}, Z_2, \bar{Y})$$

is inductive and by the first assertion once more, applied to $\neg\mathscr{Q}$,

$$\begin{aligned}\mathscr{P}_2(\bar{x}, \bar{Y}) &\Leftrightarrow \mathscr{P}_1(\bar{x}, \{\bar{y}\colon \neg\mathscr{Q}(\bar{y}, \bar{x}, \bar{Y})\}, \bar{Y})\\ &\Leftrightarrow \mathscr{R}^\dagger(\bar{x}, \{\bar{y}\colon \mathscr{Q}(\bar{y}, \bar{x}, \bar{Y})\}, \{\bar{y}\colon \neg\mathscr{Q}(\bar{y}, \bar{x}, \bar{Y})\}, \bar{Y})\end{aligned}$$

is also inductive. But trivially,

$$\mathscr{P}_2(\bar{x}, \bar{Y}) \Leftrightarrow \mathscr{R}(\bar{x}, \{\bar{y}\colon \mathscr{Q}(\bar{y}, \bar{x}, \bar{Y})\}). \qquad \dashv$$

This theorem of course applies to the special cases when $\mathscr{Q}$ has no relation arguments: If $\mathscr{R}(\bar{x}, Z, \bar{Y})$, $Q(\bar{y}, \bar{x})$ are inductive in $Q_1, \ldots, Q_m$, and $\mathscr{R}$ depends positively on Z, then

$$\mathscr{P}(\bar{x}, \bar{Y}) \Leftrightarrow \mathscr{R}(\bar{x}, \{\bar{y}\colon Q(\bar{y}, \bar{x})\}, \bar{Y})$$

is also inductive in $Q_1, \ldots, Q_m$. Similarly, if $Q(\bar{y})$ does not depend on the variables $\bar{x}$ and is inductive in $Q_1, \ldots, Q_m$ and if $\mathscr{R}(\bar{x}, Z, \bar{Y})$ is inductive in $Q_1, \ldots, Q_m$ and depends positively on Z, then

$$\mathscr{P}(\bar{x}, \bar{Y}) \Leftrightarrow \mathscr{R}(\bar{x}, Q, \bar{Y})$$

is inductive in $Q_1, \ldots, Q_m$.

For some applications it is not convenient to quote directly 6B.3, but we must combine it with Theorem 6B.2 to prove that a particular complicated substitution leads to an inductive relation.

If $\mathscr{R}(\bar{x}, Z)$ is inductive in $Q_1, \ldots, Q_m$ and depends positively on Z and if $\bar{x}$ varies over n-tuples and Z over n-ary relations, we can define an operator

$$\Gamma : Power(A^n) \to Power(A^n)$$

by

$$\Gamma(Z) = \{\bar{x} : \mathscr{R}(\bar{x}, Z)\}.$$

Such operators are called *positive, inductive in* $Q_1, \ldots, Q_m$ and it is easy to verify that they are monotone. Our next result establishes that their fixed points are also inductive in $Q_1, \ldots, Q_m$. In recursion theoretic treatments of these matters, this result is called a *Recursion Theorem*, see Exercise 6.2. It is in truth a completeness result, since it shows that the monotone operators defined by positive, elementary inductions lead to no more fixed points than the positive elementary inductions themselves.

6B.4. Positive Induction Completeness Theorem. *Let $Q_1, \ldots, Q_m$ be fixed relations on the infinite set A, let $\mathscr{R}(\bar{x}, Z)$ be inductive in $Q_1, \ldots, Q_m$ and positively dependent on Z, put*

$$\Gamma(Z) = \{\bar{x} : \mathscr{R}(\bar{x}, Z)\}.$$

Then Γ is a monotone operator, the set I_Γ built up by Γ is inductive in $Q_1, \ldots, Q_m$ and the closure ordinal of Γ is $\leqslant \kappa(A, Q_1, \ldots, Q_m)$.

Proof. By hypothesis there is a formula

$$\varphi(\bar{u}, \bar{x}, Z, S) \equiv \varphi(\bar{u}, \bar{x}, Z, S, =, Q_1, \ldots, Q_m)$$

in which $Z, S, Q_1, \ldots, Q_m$ all occur positively and constants $\bar{a}$ such that

$$\bar{x} \in \Gamma(Z) \Leftrightarrow (\bar{a}, \bar{x}) \in I_\varphi(Z).$$

Let I_Γ^ξ be the stages of Γ as in Section 1A,

$$I_\Gamma^\xi = \Gamma(\textstyle\bigcup_{\eta<\xi} I_\Gamma^\eta) = \{\bar{x} : \varphi(\bar{a}, \bar{x}, Z, \textstyle\bigcup_{\eta<\xi} I_\Gamma^\eta)\},$$
$$I_\Gamma^{<\xi} = \textstyle\bigcup_{\eta<\xi} I_\Gamma^\eta.$$

Here of course, for any Z,

$$I_\varphi^\xi(Z) = \{(\bar{u}, \bar{x}) : \varphi(\bar{u}, \bar{x}, Z, I_\varphi^{<\xi}(Z))\},$$
$$I_\varphi(Z) = \textstyle\bigcup_\xi I_\varphi^\xi(Z).$$

It is worth pointing out that each I_φ^ξ is monotone as an operator on Z,

$$Z_1 \subseteq Z_2 \Rightarrow I_\varphi^\xi(Z_1) \subseteq I_\varphi^\xi(Z_2);$$

this is easily proved by induction on ξ, using the hypothesis that Z occurs positively in φ.

Put now

$$\psi(\bar{u}, \bar{x}, S) \Leftrightarrow \varphi(\bar{u}, \bar{x}, \{\bar{x}' : (\bar{a}, \bar{x}') \in S\}, S).$$

The theorem will follow easily from the following two steps.

Step 1: *For every* $\bar{u}, \bar{x}$,

$$(\bar{u}, \bar{x}) \in I_\psi^\xi \Rightarrow (\bar{u}, \bar{x}) \in I_\varphi^\xi(I_\Gamma^{<\xi}).$$

Proof of Step 1 is by transfinite induction on ξ. Assume the induction hypothesis and notice that for every $\zeta < \xi$,

$$\begin{aligned} (\bar{a}, \bar{x}) \in I_\psi^\zeta &\Rightarrow (\bar{a}, \bar{x}) \in I_\varphi^\zeta(I_\Gamma^{<\zeta}) && \text{(by ind. hyp.)} \\ &\Rightarrow (\bar{a}, \bar{x}) \in I_\varphi(I_\Gamma^{<\zeta}) && \text{(by def. of } I_\varphi(Z)) \\ &\Rightarrow \bar{x} \in I_\Gamma^\zeta && \text{(by def. of } I_\Gamma^\zeta). \end{aligned}$$

Now compute:

$$\begin{aligned} (\bar{u}, \bar{x}) \in I_\psi^\xi &\Rightarrow \varphi(\bar{u}, \bar{x}, \{\bar{x}' : (\bar{a}, \bar{x}') \in I_\psi^{<\xi}\}, I_\psi^{<\xi}) && \text{(by def. of } I_\psi^\xi) \\ &\Rightarrow \varphi(\bar{u}, \bar{x}, I_\Gamma^{<\xi}, I_\psi^{<\xi}) && \text{(by the remark above and monotonicity)} \\ &\Rightarrow \varphi(\bar{u}, \bar{x}, I_\Gamma^{<\xi}, \textstyle\bigcup_{\eta<\xi} I_\varphi^\eta(I_\Gamma^{<\eta})) && \text{(by ind. hyp. and monotonicity)} \\ &\Rightarrow \varphi(\bar{u}, \bar{x}, I_\Gamma^{<\xi}, \textstyle\bigcup_{\eta<\xi} I_\varphi^\eta(I_\Gamma^{<\xi})) && \text{(by monotonicity of } I_\varphi^\eta(Z) \text{ in } Z) \\ &\Rightarrow (\bar{u}, \bar{x}) \in I_\varphi^\xi(I_\Gamma^{<\xi}) && \text{(by def. of } I_\varphi^\xi(Z)). \end{aligned}$$

Step 2: *For every* $\bar{u}, \bar{x}$,

$$(\bar{u}, \bar{x}) \in I_\varphi^\xi(\{\bar{x}' : (\bar{a}, \bar{x}') \in I_\psi\}) \Rightarrow (\bar{u}, \bar{x}) \in I_\psi.$$

Proof of Step 2 is also by induction on ξ. Put for convenience

$$J_\psi = \{\bar{x}' : (\bar{a}, \bar{x}') \in I_\psi\}$$

and compute:

$$\begin{aligned} (\bar{u}, \bar{x}) \in I_\varphi^\xi(J_\psi) &\Rightarrow \varphi(\bar{u}, \bar{x}, J_\psi, I_\varphi^{<\xi}(J_\psi)) && \text{(by def. of } I_\varphi^\xi(Z)), \\ &\Rightarrow \varphi(\bar{u}, \bar{x}, J_\psi, I_\psi) && \text{(by ind. hyp. and monotonicity of } \varphi \text{ in } S) \\ &\Rightarrow \varphi(\bar{u}, \bar{x}, \{\bar{x}' : (\bar{a}, \bar{x}') \in I_\psi\}, I_\psi) && \text{(by def. of } J_\psi) \\ &\Rightarrow \psi(\bar{u}, \bar{x}, I_\psi) && \text{(by def. of } \psi) \\ &\Rightarrow (\bar{u}, \bar{x}) \in I_\psi. \end{aligned}$$

Now by Step 1, using monotonicity repeatedly,

$$\begin{aligned} \bar{x} \in J_\psi &\Rightarrow (\bar{a}, \bar{x}) \in I_\psi \\ &\Rightarrow (\bar{a}, \bar{x}) \in I_\psi^\xi, \quad \text{for some } \xi, \\ &\Rightarrow (\bar{a}, \bar{x}) \in I_\varphi^\xi(I_\Gamma^{<\xi}) \\ &\Rightarrow (\bar{a}, \bar{x}) \in I_\varphi(I_\Gamma) \\ &\Rightarrow \bar{x} \in I_\Gamma, \end{aligned}$$

and by Step 2

$$\begin{aligned} \bar{x} \in \Gamma(J_\psi) &\Rightarrow (\bar{a}, \bar{x}) \in I_\varphi(J_\psi) \\ &\Rightarrow (\bar{a}, \bar{x}) \in I_\psi \\ &\Rightarrow \bar{x} \in J_\psi; \end{aligned}$$

i.e. $J_\psi \subseteq I_\Gamma$ and $\Gamma(J_\psi) \subseteq J_\psi$, so that $J_\psi = I_\Gamma$, and I_Γ is therefore inductive in $Q_1, \ldots, Q_m$.

Also, if $\lambda = \|\psi\|$ is the closure ordinal of ψ, then the equation $I_\Gamma = J_\psi$ implies

$$\begin{aligned} \bar{x} \in I_\Gamma &\Rightarrow (\bar{a}, \bar{x}) \in I_\psi^\xi, \quad \text{for some } \xi < \lambda \\ &\Rightarrow (\bar{a}, \bar{x}) \in I_\varphi^\xi(I_\Gamma^{<\xi}) \qquad \text{(by Step 1)} \\ &\Rightarrow (\bar{a}, \bar{x}) \in I_\varphi(I_\Gamma^{<\xi}) \\ &\Rightarrow \bar{x} \in I_\Gamma^\xi, \end{aligned}$$

so that the closure ordinal of Γ is $\leqslant \lambda \leqslant \kappa(A, Q_1, \ldots, Q_m)$. ⊣

We now collect in one theorem the simple properties of inductive second order relations on an infinite structure. Proofs are immediate from the preceding results and the methods of Section 1D.

6B.5. Theorem. *The class of inductive second order relations on an infinite structure $\mathfrak{A}$ is closed under the positive operations* &, ∨, ∃, ∀, *it is closed under A-valued hyperelementary substitution*

$$\mathscr{P}(\bar{x}, \bar{Y}) \Leftrightarrow \mathscr{R}(f_1(\bar{x}, \bar{Y}), \ldots, f_m(\bar{x}, \bar{Y}), \bar{Y})$$

and it is closed under the operation of relation substitution

$$(*) \qquad \mathscr{P}(\bar{x}, \bar{Y}) \Leftrightarrow \mathscr{R}(\bar{x}, \{\bar{y}: \mathscr{Q}(\bar{y}, \bar{x}, \bar{Y})\}, \bar{Y}),$$

whenever $\mathscr{R}(\bar{x}, Z, \bar{Y})$ depends positively on Z or $\mathscr{Q}$ is hyperelementary.

The class of hyperelementary second order relations on $\mathfrak{A}$ includes all elementary second order relations and is closed under all the elementary *operations* $\neg$, &, $\vee$, $\exists$, $\forall$ *and A-valued hyperelementary substitution as well as* (*).

If $Q_1, \ldots, Q_m$ *are hyperelementary on* $\mathfrak{A}$, *then the structure* $(\mathfrak{A}, Q_1, \ldots, Q_m)$ *has the same inductive second order relations as the structure* $\mathfrak{A}$.

If $\mathscr{R}(\bar{x}, Z)$ *is inductive on* $\mathfrak{A}$ *and positively dependent on* Z, *then the operator*

$$\Gamma(Z) = \{\bar{x} : \mathscr{R}(\bar{x}, Z)\}$$

is monotone, the set I_Γ *built up by it is inductive on* $\mathfrak{A}$ *and its closure ordinal is* $\leqslant \kappa^{\mathfrak{A}}$.

Every inductive second order relation is Π^1_1 *on* $\mathfrak{A}$ *and hence every hyperelementary second order relation is* Δ^1_1 *on* $\mathfrak{A}$. ⊣

Here of course Π^1_1 second order relations are those of the form

$$\mathscr{P}(\bar{x}, \bar{Y}) \Leftrightarrow (\forall \bar{Z})\mathscr{R}(\bar{x}, \bar{Y}, \bar{Z}),$$

where $\mathscr{R}$ is elementary.

6C. Extension of the theory to second order relations

We list here versions for second order relations of the most significant theorems in Chapters 1–5. In most cases the old proofs work with very minor modifications (mostly notational) and we shall omit them.

If $(\bar{x}, \bar{Y}) \in \mathscr{I}_\varphi$ for some fixed point $\mathscr{I}_\varphi$, we naturally put

$$\begin{aligned} |\bar{x}, \bar{Y}|_\varphi &= \textit{least } \xi \textit{ such that } (\bar{x}, \bar{Y}) \in \mathscr{I}^\xi_\varphi \\ &= \textit{least } \xi \textit{ such that } \bar{x} \in I^\xi_\varphi(\bar{Y}). \end{aligned}$$

Notice that if $\varphi(\bar{x}, \bar{Y}, S)$ is in the language of an infinite structure $\mathfrak{A}$ and $(\bar{x}, \bar{Y}) \in \mathscr{I}_\varphi$, then

$$|\bar{x}, \bar{Y}|_\varphi < \kappa^{(\mathfrak{A}, \bar{Y})}.$$

This is immediate from the definition, since for each fixed $\bar{Y}$ the sequence of stages $\{I^\xi_\varphi(\bar{Y})\}$ is defined by an ordinary first order induction on the structure $(\mathfrak{A}, \bar{Y})$.

The Stage Comparison Theorem follows exactly as before.

6C.1. STAGE COMPARISON THEOREM. *Let* $\varphi(\bar{x}, \bar{Y}, S)$, $\psi(\bar{z}, \bar{W}, T)$ *be formulas in the language of an infinite structure* $\mathfrak{A}$, *respectively positive in* S, T. *Define the relations* $\leqslant^*_{\varphi,\psi}$, $<^*_{\varphi,\psi}$ by

$$(\bar{x}, \bar{Y}) \leqslant^*_{\varphi,\psi} (\bar{z}, \bar{W}) \Leftrightarrow (\bar{x}, \bar{Y}) \in \mathscr{I}_\varphi \mathbin{\&} [(\bar{z}, \bar{W}) \notin \mathscr{I}_\psi \vee |\bar{x}, \bar{Y}|_\varphi \leqslant |\bar{z}, \bar{W}|_\psi],$$

$$(\bar{x}, \bar{Y}) <^*_{\varphi,\psi} (\bar{z}, \bar{W}) \Leftrightarrow (\bar{x}, \bar{Y}) \in \mathscr{I}_\varphi \mathbin{\&} [(\bar{z}, \bar{W}) \notin \mathscr{I}_\psi \vee |\bar{x}, \bar{Y}|_\varphi < |\bar{z}, \bar{W}|_\psi].$$

Then both $\leqslant^*_{\varphi,\psi}$, $<^*_{\varphi,\psi}$ *are fixed points of the structure* $\mathfrak{A}$. ⊣

From this we get immediately the extended version of Theorem 2B.1.

6C.2. THEOREM. *Let $\varphi(\bar{x}, \bar{Y}, S)$ be S-positive in the language of an infinite structure $\mathfrak{A}$. For each $\lambda < \kappa^{\mathfrak{A}}$, the set $\mathscr{I}_\varphi^\lambda$ is hyperelementary on $\mathfrak{A}$. In particular, if $\|\varphi\| < \kappa^{\mathfrak{A}}$, then $\mathscr{I}_\varphi$ is hyperelementary on $\mathfrak{A}$.* ⊣

Here

$$\|\varphi\| = \textit{least } \xi \textit{ such that } \mathscr{I}_\varphi^\xi = \bigcup_{\eta<\xi} \mathscr{I}_\varphi^\eta = \mathscr{I}_\varphi^{<\xi}$$
$$= \textit{supremum}\{|\bar{x}, \bar{Y}|_\varphi + 1 \colon (\bar{x}, \bar{Y}) \in \mathscr{I}_\varphi\}.$$

The supremum in this definition of $\|\varphi\|$ is over a set of cardinality $2^{|\mathfrak{A}|}$ and we may very well have $\|\varphi\| = |\mathfrak{A}|^+$. Thus for these second order inductions the Closure Theorem 2B.4 may fail. It is worth stating as a theorem the existence of such examples since they will be useful later on.

6C.3. THEOREM. *There are structures $\mathfrak{A}$ in which the class $\mathscr{WF}^1$ of well-founded binary relations is elementary. On such structures there are formulas $\varphi(\bar{x}, \bar{Y}, S)$ such that $\mathscr{I}_\varphi$ is elementary but $\|\varphi\| = |\mathfrak{A}|^+ > \kappa^{\mathfrak{A}}$.*

PROOF. Take $\mathfrak{A} = \langle V, \in \upharpoonright V_\lambda\rangle$, where λ is a limit ordinal with *cofinality*(λ) $> \omega$ and V_λ is the set of sets of rank less than λ. We then have

$$Y \in \mathscr{WF}^1 \Leftrightarrow Y \textit{ has no infinite descending chains}$$
$$\Leftrightarrow (\forall f)\{[f \textit{ is a function } \& \textit{ Domain}(f) = \omega$$
$$\& \textit{ Range}(f) \subseteq \textit{Field}(Y)] \rightarrow (\exists i)(f(i+1), f(i)) \notin Y\}.$$

Following the idea of the proof of Theorem 6A.1, put

$$\varphi(x, Y, S) \Leftrightarrow x \in \textit{Field}(Y) \;\&\; (\forall y)[(y, x) \in Y \Rightarrow y \in S].$$

It is easy to prove by induction on ξ that

(*) $\quad (x, Y) \in \mathscr{I}_\varphi^\xi \Leftrightarrow x \in \textit{Field}(Y) \;\&\; Y \upharpoonright x \textit{ is wellfounded of rank} \leqslant \xi,$

where

$$Y \upharpoonright x = \{(u, v) \in Y \colon (v, x) \in Y\}.$$

From this it follows immediately that $\|\varphi\| = |\mathfrak{A}|^+$. But if $\mathscr{WF}^1$ is elementary on $\mathfrak{A}$, surely $\mathscr{I}_\varphi$ is elementary on $\mathfrak{A}$,

$$(x, Y) \in \mathscr{I}_\varphi \Leftrightarrow Y \upharpoonright x \textit{ is wellfounded.}$$ ⊣

We will prove later that the extended Closure Theorem for second order inductions does hold for *countable, acceptable structures*. This depends on the fact that for such structures all Π_1^1 relations are inductive, which fails for structures in which $\mathscr{WF}^1$ is elementary.

Going on to the structure results of Chapter 3, we call a norm

$$\sigma : \mathscr{P} \twoheadrightarrow \lambda$$

on a second order relation $\mathscr{P}$ *inductive* if there exist relations $\mathscr{I}_\sigma$, $\check{\mathscr{I}}_\sigma$, inductive and coinductive respectively such that

$$\mathscr{P}(\bar{x}', \bar{Y}') \Rightarrow (\forall \bar{x}, \bar{Y})\{[\mathscr{P}(\bar{x}, \bar{Y}) \,\&\, \sigma(\bar{x}, \bar{Y}) \leqslant \sigma(\bar{x}', \bar{Y}')] \Leftrightarrow \mathscr{I}_\sigma(\bar{x}, \bar{Y}, \bar{x}', \bar{Y}') \Leftrightarrow \check{\mathscr{I}}_\sigma(\bar{x}, \bar{Y}, \bar{x}', \bar{Y}')\}.$$

We associate with a norm σ the relations $\leqslant^*_\sigma$, $<^*_\sigma$ defined by

$$(\bar{x}, \bar{Y}) \leqslant^*_\sigma (\bar{x}', \bar{Y}') \Leftrightarrow \mathscr{P}(\bar{x}, \bar{Y}) \,\&\, [\neg\mathscr{P}(\bar{x}', \bar{Y}') \vee \sigma(\bar{x}, \bar{Y}) \leqslant \sigma(\bar{x}', \bar{Y}')],$$

$$(\bar{x}, \bar{Y}) <^*_\sigma (\bar{x}', \bar{Y}') \Leftrightarrow \mathscr{P}(\bar{x}, \bar{Y}) \,\&\, [\neg\mathscr{P}(\bar{x}', \bar{Y}') \vee \sigma(\bar{x}, \bar{Y}) < \sigma(\bar{x}', \bar{Y}')].$$

The same proof as in Theorem 3A.1 shows that if $\mathscr{P}$ is inductive, then a norm σ on $\mathscr{P}$ is inductive if and only if $\leqslant^*_\sigma$, $<^*_\sigma$ are inductive. Theorem 3A.2 extends directly and leads immediately to the extended version of the Prewellordering Theorem.

6C.4. Prewellordering Theorem. *Every inductive second order relation on an infinite structure $\mathfrak{A}$ admits an inductive norm.* ⊣

The Reduction Theorem for inductive second order relations and the Separation Theorem for coinductive second order relations follow as in 3A.4, 3A.5 and we will not bother to state them explicitly.

There are partial (but significant) extensions of 3B.1, 3B.2, 3B.3.

6C.5. Hyperelementary Selection Theorem. *Suppose $\mathscr{P}(\bar{x}, \bar{Y}, \bar{z})$ is an inductive second order relation on an infinite structure $\mathfrak{A}$, where $\bar{z}$ varies over m-tuples from A. There are inductive relations $\mathscr{P}^*(\bar{x}, \bar{Y}, \bar{z})$, $\mathscr{P}^{**}(\bar{x}, \bar{Y}, \bar{z})$ such that*

(1) $$\mathscr{P}^* \subseteq \mathscr{P},$$

(2) $$(\exists \bar{z})\mathscr{P}(\bar{x}, \bar{Y}, \bar{z}) \Rightarrow (\exists \bar{z})\mathscr{P}^*(\bar{x}, \bar{Y}, \bar{z}),$$

(3) $$(\exists \bar{z})\mathscr{P}(\bar{x}, \bar{Y}, \bar{z}) \Rightarrow (\forall \bar{z})[\mathscr{P}^*(\bar{x}, \bar{Y}, \bar{z}) \Leftrightarrow \neg\mathscr{P}^{**}(\bar{x}, \bar{Y}, \bar{z})].$$

In particular, if $\mathscr{P}(\bar{x}, \bar{Y}, \bar{z})$ is inductive and $(\forall \bar{x}, \bar{Y})(\exists \bar{z})\mathscr{P}(\bar{x}, \bar{Y}, \bar{z})$, then there exists a hyperelementary $\mathscr{P}^* \subseteq \mathscr{P}$ such that $(\forall \bar{x}, \bar{Y})(\exists \bar{z})\mathscr{P}^*(\bar{x}, \bar{Y}, \bar{z})$. ⊣

The reason why 3B.1 does not extend to the case where $\bar{z}$ may vary over relations is that $\bar{z}$ is quantified in the proof of 3B.1.

The Rank Comparison Theorem 3B.4 does not extend to wellfounded relations on sets because of a similar quantification over the field of the relations in the formula φ of the proof.

The Boundedness and Covering Theorems also do not extend directly to second order relations since they are generalizations of the Closure Theorem.

The various extensions of the theory discussed in Section 3D relativize directly and yield extensions of the second order theory. No great ingenuity is needed for this.

From Chapter 4, it is worth putting down the relativized version of 4C.2.

6C.6. Fixed Point Normal Form Theorem. *A second order relation $\mathscr{P}$ on an infinite structure $\mathfrak{A}$ is a fixed point if and only if there is an elementary formula $\theta(\bar{x}, \bar{Y}, \bar{z}, \bar{u})$ in the language of $\mathfrak{A}$ and a string of quantifiers $\bar{Q}$ such that*

$$(*)\quad \mathscr{P}(\bar{x}, \bar{Y}) \Leftrightarrow \{(\bar{Q}\bar{z}_1)(\forall \bar{u}_1)(\bar{Q}\bar{z}_2)(\forall \bar{u}_2)\ldots\}[\theta(\bar{x}, \bar{Y}, \bar{z}_1, \bar{u}_1) \vee \theta(\bar{u}_1, \bar{Y}, \bar{z}_2, \bar{u}_2) \vee \ldots]$$

$$\Leftrightarrow \{(\bar{Q}\bar{z}_1)(\forall \bar{u}_1)(\bar{Q}\bar{z}_2)(\forall \bar{u}_2)\ldots\}[\theta(\bar{x}, \bar{Y}, \bar{z}_1, \bar{u}_1) \vee \textstyle\bigvee_{i \geqslant 1} \theta(\bar{u}_i, \bar{Y}, \bar{z}_{i+1}, \bar{u}_{i+1})].$$

In fact, every fixed point on $\mathfrak{A}$ satisfies (*) with a quantifier free θ.

Moreover, $\mathscr{P}$ depends positively on Y_j if and only if (*) holds with some $\theta(\bar{x}, \bar{Y}, \bar{z}, \bar{u})$ in which Y_j occurs positively. ⊣

This again gives a normal form for inductive second order relations which we will not bother to put down.

Suppose now that $\mathfrak{A}$ is acceptable. The results of 5A and 5B extend trivially to second order relations and we will not bother to list them explicitly. It is worth, however, putting down the extensions of 5C.1, 5C.2 and the Parametrization Theorems 5D.1, 5D.2.

6C.7. Theorem. *Let $\mathscr{C}$ be an elementary coding scheme on $\mathfrak{A}$, let $\mathsf{G} = \mathsf{G}^{\mathscr{C}}$ be the game quantifier associated with $\mathscr{C}$ by (*) of 5C. A second order relation $\mathscr{P}(\bar{x}, \bar{Y})$ is inductive on $\mathfrak{A}$ if and only if there is an elementary relation $\mathscr{R}(\bar{z}, \bar{x}, \bar{Y})$ such that*

$$(*)\quad \mathscr{P}(\bar{x}, \bar{Y}) \Leftrightarrow (\mathsf{G}z)\mathscr{R}(z, \bar{x}, \bar{Y})$$

$$\Leftrightarrow \{(\forall s_1)(\exists t_1)(\forall s_2)(\exists t_2)\ldots\} \textstyle\bigvee_{m \in \omega} \mathscr{R}(\langle s_1, t_1, \ldots, s_m, t_m\rangle, \bar{x}, \bar{Y}).$$

Moreover, $\mathscr{P}$ depends positively on Y_j if and only if (*) holds with some $\mathscr{R}(z, \bar{x}, \bar{Y})$ which is definable by a formula $\theta(z, \bar{x}, \bar{Y})$ in which Y_j occurs positively. ⊣

If

$$\mathscr{P}(\bar{x}, \bar{Y}) \Leftrightarrow \mathscr{P}(x_1, \ldots, x_n, Y_1, \ldots, Y_k)$$

is a second order relation, where Y_j is r_j-ary, we call the sequence

$$\nu = (n, r_1, \ldots, r_k)$$

the *signature of* $\mathscr{P}$; we then call $\mathscr{P}$ ν-ary. We also call $(r_1, \ldots, r_k)$ the signature of the sequence of variables $Y_1, \ldots, Y_k$.

6C.8. THEOREM. *Let $\mathfrak{A}$ be acceptable. For each signature $\nu = (n, r_1, \ldots, r_k)$ there is a hyperelementary relation $\mathscr{E}^\nu$ of signature $(n+1, r_1, \ldots, r_k)$ which parametrizes the elementary second order relations of signature ν, i.e. for each $\mathscr{R}$ of signature ν,*

$$\mathscr{R} \text{ is elementary} \Leftrightarrow \text{for some } a \in A,$$

$$\mathscr{R} = \mathscr{E}^\nu_a = \{(\bar{x}, \bar{Y}) : (a, \bar{x}, \bar{Y}) \in \mathscr{E}^\nu\}.$$

For each ν, there is a universal inductive relation $\mathscr{U}^\nu$ of signature $(n+1, r_1, \ldots, r_k)$, i.e. an inductive $\mathscr{U}^\nu$ such that for each $\mathscr{R}$ of signature ν,

$$\mathscr{R} \text{ is inductive} \Leftrightarrow \text{for some } a \in A,$$

$$\mathscr{R} = \mathscr{U}^\nu_a = \{(\bar{x}, \bar{Y}) : (a, \bar{x}, \bar{Y}) \in \mathscr{U}^\nu_{\mathfrak{i}}\}.$$

Similarly, there is a universal Π^1_1 relation $\mathscr{P}^\nu$ of signature $(n+1, r_1, \ldots, r_k)$ i.e. a Π^1_1 relation $\mathscr{P}^\nu$ such that for each $\mathscr{R}$ of signature ν,

$$\mathscr{R} \text{ is } \Pi^1_1 \Leftrightarrow \text{for some } a \in A,$$

$$\mathscr{R} = \mathscr{P}^\nu_a = \{(\bar{x}, \bar{Y}) : (a, \bar{x}, \bar{Y}) \in \mathscr{P}^\nu\}.$$

PROOF is by direct relativization of the proofs of 5D.1 and 5D.2. To define $\mathscr{E}^\nu$ for a signature $\nu = (n, r_1, \ldots, r_k)$, we look at the definition of E^n as given in the proof of 5D.2 relative to the structure $(\mathfrak{A}, \bar{Y})$, for any fixed $\bar{Y}$ of signature $(r_1, \ldots, r_k)$, i.e. we put

$$\mathscr{E}^\nu(a, \bar{x}, \bar{Y}) \Leftrightarrow Sat^{\mathscr{C}}(a, \langle\bar{x}\rangle) \text{ is true, for } (\mathfrak{A}, \bar{Y}).$$

Looking back to the inductive definition of $Sat^{\mathscr{C}}$ in 5B.2, it is easy to see that it is uniform in the particular relations of the structure involved and depends only on the number of arguments they have, i.e. the signature of the sequence $R_1, \ldots, R_l, Y_1, \ldots, Y_k$, if $\mathfrak{A} = \langle A, R_1, \ldots, R_l\rangle$. By this we mean that the formula χ in the proof of 5B.2 written down for the structure $(\mathfrak{A}, \bar{Y})$ will simply carry $\bar{Y}$ as parameters and hence

$$Sat^{\mathscr{C}}(a, \langle\bar{x}\rangle, \bar{Y}) \Leftrightarrow Sat^{\mathscr{C}}(a, \langle\bar{x}\rangle) \text{ holds for } (\mathfrak{A}, \bar{Y})$$

is hyperelementary and so is $\mathscr{E}^\nu$.

From this the result about $\mathscr{U}^\nu$ follows immediately as before, by using the quantifier G.

We can use the same method with the quantifier $\forall$ to get the result for Π_1^1 if we show that every Π_1^1 second order relation of signature ν is of the form

$$(\forall Z)\mathscr{Q}(Z, \bar{x}, \bar{Y}),$$

where $\mathscr{Q}$ is elementary and Z varies over subsets of A, i.e. if we show how to reduce a prefix of the form

$$(\forall Z_1)(\forall Z_2) \ldots (\forall Z_m),$$

where Z_i varies over k_i-ary relations to just $(\forall Z)$ keeping the matrix elementary. But this follows immediately from the equivalences

$$(\forall Z \subseteq A^r)\mathscr{Q}(Z) \Leftrightarrow (\forall Y \subseteq A)\mathscr{Q}(\{(t_1, \ldots, t_r): \langle t_1, \ldots, t_r\rangle \in Y\}),$$

$$(\forall X \subseteq A)(\forall Y \subseteq A)\mathscr{Q}(X, Y) \Leftrightarrow (\forall Z \subseteq A)\mathscr{Q}(\{t: \langle c_0, t\rangle \in Z\}, \{t: \langle c_1, t\rangle \in Z\}).$$

⊣

A trivial relativization of the proof of 5D.4 yields a version of that theorem which is uniform in relation parameters.

6C.9. Parametrization Theorem (for relatively hyperelementary relations). *Let $\mathfrak{A}$ be acceptable. For each $\nu = (n, r_1, \ldots, r_k)$ there is an inductive second order relation*

$$\mathscr{H}^\nu(a, \bar{x}, \bar{Y}) \Leftrightarrow \mathscr{H}^\nu(a, x_1, \ldots, x_n, Y_1, \ldots, Y_k)$$

such that for each $\bar{Y}$ the $n+1$-ary relation

$$H^\nu(\bar{Y}) = \{(a, \bar{x}): (a, \bar{x}, \bar{Y}) \in \mathscr{H}^\nu\}$$

parametrizes the n-ary relations which are hyperelementary on $(\mathfrak{A}, \bar{Y})$. Moreover, there is an inductive $\mathscr{I}^\nu(a, \bar{Y})$ and a coinductive $\check{\mathscr{H}}^\nu(a, \bar{x}, \bar{Y})$ such that for all $\bar{Y}$:

(i) *If $R \subseteq A^n$ is hyperelementary on $(\mathfrak{A}, \bar{Y})$, then for some $a \in A$ such that $\mathscr{I}^\nu(a, \bar{Y})$,*

$$R(\bar{x}) \Leftrightarrow \mathscr{H}^\nu(a, \bar{x}, \bar{Y}).$$

(ii) *If $\mathscr{I}^\nu(a, \bar{Y})$, then for every $\bar{x}$,*

$$\mathscr{H}^\nu(a, \bar{x}, \bar{Y}) \Leftrightarrow \check{\mathscr{H}}^\nu(a, \bar{x}, \bar{Y}).$$ ⊣

The proof of 5D.4 does not relativize to yield a parametrization of the second order hyperelementary relations because it depends on the Covering Theorem. There is, however, a representation theorem which appears weak but can often substitute for a parametrization theorem in applications. Its proof depends on the following covering-type result.

6C.10. THEOREM. *Let* $\mathfrak{A}$ *be acceptable, let* $\mathscr{U}^\nu$ *be universal for the* ν*-ary inductive relations on* $\mathfrak{A}$*, where* $\nu = (n, r_1, \ldots, r_k)$ *is a signature with* $n \geqslant 1$*, and suppose* $\sigma\colon \mathscr{U}^\nu \twoheadrightarrow \lambda$ *is an inductive norm on* $\mathscr{U}^\nu$*. If for some a the relation*

$$\mathscr{R}(\bar{x}, \bar{Y}) \Leftrightarrow (\bar{x}, \bar{Y}) \in \mathscr{U}^\nu_a \Leftrightarrow (a, \bar{x}, \bar{Y}) \in \mathscr{U}^\nu$$

is hyperelementary, then there is a sequence of constants $c, \bar{x}^*$ *such that for every* $\bar{Y}$*,* $(c, \bar{x}^*, \bar{Y}) \in \mathscr{U}^\nu$ *and for every* $\bar{x}, \bar{Y}$*,*

$$\mathscr{R}(\bar{x}, \bar{Y}) \Leftrightarrow [(a, \bar{x}, \bar{Y}) \in \mathscr{U}^\nu \;\&\; \sigma(a, \bar{x}, \bar{Y}) \leqslant \sigma(c, \bar{x}^*, \bar{Y})].$$

PROOF. Put

$$\mathscr{Q}(\bar{x}, \bar{Y}) \Leftrightarrow (\exists \bar{x}')\{(a, \bar{x}', \bar{Y}) \in \mathscr{U}^\nu \;\&\; [(x_1, \bar{x}, \bar{Y}) \in \mathscr{U}^\nu \;\&\; \sigma(x_1, \bar{x}, \bar{Y}) < \sigma(a, \bar{x}', \bar{Y})]\}$$

$$\Leftrightarrow (\exists \bar{x}')\{\mathscr{R}(\bar{x}', \bar{Y}) \;\&\; \neg (a, \bar{x}', \bar{Y}) \leqslant^*_\sigma (x_1, \bar{x}, \bar{Y})\},$$

where $\bar{x} = x_1, \ldots, x_n$. Since $\mathscr{R}$ is hyperelementary, $\mathscr{Q}$ is coinductive (in fact it is hyperelementary), hence there is a fixed c such that

$$\mathscr{Q}(\bar{x}, \bar{Y}) \Leftrightarrow (c, \bar{x}, \bar{Y}) \notin \mathscr{U}^\nu.$$

Now choose $\bar{x}^* = x_1^*, \ldots, x_n^*$ so that $x_1^* = c$.

We prove that for every $\bar{Y}$, $(c, \bar{x}^*, \bar{Y}) \in \mathscr{U}^\nu$ by contradiction. If not, then by the choice of c we would have $\mathscr{Q}(\bar{x}^*, \bar{Y})$ and this implies immediately that $(c, \bar{x}^*, \bar{Y}) \in \mathscr{U}^\nu$ contrary to hypothesis.

Now that we know $(c, \bar{x}^*, \bar{Y}) \in \mathscr{U}^\nu$, we have $\neg \mathscr{Q}(\bar{x}^*, \bar{Y})$ by the choice of c, and this means

$$(\forall \bar{x}')[(a, \bar{x}', \bar{Y}) \in \mathscr{U}^\nu \Rightarrow \sigma(a, \bar{x}', \bar{Y}) \leqslant \sigma(c, \bar{x}^*, \bar{Y})]$$

which is what we wanted to show. ⊣

The result can be considered as a *Uniform Covering Theorem* in the following sense. Suppose $\mathscr{R}$ is hyperelementary, $\mathscr{U}$ is inductive but not hyperelementary, σ is an inductive norm on $\mathscr{U}$ and $f(\bar{x})$ is a hyperelementary function such that

$$(\bar{x}, \bar{Y}) \in \mathscr{R} \Rightarrow (f(\bar{x}), \bar{Y}) \in \mathscr{U}.$$

The Covering Theorem 3C.2 easily implies in these circumstances that for each $\bar{Y}$ such that $\{\bar{u}\colon \mathscr{U}(\bar{u}, \bar{Y})\}$ is not hyperelementary on $(\mathfrak{A}, \bar{Y})$, there exists some $\bar{c}_{\bar{Y}}$ such that

$$(\bar{x}, \bar{Y}) \in \mathscr{R} \Rightarrow \sigma(f(\bar{x}), \bar{Y}) \leqslant \sigma(\bar{c}_{\bar{Y}}, \bar{Y}).$$

Now Theorem 6C.10 says that in the special case when $\mathscr{U}$ is universal inductive and

$$f(\bar{x}) = (a, \bar{x}),$$

we can find a *single* $\bar{c}$, independent of $\bar{Y}$, such that

$$(\bar{x}, \bar{Y}) \in \mathscr{R} \Rightarrow \sigma(f(\bar{x}), \bar{Y}) \leqslant \sigma(\bar{c}, \bar{Y}).$$

6C.11. THEOREM. *Let* $\mathfrak{A}$ *be an acceptable structure and* $\mathscr{U}$ *a universal inductive set for inductive second order relations of signature* $(n+1, r_1, \ldots, r_k)$, *let* σ *be an inductive norm on* $\mathscr{U}$. *A relation* $\mathscr{R}(\bar{x}, \bar{Y})$ *of signature* $\nu = (n, r_1, \ldots, r_k)$ *is hyperelementary on* $\mathfrak{A}$ *if and only if there exist constants* $a_1, a_2, c_1, c_2, \bar{x}^*$ *such that*

$$\textit{for all } \bar{Y}, \quad (c_1, c_2, \bar{x}^*, \bar{Y}) \in \mathscr{U}$$

and

$$\begin{aligned} \mathscr{R}(\bar{x}, \bar{Y}) &\Leftrightarrow (a_1, a_2, \bar{x}, \bar{Y}) \in \mathscr{U} \\ &\Leftrightarrow \{(a_1, a_2, \bar{x}, \bar{Y}) \in \mathscr{U} \;\&\; \sigma(a_1, a_2, \bar{x}, \bar{Y}) \leqslant \sigma(c_1, c_2, \bar{x}^*, \bar{Y})\}. \end{aligned}$$

PROOF. Pick some a_2 and put

$$\mathscr{R}'(t, \bar{x}, \bar{Y}) \Leftrightarrow t = a_2 \;\&\; \mathscr{R}(\bar{x}, \bar{Y}),$$

choose a_1 so that

$$\mathscr{R}'(t, \bar{x}, \bar{Y}) \Leftrightarrow (a_1, t, \bar{x}, \bar{Y}) \in \mathscr{U}$$

and apply Theorem 6C.10. This proves the "only if" part, the "if" part being Theorem 6C.3. ⊣

6D. The class of hyperelementary relations

We collect in this section some interesting structure properties of the class of hyperelementary relations on an acceptable structure which involve inductive second order relations in their statements or proofs. Some of these are "the easy halves" of deeper and harder results that we will establish in the next chapter.

The key tool is the following computation estimate which for the case $\mathfrak{A} = \mathbb{N}$ is due to Kleene [1959a].

For each $m \geqslant 1$, put

$$\mathscr{H}\mathscr{E}^m(\mathfrak{A}) = \{Y \subseteq A^m \colon Y \textit{ is hyperelementary on } \mathfrak{A}\},$$

$$\mathscr{H}\mathscr{E}(\mathfrak{A}) = \textstyle\bigcup_m \mathscr{H}\mathscr{E}^m(\mathfrak{A}),$$

where we will simply write $\mathscr{H}\mathscr{E}^m$ or $\mathscr{H}\mathscr{E}$ when $\mathfrak{A}$ is clearly determined by the context.

6D.1. THEOREM. *Suppose*

$$\mathscr{P}(Z, \bar{x}, \bar{Y}) \Leftrightarrow \mathscr{P}(Z, \bar{x}, Y_1, \ldots, Y_k)$$

is inductive on the acceptable structure $\mathfrak{A}$, *where* Z *varies over the m-ary relations on* $\mathfrak{A}$, *let* $j \leqslant k$ *and put*

$$\mathscr{R}(\bar{x}, \bar{Y}) \Leftrightarrow (\exists Z)[Z \in \mathscr{HE}(\mathfrak{A}, Y_1, \ldots, Y_j) \,\&\, \mathscr{P}(Z, \bar{x}, \bar{Y})].$$

Then $\mathscr{R}$ *is inductive on* $\mathfrak{A}$.

PROOF. Letting $\nu = (m, r_1, \ldots, r_j)$, by the Parametrization Theorem 6C.9 we have

$$\mathscr{R}(\bar{x}, \bar{Y}) \Leftrightarrow (\exists a)\{(a, Y_1, \ldots, Y_j) \in \mathscr{I}^\nu \\ \& \mathscr{P}(\{\bar{u}: (a, \bar{u}, Y_1, \ldots, Y_j) \in \mathscr{H}^\nu\}, \bar{x}, \bar{Y})\}.$$

By 6B.2 there is an inductive $\mathscr{P}^\dagger(Z_1, Z_2, \bar{x}, \bar{Y})$, positively dependent on Z_1, Z_2, such that

$$\mathscr{P}(Z, \bar{x}, \bar{Y}) \Leftrightarrow \mathscr{P}^\dagger(Z, \neg Z, \bar{x}, \bar{Y});$$

using 6C.9 again,

$$\mathscr{R}(\bar{x}, \bar{Y}) \Leftrightarrow (\exists a)\{(a, Y_1, \ldots, Y_j) \in \mathscr{I}^\nu \\ \& \mathscr{P}^\dagger(\{\bar{u}: (a, \bar{u}, Y_1, \ldots, Y_j) \in \mathscr{H}^\nu\}, \\ \{\bar{u}: (a, \bar{u}, Y_1, \ldots, Y_j) \notin \check{\mathscr{H}}^\nu\}, \bar{x}, \bar{Y})\},$$

so that $\mathscr{R}$ is inductive by the Substitution Theorem 6B.3. ⊣

Of course, in the case $j = 0$ the result says that the relation

$$\mathscr{R}(\bar{x}, \bar{Y}) \Leftrightarrow (\exists Z \in \mathscr{HE})\mathscr{P}(Z, \bar{x}, \bar{Y})$$

is inductive and we only need appeal to 5D.4 for the proof.

From this follows immediately:

6D.2. THEOREM. *If* $\mathfrak{A}$ *is acceptable, then the class*

$$\mathscr{HE}^m(\mathfrak{A}) = \{X \subseteq A^m: X \text{ is hyperelementary, } m\text{-ary}\}$$

is inductive.

PROOF. Take

$$\mathscr{P}(Z, X) \Leftrightarrow Z = X$$

and notice that

$$X \in \mathscr{HE} \Leftrightarrow (\exists Z \in \mathscr{HE})\mathscr{P}(Z, X).$$ ⊣

Theorem 6D.1 is already interesting in the special case when $\mathscr{P}$ is elementary. One of its consequences is that if

$$R(\bar{x}) \Leftrightarrow (\exists Z \in \mathscr{HE})\mathscr{P}(Z, \bar{x})$$

with elementary $\mathscr{P}$, then R is inductive. In the next chapter we will prove that in fact every inductive R can be represented in this form.

6D.3. Collection Theorem. *Let $\mathscr{P}(t, Z, \bar{x})$ be inductive on the acceptable structure $\mathfrak{A}$, where Z varies over m-ary relations on A; then*

$$(*) \qquad (\forall t)(\exists Z \in \mathscr{HE}^m)\mathscr{P}(t, Z, \bar{x}) \Leftrightarrow (\exists Z \in \mathscr{HE}^{m+1})(\forall t)(\exists a)\mathscr{P}(t, Z_a, \bar{x}).$$

Proof. The implication from right to left is trivial, so assume the left-hand side of (*) for a fixed $\bar{x}$. By the Parametrization Theorem 5D.4 we know that

$$(2) \qquad (\forall t)(\exists a)\{a \in I^m \,\&\, \mathscr{P}(t, H^m_a, \bar{x})\},$$

using the notation of 5D.4. Keeping $\bar{x}$ fixed, put

$$(3) \qquad R(t, a) \Leftrightarrow a \in I^m \,\&\, \mathscr{P}(t, H^m_a, \bar{x}).$$

We first verify easily that R is inductive, as in the proof of 6D.1: choose an inductive $\mathscr{P}^\dagger(t, Z_1, Z_2, \bar{x})$, positively dependent on Z_1, Z_2 and such that

$$\mathscr{P}(t, Z, \bar{x}) \Leftrightarrow \mathscr{P}^\dagger(t, Z, \neg Z, \bar{x}),$$

and then apply the Substitution Theorem 6B.3 and the obvious equivalence

$$R(t, a) \Leftrightarrow a \in I^m \,\&\, \mathscr{P}^\dagger(t, \{s: H^m(a, s)\}, \{s: \neg \check{H}^m(a, s)\}, \bar{x}).$$

We know from (2) that $(\forall t)(\exists a)R(t, a)$, hence by the Hyperelementary Selection Theorem 3B.2 there is a hyperelementary R^* such that

$$(4) \qquad R^* \subseteq R, \qquad (\forall t)(\exists a)R^*(t, a).$$

Put

$$B = \{a: (\exists t)R^*(t, a)\};$$

B is hyperelementary and by the definition of R, $B \subseteq I^m$. Moreover, $(\forall t)(\exists a)R^*(t, a)$, hence by the definition of B and the fact that $R^* \subseteq R$,

$$(5) \qquad (\forall t)(\exists a)\{a \in B \,\&\, \mathscr{P}(t, H^m_a, \bar{x})\}.$$

We now put

$$\begin{aligned}(a, \bar{y}) \in Z &\Leftrightarrow a \in B \,\&\, H^m(a, \bar{y}) \\ &\Leftrightarrow a \in B \,\&\, \check{H}^m(a, \bar{y});\end{aligned}$$

Z is evidently hyperelementary and (5) implies trivially that $(\forall t)(\exists a)\mathscr{P}(t, Z_a, \bar{x})$. ⊣

Some of the most important consequences of the last three theorems can be summarized neatly in terms of *models for theories in the second order language over a structure* $\mathfrak{A}$. We can consider a collection $\mathscr{F}$ of relations on A as a structure for interpreting second order formulas over $\mathfrak{A}$ in the natural way, i.e. by interpreting the relation variables in the formulas as ranging over $\mathscr{F}$ (rather than over all the relations on A).

The schema of Σ_1^1-*Collection* on the structure $\mathfrak{A}$ is the class of all formulas of the form

$$(\Sigma_1^1\text{-}Coll) \qquad (\forall t)(\exists Z)\varphi(t, Z, \bar{Y}) \Leftrightarrow (\exists W)(\forall t)(\exists a)\varphi(t, W_a, \bar{Y})$$

where φ is in the first order language over the structure $\mathfrak{A}$, with no relation quantifiers. Similarly, the schema of Δ_1^1-*Comprehension* on $\mathfrak{A}$ is the class of all formulas of the form

$$(\Delta_1^1\text{-}Comp) \quad (\forall \bar{x})\{(\exists Z_1)\varphi(\bar{x}, Z_1, \bar{Y}) \Leftrightarrow (\forall Z_2)\psi(\bar{x}, Z_2, \bar{Y})\}$$
$$\Rightarrow (\exists W)(\forall \bar{x})[\bar{x} \in W \Leftrightarrow (\exists Z_1)\varphi(\bar{x}, Z_1, \bar{Y})],$$

where again φ, ψ are in the first order language over $\mathfrak{A}$. A class $\mathscr{F}$ of relations on A *satisfies* Σ_1^1-*Collection* on $\mathfrak{A}$ if all the formulas Σ_1^1-*Coll* are true when the variables $\bar{Y}$ are given values in $\mathscr{F}$ and the set quantifiers are interpreted as ranging over $\mathscr{F}$. Similarly for Δ_1^1-Comprehension on $\mathfrak{A}$.

6D.4. THEOREM. *If $\mathfrak{A}$ is an acceptable structure, then $\mathscr{H}\mathscr{E}(\mathfrak{A})$ is a model of both Σ_1^1-Collection and Δ_1^1-Comprehension.*

PROOF. Take Σ_1^1-Collection first. We must show that if $\bar{Y} = Y_1, \ldots, Y_k$ are hyperelementary on $\mathfrak{A}$, then

$$(\forall t)(\exists Z \in \mathscr{H}\mathscr{E})\varphi(t, Z, \bar{x}, \bar{Y}) \Leftrightarrow (\exists Z \in \mathscr{H}\mathscr{E})(\forall t)(\exists a)\varphi(t, Z_a, \bar{x}, \bar{Y});$$

but this is an immediate consequence of the Collection Theorem, if we put

$$\mathscr{P}(t, Z, \bar{x}) \Leftrightarrow \varphi(t, Z, \bar{x}, \bar{Y})$$

and notice that $\mathscr{P}$ is hyperelementary by 6B.5.

For Δ_1^1-Comprehension, assume the hypothesis, i.e. that $\bar{Y}$ are hyperelementary and for all $\bar{x}$,

$$(\exists Z_1 \in \mathscr{H}\mathscr{E})\varphi(\bar{x}, Z_1, \bar{Y}) \Leftrightarrow (\forall Z_2 \in \mathscr{H}\mathscr{E})\psi(\bar{x}, Z_2, \bar{Y}),$$

and put

$$R(\bar{x}) \Leftrightarrow (\exists Z_1 \in \mathscr{H}\mathscr{E})\varphi(\bar{x}, Z_1, \bar{Y}).$$

Now 6D.1 implies immediately that R is both inductive and coinductive, hence the conclusion. ⊣

Exercises for Chapter 6

6.1. Let X vary over A^{2n} and Y vary over A^{2m}. Prove that there are relations $\mathscr{P}(X, Y)$, $\mathscr{Q}(X, Y)$, inductive and coinductive, respectively, such that whenever Y is wellfounded,

$$X \text{ is wellfounded } \& \ rank(X) \leqslant rank(Y) \Leftrightarrow \mathscr{P}(X, Y)$$
$$\Leftrightarrow \mathscr{Q}(X, Y). \qquad \dashv$$

6.2. Let

$$G(f, \bar{x}) = G(f, x_1, \ldots, x_n)$$

be a partial function with arguments n-ary partial functions and n-tuples from A and values in A. We call G *positive inductive on* $\mathfrak{A} = \langle A, R_1, \ldots, R_l \rangle$ if there is an inductive relation $\mathscr{R}(Z, \bar{x}, y)$, positively dependent on Z, such that for every partial function f,

$$G(f, \bar{x}) = y \Leftrightarrow \mathscr{R}(G_f, \bar{x}, y),$$

where G_f is the graph of f.

Prove that each positive inductive $G(f, \bar{x})$ is *monotone*, i.e.

$$[f_1 \subseteq f_2 \ \& \ G(f_1, \bar{x}) = y] \Rightarrow G(f_2, \bar{x}) = y,$$

and that it has *a least fixed point*, i.e. a partial function f^* such that

$$(\forall \bar{x})[G(f^*, \bar{x}) = f^*(\bar{x})],$$
$$(\forall \bar{x})[G(f, \bar{x}) = f(\bar{x})] \Rightarrow f^* \subseteq f.$$

Prove that this least fixed point f^* has inductive graph. (The *First Recursion Theorem* for positive inductive definability.) $\dashv$

6.3. Suppose $\prec$ is a wellfounded relation with field some $B \subseteq A^n$ which is hyperelementary on $\mathfrak{A}$, let $f(\bar{x}, \bar{z})$ be a function defined for $\bar{x} \in B$ with values in some A^m such that for some inductive relation $\mathscr{R}$,

$$f(\bar{x}, \bar{z}) = \bar{y} \Leftrightarrow \mathscr{R}(\{(\bar{x}', \bar{z}', f(\bar{x}', \bar{z}')) : \bar{x}' \prec \bar{x}\}, \bar{x}, \bar{z}, \bar{y}).$$

Prove that f is hyperelementary. $\dashv$

6.4. Suppose $\prec$ is a wellfounded relation with field some $B \subseteq A^n$ which is hyperelementary on $\mathfrak{A}$, let $f(\bar{x}, \bar{z})$ be a function defined for $\bar{x} \in B$ with values subsets of some A^m and such that for some hyperelementary $\mathscr{R}$,

$$\bar{y} \in f(\bar{x}, \bar{z}) \Leftrightarrow \mathscr{R}(\{(\bar{x}', \bar{z}', \bar{y}') : \bar{x}' \prec \bar{x} \ \& \ \bar{y}' \in f(\bar{x}', \bar{z}')\}, \bar{x}, \bar{z}, \bar{y}).$$

Prove that each $f(\bar{x}, \bar{z})$ is hyperelementary and that the relation

$$\bar{y} \in f(\bar{x}, \bar{z})$$

is hyperelementary. ⊣

6.5. Suppose $\mathfrak{A}$ is acceptable, $\mathscr{R}(X, Y)$ is inductive with X, Y varying over n-ary relations and X_0 is a fixed hyperelementary n-ary relation. Prove that

$$(\forall X \in \mathscr{HE})(\exists Y \in \mathscr{HE})\mathscr{R}(X, Y)$$
$$\Rightarrow (\exists W \in \mathscr{HE})\{W_{\langle 0,0\rangle} = X_0 \ \&\ (\forall n)(\forall a)(\forall b)\mathscr{R}(W_{\langle n,a\rangle}, W_{\langle n+1,b\rangle})\}.$$

(*Dependent Collection Theorem.*) ⊣

CHAPTER 7

SECOND ORDER CHARACTERIZATIONS

In the last section of Chapter 6 we showed that the class $\mathcal{HE}(\mathfrak{A})$ of hyperelementary relations on an acceptable structure $\mathfrak{A}$ satisfies certain properties which are formally expressible in the second order language over $\mathfrak{A}$. We show here that these results can be turned around to give elegant, model theoretic representation theorems and characterizations of the inductive and hyperelementary relations. Except for the trivial, folklore observations of Section 7A, the theorems of this chapter are essentially from Moschovakis [1969b], [1969c] and they extend to all acceptable structures work of Spector [1960], Gandy [1960], Kreisel [1961] and particularly Kleene [1959a] for the structure of arithmetic $\mathbb{N}$. The proofs, however, are from a very different point of view than that of Moschovakis [1969b], [1969c] and the classical work, the main new ingredient being the systematic exploitation of the game theoretic ideas introduced in Chapter 4.

7A. Inductive and Σ_1^1 relations

We have already observed that inductive relations are Π_1^1 and we will show in Chapter 8 that on *countable* acceptable structures the converse holds. On certain uncountable structures, however, it may be that inductive relations are also Σ_1^1 and in this case Theorem 6C.8 easily implies that not all Π_1^1 relations are inductive. The key computation is in the next easy result.

For $n \geqslant 1$, let Y vary over $2n$-ary relations on A (which we view as binary relations on A^n) and put

$$\mathcal{WF}^n(Y) \Leftrightarrow Y \textit{ is wellfounded.}$$

The proof of 6A.1 easily extends to show that on an infinite $\mathfrak{A}$ each $\mathcal{WF}^n$ is an inductive class of relations.

7A.1. THEOREM. *If $\varphi(\bar{x}, S)$ is S-positive in the language of an infinite structure $\mathfrak{A}$, where $\bar{x}$ varies over n-tuples and S over n-ary relations, then there is an elementary formula $\theta(Y, \bar{x})$ such that*

$$\begin{aligned}\bar{x} \in I_\varphi &\Leftrightarrow (\exists Y)\{\mathcal{WF}^n(Y) \;\&\; \theta(Y, \bar{x})\}\\ &\Leftrightarrow (\exists Y)\{Y \in \mathcal{HE}(\mathfrak{A}) \;\&\; \mathcal{WF}^n(Y) \;\&\; \theta(Y, \bar{x})\}.\end{aligned}$$

PROOF. Put

$$\theta(Y, \bar{x}) \Leftrightarrow \bar{x} \in Field(Y) \,\&\, (\forall \bar{z})[\bar{z} \in Field(Y) \Rightarrow \varphi(\bar{z}, \{\bar{x}' : (\bar{x}', \bar{z}) \in Y\})].$$

To prove

$$\bar{x} \in I_\varphi \Rightarrow (\exists Y)\{Y \in \mathscr{H}\mathscr{E}(\mathfrak{A}) \,\&\, \mathscr{W}\mathscr{F}^n(Y) \,\&\, \theta(Y, \bar{x})\},$$

we take

$$Y = \{(\bar{x}_1, \bar{x}_2) : \bar{x}_1, \bar{x}_2 \in I_\varphi \,\&\, |\bar{x}_1|_\varphi < |\bar{x}_2|_\varphi \leqslant |\bar{x}|_\varphi\}$$

and apply the Stage Comparison Theorem 2A.2.

To prove

$$\mathscr{W}\mathscr{F}^n(Y) \,\&\, \theta(Y, \bar{x}) \Rightarrow \bar{x} \in I_\varphi,$$

we verify by an easy transfinite induction on $\rho^Y(\bar{z})$ (= the rank of $\bar{z}$ for $\bar{z} \in Field(Y)$) that

$$\bar{z} \in Field(Y) \Rightarrow \bar{z} \in I_\varphi. \quad \dashv$$

From this and the version of 6A.1 for $\mathscr{W}\mathscr{F}^n$, all $n \geqslant 1$, we get immediately:

7A.2. THEOREM. *The relations $\mathscr{W}\mathscr{F}^n$ on an infinite structure $\mathfrak{A}$ are Σ_1^1 if and only if all inductive relations on $\mathfrak{A}$ are Σ_1^1.* $\dashv$

Theorem 6C.3 constructs examples of such structures.

7B. Quasistrategies

Recall from Section 4A that a game is determined by an infinite string

(1) $$\bar{Q} = Q_0, Q_1, Q_2, \ldots$$

of quantifiers and a relation $R(f)$ on infinite sequences on the relevant set A. Moreover, the game is open if

(2) $$R(x_0, x_1, x_2, \ldots) \Leftrightarrow \bigvee_i R_i(x_0, \ldots, x_i),$$

and the game is closed if

(3) $$R(x_0, x_1, x_2, \ldots) \Leftrightarrow \bigwedge_i R_i(x_0, \ldots, x_i)$$

for suitable finitary relations $R_0, R_1, R_2, \ldots$.

Suppose $\mathfrak{A}$ is acceptable and $\mathscr{C}$ is a fixed elementary coding scheme on $\mathfrak{A}$. We call an open or closed game $G = G(\bar{Q}, R)$ *elementary* (or *hyperelementary*) if (1) and (2) or (3) hold with a string $\bar{Q}$ and relations $R_0, R_1, R_2, \ldots$ such that the function

$$q_{\bar{Q}}(i) = \begin{cases} 0 & \text{if } Q_i = \exists, \\ 1 & \text{if } Q_i = \forall, \end{cases}$$

and the relation

$$R_G(z) \Leftrightarrow Seq(z) \;\&\; (\exists i)[lh(z) = i+1 \;\&\; R_i((z)_1, (z)_2, \ldots, (z)_{i+1})]$$

are elementary (or hyperelementary). It is easy to verify that this notion is independent of the particular elementary coding scheme chosen to define it, see Exercise 5.4.

If G is an elementary game on $\mathfrak{A}$ and we know that $(\exists)$ wins G, it is natural to ask whether $(\exists)$ has a definable (in particular hyperelementary) winning strategy. It may be the case that this fails, even though $(\exists)$ can clearly win in a more or less trivial manner—this is roughly because $(\exists)$ must choose each time one move out of many equally good moves and he may lack a definable choice function for $\mathfrak{A}$, see Exercise 7.2. It is more natural to reformulate the question *whether* $(\exists)$ *can win definably* in the context of quasistrategies, multiple-valued strategies which avoid the issue of choices.

A *quasistrategy* Σ for $(\exists)$ in the game $G(\bar{Q}, R)$ is a set of finite sequences such that

$$(*)\quad \begin{cases} \emptyset \in \Sigma, \\ \textit{if } Q_i = \exists \textit{ and } (x_0, \ldots, x_{i-1}) \in \Sigma, \textit{ then } (\exists x)(x_0, \ldots, x_{i-1}, x) \in \Sigma, \\ \textit{if } Q_i = \forall \textit{ and } (x_0, \ldots, x_{i-1}) \in \Sigma, \textit{ then } (\forall x)(x_0, \ldots, x_{i-1}, x) \in \Sigma. \end{cases}$$

We call Σ *winning for* $(\exists)$ if for all $f: \omega \to A$,

$$(\forall i)[(f(0), f(1), \ldots, f(i)) \in \Sigma] \Rightarrow R(f).$$

Quasistrategies and winning quasistrategies for $(\forall)$ are defined in the obvious dual manner. The idea is that a winning quasistrategy gives the player *some good moves* each time it is his turn, perhaps more than one of them. If he plays one of the good moves each time (and he must make choices to do this), then he is certain to win.

7B.1. THEOREM. *Let* $G = G(\bar{Q}, R)$ *be a game on some set* A. *Player* $(\exists)$ *wins* G *if and only if* $(\exists)$ *has a winning quasistrategy for* G, *and similarly for* $(\forall)$.

PROOF. If $(\exists)$ wins G, let

$$\mathscr{S} = \{f_i : Q_i = \exists\}$$

be a winning strategy for $(\exists)$ in the notation of Section 4A and put

$$\Sigma = \{(x_0, x_1, \ldots, x_n) : \textit{if } i \leqslant n \textit{ and } Q_i = \exists, \textit{ then } x_i = f_i(x_0, \ldots, x_{i-1})\}.$$

It is trivial to verify that Σ is a winning quasistrategy for $(\exists)$.

Conversely, suppose Σ is a winning quasistrategy for $(\exists)$ and let

$$h\colon Power(A) - \{\emptyset\} \to A$$

be a choice function which assigns to each nonempty subset of A one of its elements,

$$X \subseteq A \;\&\; X \neq \emptyset \Rightarrow h(X) \in X.$$

We define a sequence of functions f_i, one for each i, such that $Q_i = \exists$ using the choice function,

$$f_i(x_0, \ldots, x_{i-1}) = h(\{t\colon (x_0, x_1, \ldots, x_{i-1}, t) \in \Sigma\}),$$

and it is trivial to verify that $\{f_i\colon Q_i = \exists\}$ is a winning strategy for $(\exists)$. ⊣

To measure the definability of a quasistrategy Σ relative to a coding scheme $\mathscr{C}$, we put

$$\langle \Sigma \rangle^{\mathscr{C}} = \{\langle x_1, \ldots, x_n \rangle\colon (x_1, \ldots, x_n) \in \Sigma\}.$$

We call Σ *elementary*, *hyperelementary*, *inductive* or *coinductive* according as $\langle \Sigma \rangle^{\mathscr{C}}$ is, where $\mathscr{C}$ is any elementary coding scheme. As with games, those notions are independent of the particular elementary coding scheme $\mathscr{C}$ used to define them, see Exercise 5.4.

7B.2. THEOREM. *If* $(\exists)$ *wins a closed hyperelementary game on an acceptable structure* $\mathfrak{A}$, *then* $(\exists)$ *has a coinductive winning quasistrategy.*

PROOF. If the game is $G = G(\bar{Q}, R)$ with

$$\bar{Q} = Q_0, Q_1, Q_2, \ldots,$$

$$R(x_0, x_1, x_2, \ldots) \Leftrightarrow \bigwedge_i R_i(x_0, \ldots, x_i),$$

we clearly want to put

$$\Sigma = \{\emptyset\} \cup \{(x_0, \ldots, x_n)\colon \{(Q_{n+1}x_{n+1})(Q_{n+2}x_{n+2}) \ldots\} \bigwedge_i R_i(x_0, \ldots, x_i)\}.$$

To see that this is a coinductive quasistrategy, put

$$\varphi(x, S) \Leftrightarrow Seq(x) \;\&\; \{\neg R_G(x) \vee [q_{\bar{Q}}(lh(x)) = 0 \;\&\; (\forall t)[x^\frown\langle t\rangle \in S]]$$
$$\vee [q_{\bar{Q}}(lh(x)) = 1 \;\&\; (\exists t)[x^\frown\langle t\rangle \in S]]\}.$$

It is now very easy to show by the customary game-theoretic argument that

$$\langle x_0, \ldots, x_n \rangle \in I_\varphi \Leftrightarrow \{(Q_{n+1}x_{n+1})(Q_{n+2}x_{n+2}) \ldots\} \bigwedge_i R_i(x_0, \ldots, x_i),$$

so that

$$\langle \Sigma \rangle = \{\emptyset\} \cup \neg I_\varphi$$

and Σ is coinductive. ⊣

We will see in the exercises that this is the best possible for closed hyperelementary games, but that if (∃) wins an open hyperelementary game, then (∃) has a hyperelementary winning quasistrategy.

The next result is easy, but it is very useful for the applications of closed games that we give in this chapter.

7B.3. THEOREM. *Let* $\bar{\mathrm{Q}} = \mathrm{Q}_0, \mathrm{Q}_1, \ldots$ *be an infinite string of quantifiers such that the function* $q_{\bar{\mathrm{Q}}}$ *describing it is elementary on a structure* $\mathfrak{A}$, *let* $\mathscr{C}$ *be an elementary coding scheme on* $\mathfrak{A}$, *let* $R(x, \bar{y})$ *be elementary and for each* $\bar{y}$ *define the closed game*

$$G_{\bar{y}} = G(\bar{\mathrm{Q}}, R_{\bar{y}}),$$

where

$$R_{\bar{y}}(x_0, x_1, \ldots) \Leftrightarrow \bigwedge_i R(\langle x_0, \ldots, x_i \rangle, \bar{y}).$$

Then the second order relation

$$\mathscr{P}(X, \bar{y}) \Leftrightarrow X = \langle \Sigma \rangle, \textit{for some quasistrategy } \Sigma, \textit{ winning for } (\exists) \textit{ in } G_{\bar{y}}$$

is elementary. Hence the relation

$$Q(\bar{y}) \Leftrightarrow (\exists) \textit{ wins } G_{\bar{y}}$$

is Σ^1_1.

PROOF. We compute:

$$\begin{aligned} \mathscr{P}(X, \bar{y}) \Leftrightarrow\ & (\forall x)[x \in X \Rightarrow Seq(x)] \\ & \& \langle \emptyset \rangle \in X \\ & \& (\forall x)[(x \in X \ \& \ q_{\bar{\mathrm{Q}}}(lh(x)) = 0) \Rightarrow (\exists t)[x^\frown\langle t \rangle \in X]] \\ & \& (\forall x)[(x \in X \ \& \ q_{\bar{\mathrm{Q}}}(lh(x)) = 1) \Rightarrow (\forall t)[x^\frown\langle t \rangle \in X]] \\ & \& (\forall x)[x \in X \Rightarrow R(x, \bar{y})]. \end{aligned}$$

⊣

Again we will see in the exercises that this result fails for open games.

It is worth putting down explicitly a useful, immediate corollary of Theorems 7B.2 and 7B.3. For $\mathfrak{A} = \mathbb{N}$ this is due to Kleene.

7B.4. THEOREM. *If* $R(\bar{x})$ *is a coinductive relation on an acceptable structure* $\mathfrak{A}$, *then there exists an elementary* $\mathscr{P}(Y, \bar{x})$ *such that*

$$\begin{aligned} R(\bar{x}) & \Leftrightarrow (\exists Y)\mathscr{P}(Y, \bar{x}) \\ & \Leftrightarrow (\exists Y)\{Y \textit{ is coinductive } \& \ \mathscr{P}(Y, \bar{x})\}. \end{aligned}$$

PROOF. By the Normal Form Theorem 5C.2, if $R(\bar{x})$ is coinductive, then there is an elementary $P(\bar{z}, \bar{x})$ such that

$$R(\bar{x}) \Leftrightarrow (\mathrm{G}^{\cup}z)P(z, \bar{x})$$

$$\Leftrightarrow \{(\exists s_1)(\forall t_1)\ldots\}(\forall i)P(\langle s_1, t_1, \ldots, s_i, t_i\rangle, \bar{x}).$$

One easily constructs a $P^*(z, \bar{x})$ such that with the elementary string

$$\bar{\mathrm{Q}} = \exists, \forall, \exists, \ldots$$

we have in the notation of 7B.3,

$$R(\bar{x}) \Leftrightarrow (\exists) \textit{ wins the closed game } G(\bar{\mathrm{Q}}, P^*_{\bar{x}}).$$

By 7B.3 then, we have an elementary $\mathscr{P}(Y, \bar{x})$ such that

$$\mathscr{P}(Y, \bar{x}) \Leftrightarrow Y \textit{ codes a winning quasistrategy for } (\exists) \textit{ in } G(\bar{\mathrm{Q}}, P^*_{\bar{x}}).$$

Hence

$$R(\bar{x}) \Leftrightarrow (\exists Y)\mathscr{P}(Y, \bar{x}),$$

and by 7B.2,

$$R(\bar{x}) \Leftrightarrow (\exists Y)\{Y \textit{ is coinductive } \& \mathscr{P}(Y, \bar{x})\}. \qquad \dashv$$

7C. The Second Stage Comparison Theorem

The result of this section is the key computational estimate of the chapter. It is essentially the same as Lemma 17 of Moschovakis [1969c], where the proof is quite different.

Going back to the notation of the Stage Comparison Theorem 2A.2, suppose $\varphi(\bar{x}, S)$, $\psi(\bar{y}, T)$ are respectively S and T positive in the language of some structure $\mathfrak{A}$. For each $\xi < \kappa = \kappa^{\mathfrak{A}}$, put

(1) $$(\bar{x}, \bar{y}) \in H^{\xi}_{\varphi,\psi} \Leftrightarrow \bar{x} \in I_{\varphi} \;\&\; \bar{y} \in I_{\psi} \;\&\; |\bar{x}|_{\varphi} \leqslant |\bar{y}|_{\psi} \leqslant \xi.$$

The relation $H^{\xi}_{\varphi,\psi}$ is the restriction of the relation $\leqslant^*_{\varphi,\psi}$ to those $\bar{x}, \bar{y}$ with ordinals $|\bar{x}|_{\varphi}$, $|\bar{y}|_{\psi}$ no greater than ξ. We let

(2) $$\bar{H}^{\xi}_{\varphi,\psi} = \langle H^{\xi}_{\varphi,\psi}, H^{\xi}_{\varphi,\varphi}, H^{\xi}_{\psi,\varphi}, H^{\xi}_{\psi,\psi}\rangle$$

be the quadruple of these relations that come by considering all ordered pairs from the doubleton $\{\varphi, \psi\}$.

Notice that many complicated relations involving the ordinals $|\bar{x}|_{\varphi}$, $|\bar{y}|_{\psi}$ are elementary in the expanded structure $(\mathfrak{A}, \bar{H}^{\xi}_{\varphi,\psi})$. E.g., if $\xi < \|\varphi\|$, then

$$\bar{x} \in I_{\varphi} \;\&\; |\bar{x}|_{\varphi} < \xi \Leftrightarrow (\exists \bar{x}')[(\bar{x}, \bar{x}') \in H^{\xi}_{\varphi,\varphi} \;\&\; \neg(\bar{x}', \bar{x}) \in H^{\xi}_{\varphi,\varphi}].$$

When the structure $\mathfrak{A}$ is fixed, we call a relation P *elementary in* $\bar{H}^{\xi}_{\varphi,\psi}$ if P is elementary on $(\mathfrak{A}, \bar{H}^{\xi}_{\varphi,\psi})$.

7C.1. SECOND STAGE COMPARISON THEOREM. *Let $\varphi(\bar{x}, S)$, $\psi(\bar{y}, T)$ be formulas in the language of an acceptable structure $\mathfrak{A}$ which are respectively positive in S, T, let*

$$\bar{H}^{\xi} = \bar{H}^{\xi}_{\varphi,\psi}$$

be the quadruple of relations assigned to each $\xi < \kappa = \kappa^{\mathfrak{A}}$ by (1) *and* (2) *above.*

There is an elementary second order relation $\mathscr{P}(Z, \bar{x}, \bar{y})$ such that

$$\text{(3)} \qquad \bar{y} <^*_{\psi,\varphi} \bar{x} \Rightarrow (\forall Z)\neg\mathscr{P}(Z, \bar{x}, \bar{y}),$$

$$\text{(4)} \qquad \bar{x} \leqslant^*_{\varphi,\psi} \bar{y} \;\&\; |\bar{x}|_{\varphi} = \xi \Rightarrow (\exists Z)\{Z \textit{ is elementary in } \bar{H}^{\xi} \;\&\; \mathscr{P}(Z, \bar{x}, \bar{y})\}.$$

Similarly, there is an elementary second order relation $\mathscr{Q}(Z, \bar{x}, \bar{y})$ such that

$$\text{(5)} \qquad \bar{y} \leqslant^*_{\psi,\varphi} \bar{x} \Rightarrow (\forall Z)\neg\mathscr{Q}(Z, \bar{x}, \bar{y}),$$

$$\text{(6)} \qquad \bar{x} <^*_{\varphi,\psi} \bar{y} \;\&\; |\bar{x}|_{\varphi} = \xi \Rightarrow (\exists Z)\{Z \textit{ is elementary in } \bar{H}^{\xi} \;\&\; \mathscr{Q}(Z, \bar{x}, \bar{y})\}.$$

Moreover, for each $\xi < \kappa$ the relations in the quadruple $\bar{H}^{\xi}$ are hyperelementary on $\mathfrak{A}$, so that we also have

$$\text{(7)} \qquad \bar{x} \leqslant^*_{\varphi,\psi} \bar{y} \Rightarrow (\exists Z \in \mathscr{H}\mathscr{E})\mathscr{P}(Z, \bar{x}, \bar{y}),$$

$$\text{(8)} \qquad \bar{x} <^*_{\varphi,\psi} \bar{y} \Rightarrow (\exists Z \in \mathscr{H}\mathscr{E})\mathscr{Q}(Z, \bar{x}, \bar{y}).$$

PROOF. Let us first dispense with the last part of the theorem, which follows trivially from the first Stage Comparison Theorem 2A.2. If $\xi < \kappa$, choose some χ and some $\bar{z}^* \in I_{\chi}$ such that $|\bar{z}^*|_{\chi} = \xi$ and notice that

$$(\bar{x}, \bar{y}) \in H^{\xi}_{\varphi,\psi} \Leftrightarrow \bar{y} \leqslant^*_{\psi,\chi} \bar{z}^* \;\&\; \bar{x} \leqslant^*_{\varphi,\psi} \bar{y};$$

now two applications of 2A.2 imply easily that $H^{\xi}_{\varphi,\psi}$ is hyperelementary.

We proceed to define $\mathscr{P}$ and prove (3), (4), the second part of the theorem having a very similar proof.

The idea is to assign to each $\bar{x}$, $\bar{y}$ a closed game $G(\bar{x}, \bar{y})$ and put

$$\mathscr{P}(Z, \bar{x}, \bar{y}) \Leftrightarrow Z = \langle\Sigma\rangle \textit{ for some quasistrategy } \Sigma \textit{ which is winning for } (\exists) \textit{ in } G(\bar{x}, \bar{y}).$$

We will have to show that

$$G(\bar{x}, \bar{y}) = G(\bar{Q}, R_{\bar{x},\bar{y}}),$$

in the notation of 7B.3, for some elementary string $\bar{Q}$ and an elementary relation $R(z, \bar{x}, \bar{y})$ and also that

$$\text{(7)} \qquad \bar{y} <^*_{\psi,\varphi} \bar{x} \Rightarrow (\forall) \textit{ wins } G(\bar{x}, \bar{y}),$$

$$\text{(8)} \qquad \bar{x} \leqslant^*_{\varphi,\psi} \bar{y} \;\&\; |\bar{x}|^{\varphi} = \xi \Rightarrow (\exists) \textit{ has a winning quasistrategy in } G(\bar{x}, \bar{y}) \textit{ which is elementary in } \bar{H}^{\xi}.$$

In view of the game theoretic analysis of fixed points in Chapter 4, we must have done something similar in the proof of the Stage Comparison Theorem 2A.2. Because to prove that $\bar{x} \leqslant^*_{\varphi,\psi} \bar{y}$ is a fixed point means precisely to associate with each $\bar{x}, \bar{y}$ an *open* game $G^*(\bar{x}, \bar{y})$ such that

(9) $\quad \bar{x} \leqslant^*_{\varphi,\psi} \bar{y} \Leftrightarrow (\exists)$ *wins* $G^*(\bar{x}, \bar{y})$.

Notice that (9) cannot hold with (canonically assigned) *closed* games $G^*(\bar{x}, \bar{y})$, because it would imply that $\leqslant^*_{\varphi,\psi}$ is coinductive, which in general it is not. Instead we will prove implications (7), (8), which together are a bit weaker than equivalence (9), except that we can find a simple bound for the complexity of the winning quasistrategy in (8).

We may assume that φ, ψ are in canonical form,

$$\varphi(\bar{x}, S) \equiv (\bar{Q}\bar{z})(\forall \bar{u})[\theta(\bar{x}, \bar{z}, \bar{u}) \vee S(\bar{u})],$$

$$\psi(\bar{y}, T) \equiv (\bar{P}\bar{w})(\forall \bar{v})[\tau(\bar{y}, \bar{w}, \bar{v}) \vee T(\bar{v})].$$

Take $G(\bar{x}, \bar{y})$ to be the closed game corresponding to the infinitary formula

(10) $$\{(\bar{Q}\bar{z}_1)(\forall \bar{u}_1)(\bar{P}^{\cup}\bar{w}_1)(\exists \bar{v}_1)(\bar{Q}\bar{z}_2)(\forall \bar{u}_2)(\bar{P}^{\cup}\bar{w}_2)(\exists \bar{v}_2)\ldots\}$$
$$\bigwedge_{k\in\omega} [\neg\tau(\bar{v}_k, \bar{w}_{k+1}, \bar{v}_{k+1}) \vee \bigvee_{i\leqslant k} \theta(\bar{u}_i, \bar{z}_{i+1}, \bar{u}_{i+1})],$$

where we have used the notation convention

$$\bar{u}_0 = \bar{x}, \qquad \bar{v}_0 = \bar{y}.$$

This means that we define $G(\bar{x}, \bar{y})$ so that

$$(\exists) \text{ wins } G(\bar{x}, \bar{y}) \Leftrightarrow (10) \text{ is true.}$$

To motivate this game, recall the normal forms for the fixed points I_φ, I_ψ:

$$\bar{x} \in I_\varphi \Leftrightarrow \{(\bar{Q}\bar{z}_1)(\forall \bar{u}_1)(\bar{Q}\bar{z}_2)(\forall \bar{u}_2)\ldots\} \bigvee_i \theta(\bar{u}_i, \bar{z}_{i+1}, \bar{u}_{i+1}),$$

$$\bar{y} \in I_\psi \Leftrightarrow \{(\bar{P}\bar{w}_1)(\forall \bar{u}_1)(\bar{P}\bar{w}_2)(\forall \bar{v}_2)\ldots\} \bigvee_k \tau(\bar{v}_k, \bar{w}_{k+1}, \bar{v}_{k+1}),$$

where again $\bar{u}_0 = \bar{x}$, $\bar{v}_0 = \bar{y}$. Let $G(\bar{x})$, $G(\bar{y})$ be the open games determined by these formulas. We can think of $G(\bar{x}, \bar{y})$ as a *closed combination* of these two games played by players X, Y as follows. First X, Y play in $G(\bar{x})$, with X making the moves of $(\exists)$ and Y making the moves of $(\forall)$ until $\bar{z}_1$, $\bar{u}_1$ are determined; then they play in $G(\bar{y})$ with X making the moves of $(\forall)$ and Y making the moves of $(\exists)$ until $\bar{w}_1$, $\bar{v}_1$ are determined; then they go back to $G(\bar{x})$ and play similarly until $\bar{z}_2$, $\bar{u}_2$ are determined, etc. Since $G(\bar{x})$ and $G(\bar{y})$ are open games, if the player making $(\exists)$'s moves in one of them wins, we know this at a finite stage of the game. *Now X wins the combined game $G(\bar{x}, \bar{y})$ if either he wins $G(\bar{y})$, where he makes $(\forall)$'s moves, or if he wins $G(\bar{x})$ no later than the stage at which he loses $G(\bar{y})$.*

There is a rather obvious modification of this idea which yields an *open combination* $G^*(\bar{x}, \bar{y})$ of $G(\bar{x})$ and $G(\bar{y})$, where X wins if he actually wins $G(\bar{x})$, and no later than any stage at which he loses $G(\bar{y})$. This is the $G^*(\bar{x}, \bar{y})$ for which the exact equivalence (9) holds, see Exercise 7.6. Here we are interested in a closed game, so we can apply 7B.3, and we are also interested in estimating the complexity of the winning quasistrategies—the fact that we lack the full force of the exact equivalence (9) can be circumvented in the applications at the expense of some technical complications.

It is easy to see that the infinite string $\bar{\mathrm{Q}}^*$ in (10) is elementary and that $G(\bar{x}, \bar{y}) = G(\bar{\mathrm{Q}}^*, R_{\bar{x},\bar{y}})$ in the notation of 7B.3 for some elementary relation R. All that is needed is a routine computation coding these games in terms of a fixed elementary coding scheme on $\mathfrak{A}$. It remains to prove (7) and (8).

Proof of (7). Although it is not necessary to take cases for this proof, it will help understand the game if we first consider the possibility

$$\bar{y} \in I_\psi \ \&\ \bar{x} \notin I_\varphi;$$

now the hypothesis of (7) holds, so we must show that $(\forall)$ wins $G(\bar{x}, \bar{y})$. Using the canonical expressions for I_φ, I_ψ, it is obvious that $(\forall)$ can play in this case so that on the one hand

$$\bigvee_k \tau(\bar{v}_k, \bar{w}_{k+1}, \bar{v}_{k+1})$$

since $\bar{y} = \bar{v}_0 \in I_\psi$, and on the other hand

$$\bigwedge_i \neg\theta(\bar{u}_i, \bar{z}_{i+1}, \bar{u}_{i+1})$$

since $\bar{x} = \bar{u}_0 \notin I_\varphi$. This play will surely win $G(\bar{x}, \bar{y})$ for $(\forall)$.

The other possibility, if the hypothesis of (7) holds, is that

$$\bar{x} \in I_\varphi, \quad \bar{y} \in I_\psi, \quad |\bar{y}|_\psi < |\bar{x}|_\varphi.$$

We will utilize the following two equivalences which hold for every ordinal $\xi \leqslant \kappa$ and every $\bar{u}$:

$$(11)_\varphi \qquad |\bar{u}|_\varphi \leqslant \xi \Leftrightarrow (\bar{\mathrm{Q}}\bar{z})(\forall \bar{u}')[\theta(\bar{u}, \bar{z}, \bar{u}') \vee |\bar{u}'|_\varphi < \xi],$$

$$(12)_\varphi \qquad |\bar{u}|_\varphi > \xi \Leftrightarrow (\bar{\mathrm{Q}}^\smile\bar{z})(\exists \bar{u}')[\neg\theta(\bar{u}, \bar{z}, \bar{u}') \ \&\ \xi \leqslant |\bar{u}'|_\varphi].$$

The first is immediate by the definition of $|\bar{u}|_\varphi$ and the second follows by taking negations. Let $(11)_\psi$, $(12)_\psi$ be the corresponding equivalences for ψ.

Assume towards a contradiction that $(\exists)$ wins $G(\bar{x}, \bar{y})$, consider a run of this game where $(\exists)$ plays to win and $(\forall)$ plays as follows.

Let

$$\xi_0 = |\bar{y}|_\psi.$$

Since $\xi_0 < |\bar{x}|_\varphi$, we have

$$(\bar{\mathrm{Q}}^\smile\bar{z}_1)(\exists \bar{u}_1)[\neg\theta(\bar{x}, \bar{z}_1, \bar{u}_1) \ \&\ \xi_0 \leqslant |\bar{u}_1|_\varphi]$$

by $(12)_\varphi$ and $(\forall)$ can play in $G(\bar{x}, \bar{y})$ so that when $\bar{z}_1, \bar{u}_1$ are determined we have

$$(13) \qquad \neg\theta(\bar{x}, \bar{z}_1, \bar{u}_1) \;\&\; \xi_0 \leqslant |\bar{u}_1|_\varphi.$$

Also $|\bar{y}|_\psi \leqslant \xi_0$, so by $(11)_\psi$,

$$(\mathrm{P}\bar{w}_1)(\forall \bar{v}_1)[\tau(\bar{y}, \bar{w}_1, \bar{v}_1) \vee |\bar{v}_1|_\psi < \xi_0],$$

so $(\forall)$ can continue playing until $\bar{w}_1, \bar{v}_1$ are determined and we have

$$(14) \qquad \tau(\bar{y}, \bar{w}_1, \bar{v}_1) \vee |\bar{v}_1|_\psi < \xi_0.$$

Since $(\exists)$ is playing to win, we also know that at this stage

$$(15) \qquad \neg\tau(\bar{y}, \bar{w}_1, \bar{v}_1) \vee \theta(\bar{x}, \bar{z}_1, \bar{u}_1).$$

Clearly (13), (14) and (15) yield

$$\neg\theta(\bar{x}, \bar{z}_1, \bar{u}_1), \quad \neg\tau(\bar{y}, \bar{w}_1, \bar{v}_1), \quad |\bar{v}_1|_\psi < \xi_0 \leqslant |\bar{u}_1|_\varphi.$$

Put

$$\xi_1 = |\bar{v}_1|_\psi < \xi_0$$

and repeat the argument, substituting $\bar{u}_1$ for $\bar{x}$ and $\bar{v}_1$ for $\bar{y}$. After $\bar{z}_2, \bar{u}_2, \bar{w}_2, \bar{v}_2$ are determined we will again have

$$\neg\theta(\bar{u}_1, \bar{z}_2, \bar{u}_2), \quad \neg\tau(\bar{v}_1, \bar{w}_2, \bar{v}_2), \quad |\bar{v}_2|_\psi < \xi_1 \leqslant |\bar{u}_2|_\varphi,$$

and we can put

$$\xi_2 = |\bar{v}_2|_\psi < \xi_1$$

and continue repeating indefinitely. The process yields an infinite descending sequence $\xi_0 > \xi_1 > \ldots$ of ordinals, which is absurd.

Proof of (8). We assume that

$$\bar{x} \leqslant^*_{\varphi,\psi} \bar{y} \;\&\; |\bar{x}|_\varphi = \xi$$

and we describe informally how $(\exists)$ can win $G(\bar{x}, \bar{y})$. Afterwards we will argue that the quasistrategy we are defining is elementary in $\overline{H}^\xi$.

Since $|\bar{x}|_\varphi \leqslant \xi$, the right-hand side of $(11)_\varphi$ holds with $\bar{u} = \bar{x}$ and $(\exists)$ can play to produce $\bar{z}_1, \bar{u}_1$ and insure that

$$(16) \qquad \theta(\bar{x}, \bar{z}_1, \bar{u}_1) \vee |\bar{u}_1|_\varphi < \xi.$$

If the first conjunct holds, he can continue playing *trivially* since he has already won the game—i.e. no matter what $\bar{z}_i, \bar{u}_i, \bar{w}_i, \bar{v}_i$ are played after that, $\theta(\bar{x}, \bar{z}_1, \bar{u}_1)$ insures that the matrix of formula (10) will be true.

If the second disjunct of (16) holds, we know

$$|\bar{u}_1|_\varphi < |\bar{y}|_\psi;$$

so taking $\xi = |\bar{u}_1|_\varphi$ in $(12)_\psi$, (∃) can continue to play so that $\bar{w}_1$, $\bar{v}_1$ are produced and

$$(17) \qquad \neg\tau(\bar{y}, \bar{w}_1, \bar{v}_1) \;\&\; |\bar{u}_1|_\varphi \leqslant |\bar{v}_1|_\psi.$$

Now we are in a situation exactly as in the beginning of the game. We have $\bar{u}_1$, $\bar{v}_1$ determined and by (16) and (17) easily

$$\bar{u}_1 \leqslant^*_{\varphi,\psi} \bar{v}_1 \;\&\; |\bar{u}_1|_\varphi = \xi_1 < \xi,$$

$$\neg\theta(\bar{x}, \bar{z}_1, \bar{u}_1), \quad \neg\tau(\bar{y}, \bar{w}_1, \bar{v}_1).$$

The same method can be followed by (∃), until he produces $\bar{z}_2$, $\bar{u}_2$ such that $\theta(\bar{u}_1, \bar{z}_2, \bar{u}_2)$ and he can play trivially from then on, or he goes on to get $\bar{w}_2$, $\bar{v}_2$ so that $\neg\tau(\bar{v}_1, \bar{w}_2, \bar{v}_2)$ and

$$\bar{u}_2 \leqslant^*_{\varphi,\psi} \bar{v}_2 \;\&\; |\bar{u}_2|_\varphi = \xi_2 < \xi_1.$$

Continuing in this manner, (∃) eventually reaches a point from which he can win by playing trivially, since otherwise he will be defining an infinite decreasing sequence

$$\xi > \xi_1 > \xi_2 \ldots$$

of ordinals.

Let Σ be the quasistrategy for (∃) described above and consider the first part of Σ which produces $\bar{z}_1$, $\bar{u}_1$. This came out of $(11)_\varphi$, i.e. for this part (∃) was simply trying to win the finite game determined by

$$(\bar{Q}\bar{z}_1)(\forall\bar{u}_1)[\theta(\bar{x}, \bar{z}_1, \bar{u}_1) \vee |\bar{u}_1|_\varphi < \xi].$$

This finite game is easily elementary in $\bar{H}^\xi$ and surely (∃) can win it by following a quasistrategy which is elementary in $\bar{H}^\xi$. For example, if $\bar{Q} = \exists, \forall, \exists$, so that the assertion looks like

$$(\exists s)(\forall t)(\exists r)(\forall\bar{u}_1)[\theta(\bar{x}, s, t, r, \bar{u}_1) \vee |\bar{u}_1|_\varphi < \xi],$$

such a quasistrategy is given by

$$\begin{aligned}
&\{\emptyset\}\\
&\cup \{(s) : (\forall t)(\exists r)(\forall\bar{u}_1)[\theta(\bar{x}, s, t, r, \bar{u}_1) \vee |\bar{u}_1|_\varphi < \xi]\}\\
&\cup \{(s, t) : (\exists r)(\forall\bar{u}_1)[\theta(\bar{x}, s, t, r, \bar{u}_1) \vee |\bar{u}_1|_\varphi < \xi]\}\\
&\cup \{(s, t, r) : (\forall\bar{u}_1)[\theta(\bar{x}, s, t, r, \bar{u}_1) \vee |\bar{u}_1|_\varphi < \xi]\}\\
&\cup \ldots\\
&\cup \{(s, t, r, \bar{u}_1) : \theta(\bar{x}, s, t, r, \bar{u}_1) \vee |\bar{u}_1|_\varphi < \xi\}.
\end{aligned}$$

The idea is that (∃) each time can play any move from which he can go on to win, and the set of all possible moves is elementary in $\bar{H}^\xi$.

Having won this first part of the game, (∃) asks if the first or second disjunct of (16) holds. This is an elementary in $\bar{H}^\xi$ question, and then it leads him into either a trivial game or another finite game determined by (12), which is also elementary in $\bar{H}^\xi$.

To actually put down a definition of (Σ) (relative to a coding scheme) which is elementary in $\bar{H}^\xi$ requires a very messy formula, because to check whether $\langle s_0, s_1, \ldots, s_n\rangle \in \Sigma$ we must split up $s_0, s_1, \ldots, s_n$ into a sequence of blocks $\bar{z}_1, \bar{u}_1, \bar{w}_1, \bar{v}_1 \ldots$ with a tail left, and then put down the conditions guaranteeing that at each point of this sequence (∃) is on a winning quasi-strategy for the appropriate finite game determined by $(11)_\varphi$ or $(12)_\psi$. But it is clear that some formula in the language of $(\mathfrak{A}, \bar{H}^\xi)$ will do the trick, and we will omit the details. ⊣

It is clear from the proof of 7C.1 (and the proof of 7B.3 which we utilized) that the elementary formula defining $\mathscr{P}$ so that (3) and (4) hold was constructed explicitly from the given formulas φ, ψ using only a fixed elementary coding scheme on the structure $\mathfrak{A}$. This means that we may allow relation variables in φ, ψ and the same proof works, so we get the second order version of the theorem. Of course we need to relativize the definition of $\bar{H}^\xi_{\varphi,\psi}$.

Switching to the notation of the relativized Stage Comparison Theorem 6C.1, let $\varphi(\bar{x}, \bar{Y}, S)$, $\psi(\bar{z}, \bar{W}, T)$ be respectively S and T positive in the language of a structure $\mathfrak{A}$. To each $\bar{Y}, \bar{W}$ and each $\xi < \kappa^{(\mathfrak{A},\bar{Y},\bar{W})}$ we assign the relation $H^\xi_{\varphi,\psi}(\bar{Y}, \bar{W})$ defined by

$$(18)\quad (\bar{x}, \bar{z}) \in H^\xi_{\varphi,\psi}(\bar{Y}, \bar{W}) \Leftrightarrow \bar{x} \in I_\varphi(\bar{Y}) \;\&\; \bar{z} \in I_\psi(\bar{W}) \;\&\; |\bar{x}, \bar{Y}|_\varphi \leqslant |\bar{z}, \bar{W}|_\psi \leqslant \xi.$$

Put also

$$(19)\quad \bar{H}^\xi_{\varphi,\psi}(\bar{Y}, \bar{W}) = \langle H^\xi_{\varphi,\psi}(\bar{Y}, \bar{W}), H^\xi_{\varphi,\varphi}(\bar{Y}, \bar{Y}), H^\xi_{\psi,\varphi}(\bar{W}, \bar{Y}), H^\xi_{\psi,\psi}(\bar{W}, \bar{W})\rangle.$$

Extending the terminological conventions in the obvious way,

$$Z \text{ is elementary in } \bar{H}^\xi_{\varphi,\psi}(\bar{Y}, \bar{W}), \bar{Y}, \bar{W}$$
$$\Leftrightarrow Z \text{ is elementary on } (\mathfrak{A}, \bar{H}^\xi_{\varphi,\psi}(\bar{Y}, \bar{W}), \bar{Y}, \bar{W}).$$

7C.2. Second Stage Comparison Theorem (relativized). *Let $\varphi(\bar{x}, \bar{Y}, S)$, $\psi(\bar{z}, \bar{W}, T)$ be formulas in the language of an acceptable structure $\mathfrak{A}$ respectively positive in S, T, let $\bar{H}^\xi(\bar{Y}, \bar{W})$ be defined for each $\bar{Y}, \bar{W}$ and $\xi < \kappa^{(\mathfrak{A},\bar{Y},\bar{W})}$ by* (18), (19) *above.*

There is an elementary second order relation $\mathscr{P}(Z, \bar{x}, \bar{Y}, \bar{z}, \bar{W})$ such that

$$(20)\quad (\bar{z}, \bar{W}) \in \mathscr{I}_\psi \;\&\; |\bar{z}, \bar{W}|_\psi < |\bar{x}, \bar{Y}|_\varphi \Rightarrow (\forall Z)\neg\mathscr{P}(Z, \bar{x}, \bar{Y}, \bar{z}, \bar{W}),$$

$$(21)\quad (\bar{x}, \bar{Y}) \in \mathscr{I}_\varphi \;\&\; |\bar{x}, \bar{Y}|_\varphi = \xi \leqslant |\bar{z}, \bar{W}|_\psi$$
$$\Rightarrow (\exists Z)\{Z \text{ is elementary in } \bar{H}^\xi_{\varphi,\psi}(\bar{Y}, \bar{W}), \bar{Y}, \bar{W} \;\&\; \mathscr{P}(Z, \bar{x}, \bar{Y}, \bar{z}, \bar{W})\}.$$

Similarly, there is an elementary second order relation $\mathscr{Q}(Z, \bar{x}, \bar{Y}, \bar{z}, \bar{W})$ *such that*

(22) $(\bar{z}, \bar{W}) \in \mathscr{I}_\psi \;\&\; |\bar{z}, \bar{W}|_\psi \leqslant |\bar{x}, \bar{Y}|_\varphi \Rightarrow (\forall Z)\neg\mathscr{Q}(Z, \bar{x}, \bar{Y}, \bar{z}, \bar{W}),$

(23) $(\bar{x}, \bar{Y}) \in \mathscr{I}_\varphi \;\&\; |\bar{x}, \bar{Y}|_\varphi = \xi < |\bar{z}, \bar{W}|_\psi$

$\Rightarrow (\exists Z)\{Z$ *is elementary in* $\bar{H}^\xi_{\varphi,\psi}(\bar{Y}, \bar{W})\bar{Y}, \bar{W} \;\&\; \mathscr{Q}(Z, \bar{x}, \bar{Y}, \bar{z}, \bar{W})\}.$

Moreover, for each $\bar{Y}, \bar{W}$ *and each* $\xi < \kappa^{(\mathfrak{A}, \bar{Y}, \bar{W})}$ *the relations in the quadruple* $\bar{H}^\xi_{\varphi,\psi}(\bar{Y}, \bar{W})$ *are hyperelementary on* $(\mathfrak{A}, \bar{Y}, \bar{W})$, *so that we also have*

(24) $(\bar{x}, \bar{Y}) \in \mathscr{I}_\varphi \;\&\; |\bar{x}, \bar{Y}|_\varphi \leqslant |\bar{z}, \bar{W}|_\psi \Rightarrow (\exists Z)\{Z$ *is hyperelementary on* $(\mathfrak{A}, \bar{Y}, \bar{W}) \;\&\; \mathscr{P}(Z, \bar{x}, \bar{Y}, \bar{z}, \bar{W})\},$

(25) $(\bar{x}, \bar{Y}) \in \mathscr{I}_\varphi \;\&\; |\bar{x}, \bar{Y}|_\varphi < |\bar{z}, \bar{W}|_\psi \Rightarrow (\exists Z)\{Z$ *is hyperelementary on* $(\mathfrak{A}, \bar{Y}, \bar{W}) \;\&\; \mathscr{Q}(Z, \bar{x}, \bar{Y}, \bar{z}, \bar{W})\}.$ ⊣

7D. The Abstract Spector–Gandy Theorem

Let $\mathfrak{A} = \langle A, R_1, \ldots, R_l\rangle$ be a structure, $R(\bar{x})$ a relation on A, $\mathscr{F}$ a collection of relations on A. R will be called Σ^1_1-*definable* (on $\mathfrak{A}$) *with range* $\mathscr{F}$ if there exists an elementary second order relation $\mathscr{P}(Y, \bar{x})$ such that Y ranges over subsets of A and

(1) $$R(\bar{x}) \Leftrightarrow (\exists Y \in \mathscr{F})\mathscr{P}(Y, \bar{x}).$$

R is called Σ^1_1-*definable* (on $\mathfrak{A}$) *with basis* $\mathscr{F}$ if there exists an elementary second order $\mathscr{P}(Y, \bar{x})$ such that

(2) $$R(\bar{x}) \Leftrightarrow (\exists Y)\mathscr{P}(Y, \bar{x}) \Leftrightarrow (\exists Y \in \mathscr{F})\mathscr{P}(Y, \bar{x}).$$

If R is Σ^1_1-definable with basis $\mathscr{F}$, then clearly R is also Σ^1_1-definable with range $\mathscr{F}$. The converse need not hold, roughly because definitions of the form (1) allow us to utilize the structure of $\mathscr{F}$ and extract information from it. In particular, if R is Σ^1_1-definable with basis $\mathscr{F}$, then R is Σ^1_1 on $\mathfrak{A}$, no matter what $\mathscr{F}$ is; on the other hand every $R \subseteq A$ is Σ^1_1-definable with range

$$\mathscr{F}_R = \{\{x\} : x \in R\},$$

since clearly

$$R(x) \Leftrightarrow (\exists Y \in \mathscr{F})[x \in Y].$$

In this terminology the trivial Theorem 7A.1 easily implies that if the class $\mathscr{WF}^1$ of wellfounded binary relations is elementary on the acceptable structure $\mathfrak{A}$, then every inductive relation on $\mathfrak{A}$ is Σ^1_1-definable with basis $\mathscr{HE}(\mathfrak{A})$. This result fails for countable acceptable structures. We prove in this section two weaker results which hold for all acceptable structures and which are basic to the theory: that hyperelementary relations are Σ_1^1-definable with basis $\mathcal{HE}$ and that inductive relations are Σ_1^1-definable with range $\mathcal{HE}$.

Actually a stronger, relativized version of the first of these assertions holds and is easier to apply.

7D.1. THEOREM. *If* $\mathscr{R}(\bar{x}, \bar{Y})$ *is hyperelementary on an acceptable structure* $\mathfrak{A}$, *then there exists an elementary* $\mathscr{P}(Z, \bar{x}, \bar{Y})$ *such that*

$$(3) \qquad \mathscr{R}(\bar{x}, \bar{Y}) \Leftrightarrow (\exists Z)\mathscr{P}(Z, \bar{x}, \bar{Y})$$

$$\Leftrightarrow (\exists Z)\{Z \textit{ is hyperelementary on } (\mathfrak{A}, \bar{Y}) \ \& \ \mathscr{P}(Z, \bar{x}, \bar{Y})\}.$$

In particular, every hyperelementary first order relation is Σ_1^1*-definable with basis* $\mathcal{HE}$.

PROOF. By Theorem 6C.11 we know that if $\mathscr{R}(\bar{x}, \bar{Y})$ is hyperelementary, if $\mathscr{U}$ is a universal inductive set of the appropriate signature and if σ is any inductive norm on $\mathscr{U}$, then there exist $a_1, a_2, c_1, c_2, \bar{x}^*$ such that for all $\bar{Y}$, $(c_1, c_2, \bar{x}^*, \bar{Y}) \in \mathscr{U}$, and

$$\mathscr{R}(\bar{x}, \bar{Y}) \Leftrightarrow [(a_1, a_2, \bar{x}, \bar{Y}) \in \mathscr{U} \ \& \ \sigma(a_1, a_2, \bar{x}, \bar{Y}) \leqslant \sigma(c_1, c_2, \bar{x}^*, \bar{Y})].$$

Since $\mathscr{U}$ is inductive, there is a formula φ and a constant d such that

$$(\bar{u}, \bar{Y}) \in \mathscr{U} \Leftrightarrow (d, \bar{u}, \bar{Y}) \in \mathscr{I}_{\varphi},$$

and it is easy to verify (as in the proof of the Prewellordering Theorem 3A.3) that there is an inductive norm σ on $\mathscr{U}$ such that

$$\sigma(\bar{u}_1, \bar{Y}_1) \leqslant \sigma(\bar{u}_2, \bar{Y}_2) \Leftrightarrow |d, \bar{u}_1, \bar{Y}_1|_{\varphi} \leqslant |d, u_2, \bar{Y}_2|_{\varphi}.$$

Using this σ, we get the representation

$$\mathscr{R}(\bar{x}, \bar{Y}) \Leftrightarrow [(d, a_1, a_2, \bar{x}, \bar{Y}) \in \mathscr{I}_{\varphi} \ \& \ |d, a_1, a_2, \bar{x}, \bar{Y}|_{\varphi} \leqslant |d, c_1, c_2, \bar{x}^*, \bar{Y}|_{\varphi}]$$

$$\Leftrightarrow [(\bar{e}, \bar{x}, \bar{Y}) \in \mathscr{I}_{\varphi} \ \& \ |\bar{e}, \bar{x}, \bar{Y}|_{\varphi} \leqslant |\bar{f}, \bar{x}^*, \bar{Y}|_{\varphi}]$$

after renaming the constants, where

$$(4) \qquad (\forall \bar{Y})(\bar{f}, \bar{x}^*, \bar{Y}) \in \mathscr{I}_{\varphi}.$$

Now applying 7C.2 with $\varphi \equiv \psi$ we obtain an elementary $\mathscr{P}'$ such that by (24) of 7C.2,

$$\mathscr{R}(\bar{x}, \bar{Y}) \Rightarrow (\exists Z)[Z \textit{ is hyperelementary on } (\mathfrak{A}, \bar{Y})$$

$$\& \ \mathscr{P}'(Z, \bar{e}, \bar{x}, \bar{Y}, \bar{f}, \bar{x}^*, \bar{Y})]$$

$$\Rightarrow (\exists Z)[Z \textit{ is hyperelementary on } (\mathfrak{A}, \bar{Y}) \ \& \ \mathscr{P}(Z, \bar{x}, \bar{Y})]$$

with $\mathscr{P}$ defined in the obvious way. On the other hand, by the contrapositive of (20) of 7C.2,

$$(\exists Z)\mathscr{P}(Z, \bar{x}, \bar{Y}) \Rightarrow [(\bar{f}, \bar{x}^*, \bar{Y}) \notin \mathscr{I}_\varphi \vee |\bar{e}, \bar{x}, \bar{Y}|_\varphi \leqslant |\bar{f}, \bar{x}^*, \bar{Y}|_\varphi] \Rightarrow \mathscr{R}(\bar{x}, \bar{Y})$$

by (4). ⊣

We will see in the exercises that there is an almost trivial proof of this result for the structure $\mathbb{N}$ of arithmetic.

The second theorem of this section was first established for $\mathbb{N}$ by Spector [1960] and then by Gandy [1960]. The present abstract result is essentially Theorem 15 of Moschovakis [1969b].

7D.2. Abstract Spector–Gandy Theorem. *A relation $R(\bar{x})$ is inductive on an acceptable structure $\mathfrak{A}$ if and only if it is Σ_1^1-definable with range $\mathscr{HE}^1(\mathfrak{A})$, i.e. if and only if there is some elementary second order $\mathscr{P}(Y, \bar{x})$ such that*

(5) $$R(\bar{x}) \Leftrightarrow (\exists Y \in \mathscr{HE})\mathscr{P}(Y, \bar{x}).$$

Proof. The "if" part is covered by Theorem 6D.1.

Assume then that

$$R(\bar{x}) \Leftrightarrow (\bar{a}, \bar{x}) \in I_\varphi$$

for some formula $\varphi(\bar{u}, \bar{x}, S)$ and constants $\bar{a}$; we must define an elementary $\mathscr{P}(Y, \bar{x})$ so that (5) holds.

By Theorem 6D.2, the class $\mathscr{HE}^1$ of hyperelementary subsets of A is inductive. It is conjectured that $\mathscr{HE}^1$ is not coinductive, but we can only establish this for countable, acceptable $\mathfrak{A}$ at this time. So our proof will split into two cases, to cover the possibility that the natural conjecture is false and that for some $\mathfrak{A}$, $\mathscr{HE}^1$ might be hyperelementary.

Case 1: *The class $\mathscr{HE}^1$ is not coinductive.*

Proof in Case 1. Since $\mathscr{HE}^1$ is inductive, there is some fixed $\psi(\bar{v}, Y, S)$ and constants $\bar{b}$ such that

$$Y \in \mathscr{HE} \Leftrightarrow (\bar{b}, Y) \in \mathscr{I}_\psi.$$

Theorem 6C.3 implies that if $\kappa = \kappa^{\mathfrak{A}}$ is the ordinal of the structure, then

$$\textit{supremum}\{|\bar{b}, Y|_\psi : Y \in \mathscr{HE}\} \geqslant \kappa$$

or else $\mathscr{HE}$ would be hyperelementary. This means that

(6) $$R(\bar{x}) \Rightarrow (\exists Y \in \mathscr{HE})\{|\bar{a}, \bar{x}|_\varphi \leqslant |\bar{b}, \bar{Y}|_\psi\}.$$

We apply the Second Stage Comparison Theorem 7C.2 to the formulas φ, ψ where the list $\bar{Y}$ of relation variables in φ is empty and $\bar{W} = Y$. There is then an elementary $\mathscr{P}(Z, \bar{u}, \bar{x}, \bar{v}, \bar{Y})$ such that by (24)

$$(7) \qquad (\bar{u}, \bar{x}) \in I_\varphi \ \& \ |\bar{u}, \bar{x}|_\varphi \leqslant |\bar{v}, \bar{Y}|_\psi \Rightarrow (\exists Z)\{Z \textit{ is hyperelementary on } (\mathfrak{A}, Y) \ \& \ \mathscr{P}(Z, \bar{u}, \bar{x}, \bar{v}, Y)\}$$

and by the contrapositive of (20),

$$(8) \qquad (\exists Z)\mathscr{P}(Z, \bar{u}, \bar{x}, \bar{v}, \bar{Y}) \Rightarrow (\bar{v}, Y) \notin \mathscr{I}_\psi \vee |\bar{u}, \bar{x}|_\varphi \leqslant |\bar{v}, Y|_\psi.$$

We now claim that

$$(9) \qquad R(\bar{x}) \Leftrightarrow (\exists Y \in \mathscr{H}\mathscr{E})(\exists Z \in \mathscr{H}\mathscr{E})\mathscr{P}(Z, \bar{a}, \bar{x}, \bar{b}, Y).$$

If $R(\bar{x})$, then there is some $Y \in \mathscr{H}\mathscr{E}$ such that $|\bar{a}, \bar{x}|_\varphi \leqslant |\bar{b}, Y|_\psi$ by (6), and for this Y, by (7) there is some Z such that $\mathscr{P}(Z, \bar{a}, \bar{x}, \bar{b}, Y)$ and Z is hyperelementary on $(\mathfrak{A}, Y)$; but then this Z must be hyperelementary, so the right-hand side of (9) holds.

If the right-hand side of (9) holds, fix a hyperelementary Y such that $(\exists Z)\mathscr{P}(Z, \bar{a}, \bar{x}, \bar{b}, Y)$ and apply (8). Since $(\bar{b}, Y) \in \mathscr{I}_\psi$, we must have $|\bar{a}, \bar{x}_\varphi| \leqslant |\bar{b}, Y|_\psi$ which implies $(\bar{a}, \bar{x}) \in I_\varphi$, i.e. $R(\bar{x})$.

The theorem follows by contracting the set variables in (9). Choose $c_0 \neq c_1$ and put

$$\mathscr{P}'(Y, \bar{x}) \Leftrightarrow \mathscr{P}(\{t: \langle c_0, t\rangle \in Y\}, \bar{a}, \bar{x}, \bar{b}, \{t: \langle c_1, t\rangle \in Y\}).$$

Then $\mathscr{P}'$ is elementary and easily

$$(\exists Y \in \mathscr{H}\mathscr{E})\mathscr{P}'(Y, \bar{x}) \Leftrightarrow (\exists Y \in \mathscr{H}\mathscr{E})(\exists Z \in \mathscr{H}\mathscr{E})\mathscr{P}(Z, \bar{a}, \bar{x}, \bar{b}, Y) \Leftrightarrow R(\bar{x}).$$

Case 2: *The class* $\mathscr{H}\mathscr{E}^1$ *is hyperelementary.*

Proof in Case 2. In this case we will show that $R(\bar{x})$ is actually Σ^1_1-definable with basis $\mathscr{H}\mathscr{E}^1$.

Let I^1 be the inductive index set for $\mathscr{H}\mathscr{E}^1$ defined in the Parametrization Theorem 5D.4 and choose a fixed $d_0 \in I^1$. Put

$$(10) \qquad \mathscr{Q}(Y, d) \Leftrightarrow [Y \notin \mathscr{H}\mathscr{E} \ \& \ d = d_0] \vee [Y \in \mathscr{H}\mathscr{E} \ \& \ d \in I^1 \ \& \ Y = H^1_d],$$

still using the notation of 5D.4. It is easy to verify that $\mathscr{Q}(Y, d)$ is inductive, since

$$d \in I^1 \ \& \ Y = H^1_d \Leftrightarrow d \in I^1 \ \& \ (\forall t)[t \in Y \Rightarrow H^1(d, t)] \ \& \ (\forall t)[\breve{H}^1(d, t) \Rightarrow t \in Y],$$

where $\breve{H}$ is chosen coinductive by 5D.4. On the other hand,

$$(\forall Y)(\exists d)\mathscr{Q}(Y, d),$$

so by the Hyperelementary Selection Theorem 6C.5 there is a hyperelementary $\mathcal{Q}^*$ such that

$$(11) \qquad \mathcal{Q}^*(Y, d) \Rightarrow \mathcal{Q}(Y, d),$$

$$(12) \qquad (\forall Y)(\exists d)\mathcal{Q}^*(Y, d).$$

Since I^1 is inductive, there is a formula $\chi(\bar{w}, d, S)$ and constants $\bar{c}$ such that

$$(13) \qquad d \in I^1 \Leftrightarrow (\bar{c}, d) \in I_\chi.$$

We claim that for every d,

$$(14) \qquad d \in I^1 \Rightarrow (\exists Y \in \mathscr{H}\mathscr{E})(\exists d')\{\mathcal{Q}^*(Y, d') \mathbin{\&} |\bar{c}, d|_\chi \leqslant |\bar{c}, d'|_\chi\}.$$

To prove (14), notice that if it failed, then there would be a fixed $d^* \in I^1$ such that

$$(\forall Y \in \mathscr{H}\mathscr{E})(\forall d')\{\mathcal{Q}^*(Y, d') \Rightarrow |\bar{c}, d'|_\chi < |\bar{c}, d^*|_\chi\}$$

which by the definition of $\mathcal{Q}$ and (11), (12) implies

$$(15) \qquad (\forall Y \in \mathscr{H}\mathscr{E})(\exists d)\{|\bar{c}, d|_\chi < |\bar{c}, d^*|_\chi \mathbin{\&} Y = H_d^1\}.$$

This implies that the class $\mathscr{H}\mathscr{E}^1$ can be parametrized on the hyperelementary index set

$$\{d: |\bar{c}, d|_\chi < |\bar{c}, d^*|_\chi\},$$

which leads easily to a contradiction by the usual diagonal argument: the set

$$Y = \{t: |\bar{c}, t|_\chi < |\bar{c}, d^*|_\chi \mathbin{\&} t \notin H_t^1\}$$

is hyperelementary but cannot be H_d^1 for any d such that $|\bar{c}, d|_\chi < |\bar{c}, d^*|_\chi$, because then we would have

$$d \in H_d^1 \Leftrightarrow d \in Y \Leftrightarrow d \notin H_d^1.$$

The assertion (14) is the key to the proof in this case. Using it we first establish that

$$(16) \qquad R(\bar{x}) \Leftrightarrow (\exists Y)(\exists d)\{\mathcal{Q}^*(Y, d) \mathbin{\&} |\bar{a}, \bar{x}|_\varphi \leqslant |\bar{c}, d|_\chi\}$$

$$\Leftrightarrow (\exists Y \in \mathscr{H}\mathscr{E})(\exists d)\{\mathcal{Q}^*(Y, d) \mathbin{\&} |\bar{a}, \bar{x}|_\varphi \leqslant |\bar{c}, d|_\chi\}.$$

First assume $R(\bar{x})$ and let $\xi = |\bar{a}, \bar{x}|_\varphi$. Since I^1 is not hyperelementary by 5D.4, *supremum*$\{|c, d|_\chi: d \in I^1\} = \kappa$ by 2B.1, hence by (14) there is some $Y \in \mathscr{H}\mathscr{E}$ and some d' such that $\mathcal{Q}^*(Y, d')$ and $|\bar{c}, d'|_\chi > \xi$, which proves the right-hand side of (16).

On the other hand, if for some Y, d we have $\mathcal{Q}^*(Y, d) \mathbin{\&} |\bar{a}, \bar{x}|_\varphi \leqslant |\bar{c}, d|_\chi$, then $(\bar{c}, d) \in I_\chi$ and hence $(\bar{a}, \bar{x}) \in I_\varphi$, i.e. $R(\bar{x})$.

Having (16), we apply 7D.1 to get an elementary $\mathscr{P}_1(Z, Y, d)$ such that

$$(17)\quad \mathscr{Q}^*(Y, d) \Leftrightarrow (\exists Z)\mathscr{P}_1(Z, Y, d)$$
$$\Leftrightarrow (\exists Z)\{Z \textit{ is hyperelementary on } (\mathfrak{A}, Y) \ \&\ \mathscr{P}_1(Z, Y, d)\}$$

and we apply 7C.1 on the formulas φ, χ to get an elementary $\mathscr{P}_2(Z, \bar{u}, \bar{x}, \bar{w}, d)$ such that

$$(18)\qquad (\bar{u}, \bar{x}) \in I_\varphi \ \&\ |\bar{u}, \bar{x}|_\varphi \leqslant |\bar{w}, d|_\chi \Rightarrow (\exists Z \in \mathscr{H}\mathscr{E})\mathscr{P}_2(Z, \bar{u}, \bar{x}, \bar{w}, d),$$

$$(19)\qquad (\exists Z)\mathscr{P}_2(Z, \bar{u}, \bar{x}, \bar{w}, d) \Rightarrow (\bar{w}, d) \notin I_\chi \vee |\bar{u}, \bar{x}|_\varphi \leqslant |\bar{w}, d|_\chi.$$

We now claim that

$$(20)\qquad R(\bar{x}) \Rightarrow (\exists Y \in \mathscr{H}\mathscr{E})(\exists d)\{(\exists Z \in \mathscr{H}\mathscr{E})\mathscr{P}_1(Z, Y, d)$$
$$\&\ (\exists Z \in \mathscr{H}\mathscr{E})\mathscr{P}_2(Z, \bar{a}, \bar{x}, \bar{c}, d)\},$$

$$(21)\qquad (\exists Y)(\exists d)\{(\exists Z)\mathscr{P}_1(Z, Y, d) \ \&\ (\exists Z)\mathscr{P}_2(Z, \bar{a}, \bar{x}, \bar{c}, d)\} \Rightarrow R(\bar{x}).$$

Of these, (20) follows trivially from (16), (17) and (18). To prove (21), choose Y, d so that the matrix of the hypothesis holds. From $(\exists Z)\mathscr{P}_1(Z, Y, d)$ and (17) we get $\mathscr{Q}^*(Y, d)$ and hence that $d \in I^1$ by the definition of $\mathscr{Q}$, so that $(\bar{c}, d) \in I_\chi$. From $(\exists Z)\mathscr{P}_2(Z, \bar{a}, \bar{x}, \bar{c}, d)$ and (19) we get $(\bar{c}, d) \notin I_\chi \vee |\bar{a}, \bar{x}|_\varphi \leqslant |\bar{c}, d|_\chi$, so that we have altogether $\mathscr{Q}^*(Y, d) \ \&\ |\bar{a}, \bar{x}|_\varphi \leqslant |\bar{c}, d|_\chi$, i.e. $R(\bar{x})$ by (16).

To complete the proof we contract variables, i.e. we put

$$\mathscr{P}(Y, \bar{x}) \Leftrightarrow (\exists d)[\mathscr{P}_1(\{t: \langle c_1, t\rangle \in Y\}, \{t: \langle c_2, t\rangle \in Y\}, d)$$
$$\&\ \mathscr{P}_2(\{t: \langle c_3, t\rangle \in Y\}, \bar{a}, \bar{x}, \bar{c}, d)],$$

and we verify easily that

$$R(\bar{x}) \Leftrightarrow (\exists Y)\mathscr{P}(Y, \bar{x}) \Leftrightarrow (\exists Y \in \mathscr{H}\mathscr{E})\mathscr{P}(Y, \bar{x}). \qquad \dashv$$

7E. The hierarchy of hyperelementary sets

It is immediate from 7D.1 and 6D.1 that a relation $R(\bar{x})$ is hyperelementary on an acceptable structure $\mathfrak{A}$ if and only if it is Δ_1^1-*definable with basis* $\mathscr{H}\mathscr{E}(\mathfrak{A})$, i.e. if and only if both R and $\neg R$ are Σ_1^1-definable with basis $\mathscr{H}\mathscr{E}(\mathfrak{A})$. In this section we will impose a hierarchy on $\mathscr{H}\mathscr{E}(\mathfrak{A})$ by showing that *each hyperelementary relation is* Δ_1^1-*definable with basis the class of previously constructed hyperelementary relations and previously constructed relations as parameters.*

To make this precise, fix an acceptable structure $\mathfrak{A}$ and call a relation $R(\bar{x})$ *Σ_1^1-definable with basis $\mathscr{F}$ and parameters from $\mathscr{F}$* if there is an elementary relation $\mathscr{P}(Z, \bar{x}, Y_1, \ldots, Y_k)$ and relations $Y_1, \ldots, Y_k$ in $\mathscr{F}$ such that

$$R(\bar{x}) \Leftrightarrow (\exists Z)\mathscr{P}(Z, \bar{x}, Y_1, \ldots, Y_k)$$
$$\Leftrightarrow (\exists Z \in \mathscr{F})\mathscr{P}(Z, \bar{x}, Y_1, \ldots, Y_k).$$

Put

(1) $\Delta(\mathscr{F}) = \{R\colon R \textit{ is } \Delta_1^1\textit{-definable with basis } \mathscr{F} \textit{ and parameters from } \mathscr{F}\}$
$= \{R\colon \textit{both } R \textit{ and } \neg R \textit{ are } \Sigma_1^1\textit{-definable with basis } \mathscr{F} \textit{ and parameters from } \mathscr{F}\}.$

Clearly Δ is a *monotone operator* on classes of relations, i.e.

$$\mathscr{F}_1 \subseteq \mathscr{F}_2 \Rightarrow \Delta(\mathscr{F}_1) \subseteq \Delta(\mathscr{F}_2),$$

and by 6D.1, $\mathscr{H}\mathscr{E}$ is a fixed point of Δ,

$$\Delta(\mathscr{H}\mathscr{E}) = \mathscr{H}\mathscr{E}.$$

To each ordinal ξ we assign the ξth stage of Δ by the induction

(2) $\mathscr{D}^0 = \textit{the class of elementary relations on } \mathfrak{A},$

(3) $\mathscr{D}^\xi = \Delta(\bigcup_{\eta<\xi} \mathscr{D}^\eta), \textit{ if } \xi > 0,$

where the special case for $\xi = 0$ is needed to get the induction started since the usual logical conventions imply that $\Delta(\emptyset) = \emptyset$.

The main result of this section is that the sequence $\{\mathscr{D}^\xi\}_{\xi<\kappa}$ of classes of relations on A is properly increasing and closes at κ, i.e.

(4) $$\mathscr{D}^\kappa = \bigcup_{\xi<\kappa} \mathscr{D}_\xi = \mathscr{H}\mathscr{E},$$

where $\kappa = \kappa^{\mathfrak{A}}$ is the ordinal of the structure $\mathfrak{A}$. In particular, $\mathscr{H}\mathscr{E}$ is the smallest (nonempty) fixed point of Δ. In the next section we will apply this to get several elegant model theoretic characterizations of the class $\mathscr{H}\mathscr{E}$.

In Section 7 of Moschovakis [1969c] we argued that this result justifies identifying the hyperelementary relations on an acceptable structure $\mathfrak{A}$ with the *predicatively definable* relations on $\mathfrak{A}$, as we intuitively understand this notion. This is in analogy with *Church's Thesis*, the identification of the *recursive* relations on the structure $\mathbb{N}$ with the *effectively decidable* relations on $\mathbb{N}$, intuitively understood. The crux of the argument was that if a predicativist accepts every relation in some class $\mathscr{F}$, then he must also accept every relation in $\Delta(\mathscr{F})$, even if he does not understand the class $\mathscr{F}$ as a completed totality. The suggestion was not received very favourably by people who have worked on the foundational problems posed by the notion of predicative definability; in particular see Feferman's review [1971] of Moschovakis [1969c]. In any case, the present result gives a step-by-step construction of $\mathcal{HE}$, where each of the steps is presumably simpler to understand than the construction of all of $\mathcal{HE}$ in one move via inductive definitions. It would be interesting to find other natural operators which construct $\mathcal{HE}$ like Δ but which may be accepted as more "predicative."

For the structure $\mathbb{N}$ of arithmetic the results of this section are in the pioneering paper Kleene [1959a]. Some similar results which also imply the chief corollaries in the next section were proved by Kreisel [1961].

7E.1. THEOREM. *Let $\mathfrak{A}$ be an acceptable structure and let $\mathcal{D}^\xi$ be defined for each ξ by* (2), (3) *above. Then*

$$\mathcal{D}^\kappa = \bigcup_{\xi<\kappa} \mathcal{D}^\xi = \mathcal{HE}(\mathfrak{A}),$$

where $\kappa = \kappa^{\mathfrak{A}}$ is the ordinal of $\mathfrak{A}$.

PROOF. An easy induction on ξ (using 6D.1) shows that $\mathcal{D}^\xi \subseteq \mathcal{HE}$, so it will be enough to prove that every hyperelementary relation occurs in some $\mathcal{D}^\xi$ with $\xi < \kappa$. The main tools are the Normal Form Theorem 5C.2 for inductive relations on an acceptable structure in terms of the quantifier G and the Second Stage Comparison Theorem 7C.1.

Let P be any inductive set. By 5C.2 there is an elementary $R(z, \bar{x})$ such that

$$P(\bar{x}) \Leftrightarrow (\mathrm{G}z)R(z, \bar{x}),$$

where G is defined relative to a fixed elementary coding scheme on $\mathfrak{A}$. Put

$$\varphi(w, \bar{x}, S) \Leftrightarrow [Seq(w) \;\&\; lh(w) \text{ is even} \;\&\; [R(w, \bar{x}) \vee (\forall s)S(w^\frown\langle s\rangle, \bar{x})]]$$
$$\vee\ [Seq(w) \;\&\; lh(w) \text{ is odd} \;\&\; (\exists t)S(w^\frown\langle t\rangle, \bar{x})].$$

This is a different inductive analysis of the application of the quantifier G than we gave in 5C.1, but it is very simple to modify that proof and show

(5) $$(\mathrm{G}z)R(z, \bar{x}) \Leftrightarrow (\langle\emptyset\rangle, \bar{x}) \in I_\varphi.$$

For each $\xi < \kappa$, let $H_\varphi^\xi = H_{\varphi,\varphi}^\xi$ as we defined this relation in Section 7C, i.e.

$$(w_1, \bar{x}_1, w_2, \bar{x}_2) \in H_\varphi^\xi \Leftrightarrow (w_2, \bar{x}_2) \in I_\varphi^\xi \;\&\; |w_1, \bar{x}_1|_\varphi \leqslant |w_2, \bar{x}_2|_\varphi.$$

We will prove:

(*) *if $\xi = \lambda + m < \|\varphi\|$, where $\lambda = 0$ or λ is a limit ordinal, then H_φ^ξ is elementary on $(\mathfrak{A}, Q)$ for some $Q \in \mathcal{D}^\lambda$.*

Before going to the proof of (*), let us show that it will establish the theorem. We apply (*) to the case that $P = U^n$ is a universal inductive set in the sense of 5D.2, so we know that $\|\varphi\| = \kappa$. If R is hyperelementary, then for some a,

$$R(\bar{x}) \Leftrightarrow (a, \bar{x}) \in U^n,$$

and the Covering Theorem 3C.2 implies that for some $\xi < \kappa$,

$$R(\bar{x}) \Leftrightarrow (\langle\emptyset\rangle, a, \bar{x}) \in I_\varphi^\xi$$
$$\Leftrightarrow (\langle\emptyset\rangle, a, \bar{x}, \langle\emptyset\rangle, a, \bar{x}) \in H_\varphi^\xi;$$

so by (*), R is elementary on $(\mathfrak{A}, Q)$ with Q in $\mathscr{D}^\lambda$ for some $\lambda \leqslant \xi$. But then $R \in \mathscr{D}^{\lambda+1}$.

Proof of (*) is by transfinite induction on ξ.

Case 1: $\xi = 0$. In this case H_φ^0 is trivially elementary,

$$(w_1, \bar{x}_1, w_2, \bar{x}_2) \in H_\varphi^0 \Leftrightarrow \varphi(w_2, \bar{x}_2, \emptyset) \;\&\; \varphi(w_1, \bar{x}_1, \emptyset).$$

Case 2: $\xi = \lambda+m+1$ *is a successor ordinal.* We have by the definition

$$(w_1, \bar{x}_1, w_2, \bar{x}_2) \in H_\varphi^\xi \Leftrightarrow [|w_2, \bar{x}_2|_\varphi \leqslant \lambda+m \;\&\; H_\varphi^{\lambda+m}(w_1, \bar{x}_1, w_2, \bar{x}_2)]$$
$$\vee\; [|w_2, \bar{x}_2|_\varphi = \lambda+m+1 \;\&\; |w_1, \bar{x}_1|_\varphi \leqslant \lambda+m+1],$$

where we can substitute successively

$$|w_2, \bar{x}_2|_\varphi \leqslant \lambda+m \Leftrightarrow H^{\lambda+m}(w_2, \bar{x}_2, w_2, \bar{x}_2),$$
$$|w_2, \bar{x}_2|_\varphi \leqslant \lambda+m+1 \Leftrightarrow \varphi(w_2, \bar{x}_2, \{(w, \bar{x}) : H^{\lambda+m}(w, \bar{x}, w, \bar{x})\}),$$
$$|w_2, \bar{x}_2|_\varphi = \lambda+m+1 \Leftrightarrow |w_2, \bar{x}_2|_\varphi \leqslant \lambda+m+1 \;\&\; \neg(|w_2, \bar{x}_2|_\varphi \leqslant \lambda+m),$$
$$|w_1, \bar{x}_1|_\varphi \leqslant \lambda+m+1 \Leftrightarrow \varphi(w_1, \bar{x}_1, \{(w, \bar{x}) : H^{\lambda+m}(w, \bar{x}, w, \bar{x})\}).$$

Hence H_φ^ξ is elementary on $(\mathfrak{A}, H_\varphi^{\lambda+m})$, and by the induction hypothesis it is elementary on $(\mathfrak{A}, Q)$ for some $Q \in \mathscr{D}^\lambda$.

Case 3: $\xi = \lambda$ *is limit.* We know that

(6) $$\lambda = |w^*, \bar{x}^*|_\varphi$$

for some $(w^*, \bar{x}^*) \in P$. Now the structure of the formula φ is such that we must have

(7) $$Seq(w^*) \;\&\; lh(w^*) \textit{ is even} \;\&\; (\forall s)(w^* {}^\frown \langle s\rangle, \bar{x}) \in \textstyle\bigcup_{\xi<\lambda} I_\varphi^\xi;$$

because the only other possibilities for $\varphi(w^*, \bar{x}^*, \bigcup_{\xi<\lambda} I_\varphi^\xi)$ to hold are that either $R(w^*, \bar{x}^*)$, in which case $|w^*, \bar{x}^*|_\varphi = 0$, or $lh(w^*)$ *is odd* & $(\exists t)[(w^* {}^\frown \langle t\rangle, \bar{x}^*) \in \bigcup_{\xi<\lambda} I_\varphi^\xi]$ from which it follows immediately that for some $\xi < \lambda$, $|w^*, \bar{x}^*|_\varphi = \xi+1 < \lambda$.

This is where we use the very special form of φ. The assertions (6) and (7) yield

$$(8) \qquad (\forall s)[|w^{*\frown}\langle s\rangle, \bar{x}^*|_\varphi < \lambda],$$

$$(9) \qquad \lambda = \textit{supremum}\{|w^{*\frown}\langle s\rangle, \bar{x}^*|_\varphi : s \in A\},$$

so that we have a uniform way of "assigning codes" to an unbounded subset of the limit ordinal λ. Put

$$(10) \quad Q(w_1, \bar{x}_1, w_2, \bar{x}_2, s) \Leftrightarrow |w_1, \bar{x}_1|_\varphi \leqslant |w_2, \bar{x}_2|_\varphi \ \&\ |w_2, \bar{x}_2|_\varphi \leqslant |w^{*\frown}\langle s\rangle, \bar{x}^*|_\varphi.$$

We aim to show that this Q is in $\mathscr{D}^\lambda$, and from this it will be easy to complete the proof.

The Stage Comparison Theorem 7C.1 with $\psi \equiv \varphi$ guarantees an elementary $\mathscr{P}(Z, w_1, \bar{x}_1, w_2, \bar{x}_2)$ such that the following two implications hold:

$$(11) \quad (w_1, \bar{x}_1) \in I_\varphi \ \&\ |w_1, \bar{x}_1|_\varphi = \xi \leqslant |w_2, \bar{x}_2|_\varphi$$
$$\Rightarrow (\exists Z)\{Z \textit{ is elementary on } (\mathfrak{A}, H^\xi) \ \&\ \mathscr{P}(Z, w_1, \bar{x}_1, w_2, \bar{x}_2)],$$

$$(12) \quad (\exists Z)\mathscr{P}(Z, w_1, \bar{x}_1, w_2, \bar{x}_2) \Rightarrow (w_2, \bar{x}_2) \notin I_\varphi \vee [|w_1, \bar{x}_1|_\varphi \leqslant |w_2, \bar{x}_2|_\varphi < \kappa].$$

We claim that

$$(13) \quad Q(w_1, \bar{x}_1, w_2, \bar{x}_2, s) \Rightarrow (\exists Z)[Z \in \textstyle\bigcup_{\xi<\lambda} \mathscr{D}^\xi \ \&\ \mathscr{P}(Z, w_1, \bar{x}_1, w_2, \bar{x}_2)]$$
$$\&\ (\exists Z)[Z \in \textstyle\bigcup_{\xi<\lambda} \mathscr{D}^\xi \ \&\ \mathscr{P}(Z, w_2, \bar{x}_2, w^{*\frown}\langle s\rangle, \bar{x}^*)];$$

this is obvious from (10), (11), and the induction hypothesis which guarantees that

$$\xi < \lambda \ \&\ Z \textit{ is elementary on } (\mathfrak{A}, H^\xi) \Rightarrow Z \in \mathscr{D}^{\xi+1}.$$

Also,

$$(14) \qquad (\exists Z)\mathscr{P}(Z, w_1, \bar{x}_1, w_2, \bar{x}_2) \ \&\ (\exists Z)\mathscr{P}(Z, w_2, \bar{x}_2, w^{*\frown}\langle s\rangle, \bar{x}^*)$$
$$\Rightarrow Q(w_1, \bar{x}_1, w_2, \bar{x}_2);$$

this is because the second conjunct in the hypothesis with (12) and (8) yields $|w_2, \bar{x}_2|_\varphi \leqslant |w^{*\frown}\langle s\rangle, \bar{x}^*|_\varphi$ and then the first conjunct in the hypothesis with this yields $|w_1, \bar{x}_1|_\varphi \leqslant |w_2, \bar{x}_2|_\varphi$.

From (13) it follows that with $c_0 \neq c_1$,

$$Q(w_1, \bar{x}_1, w_2, \bar{x}_2, s) \Rightarrow (\exists Z \in \textstyle\bigcup_{\xi<\lambda} \mathscr{D}^\xi)[\mathscr{P}(\{t : \langle c_0, t\rangle \in Z\}, w_1, \bar{x}_1, w_2, \bar{x}_2)$$
$$\&\ \mathscr{P}(\{t : \langle c_1, t\rangle \in Z\}, w_1, \bar{x}_2, w^{*\frown}\langle s\rangle, \bar{x}^*)],$$

since obviously

$$Z_1, Z_2 \in \textstyle\bigcup_{\xi<\lambda} \mathscr{D}^\xi \Rightarrow \{\langle c_0, t\rangle : t \in Z_1\} \cup \{\langle c_1, t\rangle : t \in Z_2\} \in \textstyle\bigcup_{\xi<\lambda} \mathscr{D}^\xi.$$

Also

$$(\exists Z)[\mathscr{P}(\{t: \langle c_0, t\rangle \in Z\}, w_1, \bar{x}_1, w_2, \bar{x}_2) \;\&\; \mathscr{P}(\{t: \langle c_1, t\rangle \in Z\}, w_2, \bar{x}_2, w^*(s), x^*)] \Rightarrow Q(w_1, \bar{x}_1, w_2, \bar{x}_2, s)$$

is trivial, hence Q is Σ_1^1-definable with basis $\bigcup_{\xi<\lambda}\mathscr{D}^\xi$.

A very similar argument using the second part of 7C.1 shows that $\neg Q$ is Σ_1^1-definable with basis $\bigcup_{\xi<\lambda}\mathscr{D}^\xi$, hence Q is Δ_1^1-definable with basis $\bigcup_{\xi<\lambda}\mathscr{D}^\xi$ and $Q \in \mathscr{D}^\lambda$.

Now

$$\begin{aligned}(w_1, \bar{x}_1, w_2, \bar{x}_2) \in H^\lambda \Leftrightarrow {} & |w_1, \bar{x}_1|_\varphi \leqslant |w_2, \bar{x}_2|_\varphi < \lambda \\ & \vee [|w_1, \bar{x}_1|_\varphi \leqslant \lambda \;\&\; \neg(|w_2, \bar{x}_2|_\varphi < \lambda)],\end{aligned}$$

so to show that H^λ is elementary on $(\mathfrak{A}, Q)$ it will be enough to show that the three relations

$$|w_1, \bar{x}_1|_\varphi \leqslant |w_2, \bar{x}_2|_\varphi < \lambda, \qquad |w, \bar{x}|_\varphi < \lambda, \qquad |w, \bar{x}| \leqslant \lambda$$

are elementary on $(\mathfrak{A}, Q)$. But obviously

$$\begin{aligned}|w_1, \bar{x}_1|_\varphi \leqslant |w_2, \bar{x}_2|_\varphi < \lambda &\Leftrightarrow (\exists s)Q(w_1, \bar{x}_1, w_2, \bar{x}_2, s), \\ |w, \bar{x}|_\varphi < \lambda &\Leftrightarrow (\exists s)Q(w, \bar{x}, w, \bar{x}, s), \\ |w, \bar{x}|_\varphi \leqslant \lambda &\Leftrightarrow \varphi(w, \bar{x}, \{(w', \bar{x}'): |w'\bar{x}'|_\varphi < \lambda\}),\end{aligned}$$

so the proof is complete. ⊣

In order to prove that the sequence $\{\mathscr{D}^\xi\}_{\xi<\kappa}$ is properly increasing, i.e. that the induction defining $\{\mathscr{D}^\xi\}_{\xi<\kappa}$ closes exactly at κ, it will be convenient to establish a much stronger result. This is that *every hyperelementary set occurs in the ramified second order hierarchy over* $\mathfrak{A}$ *at a stage below* κ, *and that this hierarchy does not close below* κ. The idea of this proof is simple but there is a lot of computation, which we will only outline.

Let us specify that for each $n \geqslant 1$, the n-ary relation variables of the second order language $\mathscr{L}_2^{\mathfrak{A}}$ over a structure $\mathfrak{A}$ are

$$V_1^n, V_2^n, V_3^n, \ldots.$$

When it is convenient we can rename the individual variables $v_1, v_2, v_3, \ldots$ as $V_1^0, V_2^0, V_3^0, \ldots.$

To interpret a formula φ of $\mathscr{L}_2^{\mathfrak{A}}$ we must be given members $x_1, x_2, x_3, \ldots$ of A interpreting the individual variables $v_1, v_2, v_3, \ldots$ which are free in φ, n-ary relations $Y_1^n, Y_2^n, Y_3^n, \ldots$ for each n which interpret the n-ary relation variables $V_1^n, V_2^n, V_3^n, \ldots$ which are free, and a collection $\mathscr{F}$ of relations on A to serve as the *range* of the relation variables in φ. Of course we may have

$\mathscr{F}$ = *all relations on A*,

the standard interpretation, but we will be looking at much smaller ranges. The basic satisfaction relation for formulas of $\mathscr{L}_2^{\mathfrak{A}}$ is then

$$\mathfrak{A}, \mathscr{F}, \{Y_i^n\}_{n\geq 1, i\geq 1}, \{x_i\}_{i\geq 1} \models \varphi \Leftrightarrow \varphi \textit{ is true when interpreted on } \mathfrak{A} \textit{ with } \mathscr{F} \textit{ as the range of the relation variables and } V_i^n = Y_i^n, v_i = x_i.$$

If $\mathfrak{A}$ is acceptable, then we can code collections of relations by simple subsets of A using some elementary coding scheme. For each $Z \subseteq A$, each $a \in A$ and each integer n (in the copy of ω that is part of the coding scheme), let

$$(15) \qquad Z_a^{(n)} = \{(x_1, \ldots, x_n) : \langle n, a, x_1, \ldots, x_n \rangle \in Z\}.$$

Now every double sequence $\{Y_i^n\}_{n\geq 1, i\geq 1}$ of relations on A, Y_i^n being n-ary, can be coded into a single Y such that

$$Y_i^n = Y_i^{(n)}.$$

Also any collection of relations of the same cardinality as A is of the form

$$\{Z_a^{(n)} : n \geq 1, a \in A\}$$

for some $Z \subseteq A$.

We will not bother to define explicitly codings for the formulas of $\mathscr{L}_2^{\mathfrak{A}}$ and prove in detail that the natural syntactical and semantical relations of $\mathscr{L}_2^{\mathfrak{A}}$ are hyperelementary. However, the following lemma can be proved easily by the methods of Section 5B, relativized.

7E.2. LEMMA. *Let $\mathscr{C}$ be an elementary coding scheme on some acceptable structure $\mathfrak{A}$. We can assign codes (relative to $\mathscr{C}$) to the formulas of the second order language $\mathscr{L}_2^{\mathfrak{A}}$ over $\mathfrak{A}$ so that the following relations are hyperelementary:*

$$Fml_2(a) \Leftrightarrow a \textit{ codes a formula (of } \mathscr{L}_2^{\mathfrak{A}}),$$

$$Free_2(a, n, i) \Leftrightarrow a \textit{ codes a formula } \varphi \textit{ and } V_i^n \textit{ is free in } \varphi,$$

$$Sat_2(a, Z, Y, x) \Leftrightarrow a \textit{ codes a formula } \varphi \;\&\; \mathfrak{A}, \{Z_a^{(n)} : n \geq 1, a \in A\}, \{Y_i^{(n)}\}_{n\geq 1, i\geq 1}, \{(x)_i\}_{i\geq 1} \models \varphi. \quad \dashv$$

A relation $R(x_1, \ldots, x_n)$ on A is *second order definable with range $\mathscr{F}$ and parameters from $\mathscr{F}$* if there are relations $X_1, \ldots, X_n$ in $\mathscr{F}$ and a formula φ of $\mathscr{L}_2^{\mathfrak{A}}$ whose only free individual variables are $v_1, \ldots, v_n$ such that

$$R(x_1, \ldots, x_n) \Leftrightarrow \mathfrak{A}, \mathscr{F}, \{Y_i^n\}, x_1, \ldots, x_n \models \varphi,$$

where the double sequence of relations $\{Y_i^n\}$ assigns each X_j to a relation variable of the appropriate number of arguments and assigns the empty relation to all other relation variables.

Let

$\Gamma(\mathscr{F})$ = *all relations on* $\mathfrak{A}$ *which are second order definable with range* $\mathscr{F}$ *and parameters from* $\mathscr{F}$,

and by induction on ξ,

(16) $$\mathscr{E}_2^0 = \textit{all elementary relations on } \mathfrak{A},$$

(17) $$\mathscr{E}_2^\xi = \Gamma(\textstyle\bigcup_{\eta<\xi}\mathscr{E}_2^\eta) \quad \textit{if } \xi > 0.$$

The sequence $\{\mathscr{E}_2^\xi\}_\xi$ of classes of relations is the *ramified second order hierarchy over* $\mathfrak{A}$.

7E.3. THEOREM. *If* $\mathfrak{A}$ *is acceptable with ordinal* $\kappa = \kappa^{\mathfrak{A}}$, *then for every* $\xi < \kappa$,

(18) $$\mathscr{D}^\xi \subseteq \mathscr{E}_2^\xi \subsetneqq \mathscr{H}\mathscr{E}(\mathfrak{A});$$

hence for every $\xi < \kappa$,

(19) $$\mathscr{D}^\xi \subsetneqq \mathscr{D}^{\xi+1},$$

and the sequence $\{\mathscr{D}^\xi\}_{\xi<\kappa}$ *is properly increasing.*

PROOF. The inclusion

$$\mathscr{D}^\xi \subseteq \mathscr{E}_2^\xi$$

is proved by transfinite induction on ξ; assuming $\bigcup_{\eta<\xi}\mathscr{D}^\eta \subseteq \bigcup_{\eta<\xi}\mathscr{E}_2^\eta$, if R is in $\mathscr{D}^\xi$ then R must be Σ_1^1-definable with basis $\bigcup_{\eta<\xi}\mathscr{E}_2^\eta$ and parameters from $\bigcup_{\eta<\xi}\mathscr{D}^\eta$, hence it is Σ_1^1-definable with range $\bigcup_{\eta<\xi}\mathscr{E}_2^\eta$ and parameters in $\bigcup_{\eta<\xi}\mathscr{E}_2^\eta$, hence it is in $\mathscr{E}_2^\xi$. Then (19) follows trivially from (18), because if $\mathscr{D}^{\xi+1} = \mathscr{D}^\xi$ for some $\xi < \kappa$, then $\mathscr{D}^\kappa = \mathscr{D}^\xi$, hence $\mathscr{H}\mathscr{E} = \mathscr{D}^\kappa \subseteq \mathscr{E}_2^\xi$ contradicting (18).

The nontrivial part of the theorem is the proper inclusion

(20) $$\mathscr{E}_2^\xi \subsetneqq \mathscr{H}\mathscr{E}.$$

To simplify the computation, let us first notice that each class $\mathscr{E}_2^\xi$ is determined by the *unary relations* (sets) in it,

$$\mathscr{E}_{2,u}^\xi = \{X \subseteq A \colon X \in \mathscr{E}_2^\xi\}.$$

This is because each $\mathscr{E}_2^\xi$ is obviously closed under elementary definability, and if $R \subseteq A^n$ is in $\mathscr{E}_2^\xi$, then its unary contraction P defined by

(21) $$P(x) \Leftrightarrow R((x)_1, \ldots, (x)_n)$$

is in $\mathscr{E}_{2,u}^\xi$ and we can recover R from P by

(22) $$R(x_1, \ldots, x_n) \Leftrightarrow P(\langle x_1, \ldots, x_n\rangle).$$

Moreover, the transfinite sequence $\{\mathscr{E}^{\xi}_{2,u}\}_{\xi}$ of the unary relations in the ramified second order hierarchy over $\mathfrak{A}$ can be defined directly by the induction

$\mathscr{E}^{0}_{2,u}$ = *all unary elementary relations on* $\mathfrak{A}$,

$\mathscr{E}^{\xi}_{2,u}$ = *all unary relations which are second order definable with range* $\bigcup_{\eta<\xi} \mathscr{E}^{\eta}_{2,u}$ *and parameters from* $\bigcup_{\eta<\xi} \mathscr{E}^{\xi}_{2,u}$, *by a formula of* $\mathscr{L}^{\mathfrak{A}}_{2}$ *which has only unary relation variables, if* $\xi > 0$.

This is because we can replace n-ary variable quantification by an equivalent unary quantification using again the coding of (21), (22).

After these preliminaries, we establish the proper inclusion (20) by outlining a proof that for each $\xi < \kappa$, there is a hyperelementary set $E^{\xi} \subseteq A \times A$ which parametrizes $\mathscr{E}^{\xi}_{2,u}$, i.e.

$$Z \in \mathscr{E}^{\xi}_{2,u} \Leftrightarrow \textit{for some } a \in A, \qquad Z = \{t: (a, t) \in E^{\xi}\}.$$

This yields immediately that $\mathscr{E}^{\xi}_{2,u} \subseteq \mathscr{H}\mathscr{E}^{1}$, and the properness of the inclusion follows by the usual diagonal argument that

$$\{a: (a, a) \notin E^{\xi}\}$$

is hyperelementary but not in $\mathscr{E}^{\xi}_{2}$.

Fix any hyperelementary wellfounded relation $\prec$ and assign to each $\bar{u} \in \mathit{Field}(\prec)$ the binary relation $E^{\bar{u}}$ by the following recursion on $\rho^{\prec}(\bar{u})$:

(23) $(a, x) \in E^{\bar{u}} \Leftrightarrow$ *there exist* $b, a_1, \bar{u}_1, a_2, \bar{u}_2, \ldots, a_n, \bar{u}_n$, *such that* $a = \langle b, a_1, \bar{u}_1, a_2, \bar{u}_2, \ldots, a_n, \bar{u}_n\rangle$ *and* b *is the code of a formula with only unary relation variables in which* v_1 *is the only free individual variable and* $\bar{u}_1 \prec \bar{u}, \bar{u}_2 \prec \bar{u}, \ldots, \bar{u}_n \prec \bar{u}$ *and*

$$\mathit{Sat}_2(b, \{\langle 1, \langle \bar{u}', a'\rangle, x'\rangle: \bar{u}' \prec \bar{u} \;\&\; (a', x') \in E^{\bar{u}'}\}, \{\langle 1, i, t\rangle: 1 \leqslant i \leqslant n \;\&\; (a_i, t) \in E^{\bar{u}_i}\}, \langle x\rangle).$$

It is not hard to prove by induction on $\rho^{\prec}(\bar{u})$ that

(24) *if* $\bar{u} \in \mathit{Field}(\prec)$ & $\xi = \rho^{\prec}(\bar{u})$, *then* $E^{\bar{u}}$ *parametrizes* $\mathscr{E}^{\xi}_{2,\bar{u}}$.

Also it is not hard to construct a hyperelementary second order relation $\mathscr{R}(Z, \bar{u}, a, x)$ such that

(25) $(a, x) \in E^{\bar{u}} \Leftrightarrow \mathscr{R}(\{\langle \bar{u}', a', x'\rangle: \bar{u}' \prec \bar{u} \;\&\; (a', x') \in E^{\bar{u}'}\}, \bar{u}, a, x)$.

It then follows from Exercise 6.4 that each $E^{\bar{u}}$ is hyperelementary, and this completes the proof, since for each $\xi < \kappa$ there is a hyperelementary wellfounded $\prec$ and some $\bar{u} \in \mathit{Field}(\prec)$ so that $\xi = \rho^{\prec}(\bar{u})$. ⊣

It should be pointed out that the hierarchy $\{\mathscr{E}_2^\xi\}_\xi$ does not close at κ, since by the Abstract Spector–Gandy Theorem 7D.2 and 7E.2, if R is inductive not hyperelementary, then

$$R \in \mathscr{E}_2^\kappa - \bigcup_{\xi<\kappa} \mathscr{E}_2^\xi.$$

7F. Model theoretic characterizations

Recall the schemata of Σ_1^1-Collection and Δ_1^1-Comprehension on a structure $\mathfrak{A}$ that we introduced in Section 6D. In the same spirit, the schema of Δ_∞^0-*Comprehension* on $\mathfrak{A}$ consists of all formulas of the form

$$(\Delta_\infty^0\text{-}Comp) \qquad (\exists Z)(\forall \bar{x})[\bar{x} \in Z \Leftrightarrow \varphi(\bar{x}, \bar{Y})],$$

where φ is in the (first order) language over $\mathfrak{A}$.

7F.1. THEOREM. *If $\mathfrak{A}$ is acceptable, then $\mathscr{HE}(\mathfrak{A})$ is the smallest class of relations on A which satisfies the schema of Δ_1^1-Comprehension on $\mathfrak{A}$. Also $\mathscr{HE}(\mathfrak{A})$ is the smallest class of relations on $\mathfrak{A}$ which satisfies the schemata of Σ_1^1-Collection and Δ_∞^0-Comprehension.*

PROOF. By Theorem 6D.4, $\mathscr{HE}$ satisfies both Σ_1^1-Collection and Δ_1^1-Comprehension, and Δ_1^1-Comprehension trivially implies Δ_∞^0-Comprehension.

For the converse to the first assertion, suppose $\mathscr{F}$ is a class of relations on A which satisfies Δ_1^1-Comprehension. A trivial induction on ξ shows that $\mathscr{D}^\xi \subseteq \mathscr{F}$, hence by 7E.1, $\mathscr{HE} = \mathscr{D}^\kappa \subseteq \mathscr{F}$.

The easiest way to prove the converse to the second assertion is to prove (essentially following Kreisel [1962]) that the schemata of Σ_1^1-Collection and Δ_∞^0-Comprehension imply Δ_1^1-Comprehension. Because from the hypothesis

$$(\forall \bar{x})\{(\exists Z_1)\varphi(\bar{x}, Z_1, \bar{Y}) \Leftrightarrow (\forall Z_2)\psi(\bar{x}, Z_2, \bar{Y})\}$$

we get immediately

$$(\forall \bar{x})(\exists Z)\{\varphi(\bar{x}, Z, \bar{Y}) \vee \neg\psi(\bar{x}, Z, \bar{Y})\},$$

so by Σ_1^1-Collection there is a fixed Z such that

$$(\forall \bar{x})(\exists a)\{\varphi(\bar{x}, Z_a, \bar{Y}) \vee \neg\psi(\bar{x}, Z_a, \bar{Y})\},$$

and we can put

$$W = \{\bar{x}: (\exists a)\varphi(\bar{x}, Z_a, \bar{Y})\}$$

using Δ_∞^0-Comprehension. It is an easy exercise to verify that

$$\bar{x} \in W \Leftrightarrow (\exists Z_1)\varphi(\bar{x}, Z_1, \bar{Y}). \qquad \dashv$$

For the structure $\mathbb{N}$, these results were announced in Kreisel [1961] but at least the first of them is also implicit in Kleene [1959a]. Another result of Kreisel [1961] which is implicit in Kleene [1959a] is the following characterization of $\mathscr{HE}(\mathfrak{A})$ in terms of *invariant definability* (for $\mathfrak{A} = \mathbb{N}$).

7F.2. THEOREM. *If $\mathfrak{A}$ is acceptable, then a relation $R(\bar{x})$ is hyperelementary on $\mathfrak{A}$ if and only if there exist first order formulas $\varphi(Z, \bar{x})$, $\psi(Z, \bar{x})$ in the language of $\mathfrak{A}$ such that for every class $\mathscr{F}$ of relations which satisfies Δ^1_1-Comprehension on $\mathfrak{A}$,*

$$(1) \qquad R(\bar{x}) \Leftrightarrow (\exists Z \in \mathscr{F})\varphi(Z, \bar{x}) \Leftrightarrow (\forall Z \in \mathscr{F})\psi(Z, \bar{x});$$

i.e. if and only if R is Δ^1_1 invariantly definable over all models of Δ^1_1-Comprehension on $\mathfrak{A}$.

PROOF. If R satisfies (1) for every model $\mathscr{F}$ of Δ^1_1-Comprehension, then we can prove that R is hyperelementary by taking $\mathscr{F} = \mathscr{HE}$ and applying 6D.1. On the other hand, if R is hyperelementary, then both R and $\neg R$ are Σ^1_1-definable with basis $\mathscr{HE}$ by 7D.1 and we get the representations (1) immediately by 7F.1. ⊣

Exercises for Chapter 7

7.1. Prove that if λ is a cardinal with *cofinality*$(\lambda) > \omega$, then the relations $\mathscr{WF}^n$ are Σ^1_1 on $\boldsymbol{\lambda} = \langle \lambda, \leqslant \rangle$, so that all inductive relations are Σ^1_1 on $\boldsymbol{\lambda}$.

HINT: Notice that if $X \subseteq \lambda^{2n}$, then

$$X \textit{ is wellfounded} \Leftrightarrow (\forall \xi < \lambda)[X \cap \xi^{2n} \textit{ is wellfounded}]. \qquad ⊣$$

7.2. A game $G(\bar{Q}, R)$ is *finite* if there is a k-ary relation $P(x_0, \ldots, x_{k-1})$ such that

$$R(x_0, x_1, x_2, \ldots) \Leftrightarrow P(x_0, x_1, \ldots, x_{k-1}),$$

i.e. if the outcome of the game is known after the first k moves with a fixed, predetermined k.

Give an example of a structure $\mathfrak{A} = \langle A, R_1, \ldots, R_l \rangle$ and a finite elementary game G on A such that $(\exists)$ wins G, but $(\exists)$ has no winning strategy which is definable in the second order language over $\mathfrak{A}$. Show that $\mathfrak{A}$ can be chosen acceptable.

HINT: $(\forall)$ plays a finite set $B \subseteq A$, $(\exists)$ plays some $x \in A$ and wins if $x \notin B$. ⊣

7.3. Let $\mathfrak{A}$ be acceptable, let $G = G(\bar{Q}, R)$ be an open hyperelementary game. Prove that if $(\exists)$ wins G, then $(\exists)$ has a hyperelementary winning quasistrategy.

HINT: Assign ordinals to the set of winning positions for (∃) by the induction

$(x_0, x_1, \ldots, x_{i-1}) \in C \Leftrightarrow R_{i-1}(x_0, x_1, \ldots, x_{i-1}) \vee (Q_i x_i)[(x_0, x_1, \ldots, x_i) \in C],$

of the proof of (5C.1). Now have (∃) play so that he is always in a winning position and he continues decreasing the ordinal stage of his position in this induction. ⊣

7.4. Prove that for some open, elementary game G on the structure $\mathbb{N}$ of arithmetic the relation

$$\mathscr{P}(X) \Leftrightarrow X = \langle \Sigma \rangle, \textit{for some quasistrategy } \Sigma, \textit{ winning for } (\exists) \textit{ in } G$$

is not hyperarithmetical.

HINT: Show that the contrary hypothesis implies that every inductive relation is Σ^1_1. ⊣

7.5. Prove that if $\mathfrak{A}$ is acceptable, then there exists a closed elementary game G on $\mathfrak{A}$ such that (∃) wins G but (∃) has no hyperelementary winning quasistrategy. ⊣

7.6. In the notation of the Stage Comparison Theorems, let

$$\varphi(\bar{x}, S) \equiv (\bar{Q}\bar{z})(\forall\bar{u})[\theta(\bar{x}, \bar{z}, \bar{u}) \vee S(\bar{u})],$$
$$\psi(\bar{y}, T) \equiv (\bar{P}\bar{w})(\forall\bar{v})[\tau(\bar{y}, \bar{w}, \bar{v}) \vee T(\bar{v})]$$

be two formulas in canonical positive form. Prove that

$$\bar{x} \leqslant^*_{\varphi,\psi} \bar{y} \Leftrightarrow \{\bar{Q}(\bar{z}_1)(\forall\bar{u}_1)(\bar{P}^{\cup}\bar{w}_1)(\exists\bar{v}_1)(\bar{Q}\bar{z}_2)(\forall\bar{u}_2)(\bar{P}^{\cup}\bar{w}_2)(\exists\bar{v}_2)\ldots\}$$
$$\bigvee_i [\theta(\bar{u}_i, \bar{z}_{i+1}, \bar{u}_{i+1}) \;\&\; \bigwedge_{k<i} \neg\tau(\bar{v}_k, \bar{w}_{k+1}, \bar{v}_{k+1})].$$

Find a similar infinitary formula for $\bar{x} < ^*_{\varphi,\psi}\bar{y}$. ⊣

7.7. Give a trivial proof of Theorem 7D.1 for the structure of arithmetic $\mathbb{N}$.

HINT: By the relativized version of Exercise 3.1, if $\mathscr{R}(\bar{x}, \bar{Y})$ is Π^1_1 on $\mathbb{N}$, then

$$\mathscr{R}(\bar{x}, \bar{Y}) \Leftrightarrow \{(u, v) \colon \mathscr{P}(u, v, \bar{x}, \bar{Y})\} \textit{ is a wellordering},$$

with $\mathscr{P}$ elementary. Prove first that if $\mathscr{R}$ is also Σ^1_1, then

$$\textit{supremum}\{\textit{rank}(\{(u, v) \colon \mathscr{P}(u, v, \bar{x}, \bar{Y})\}) \colon \mathscr{R}(\bar{x}, \bar{Y})\} < \kappa^{\mathbb{N}}.$$

Then use the fact that if $\prec_1, \prec_2$ are linear orderings and $\prec_2$ is a wellordering, then

$$\prec^1 \textit{ is a wellordering of rank} < \textit{rank}(\prec_2)$$
$$\Leftrightarrow (\exists f)[f \textit{ is a similarity of } \prec_1 \textit{ with an initial segment of } \prec_2]$$
$$\Leftrightarrow (\exists ! f)[f \textit{ is a similarity of } \prec_1 \textit{ with an initial segment of } \prec_2].$$ ⊣

CHAPTER 8

COUNTABLE ACCEPTABLE STRUCTURES

There are two basic theorems about countable, acceptable structures. We prove these in Sections 8A and 8B and we also derive some of their consequences in these two sections and in Section 8C. In Section 8D we look briefly at some special properties of the structure $\mathbb{N}$ of arithmetic which do not generalize. (The method of constructing these counterexamples uses an infinitary language with game quantifiers which is interesting in itself.) The last section examines the status in the abstract theory of the Suslin–Kleene Theorem, the basic result about hyperarithmetical relations on ω.

8A. The Abstract Kleene Theorem

For the case of $\mathbb{N}$, this was the chief result of Kleene [1955a]. The strong version given there was the basis of Kleene's approach to the theory of hyperarithmetical sets. The theorem was extended to structures of the form $\langle A, \in \restriction A\rangle$ with A a countable, transitive set closed under pairing, in Barwise–Gandy–Moschovakis [1971]. For the abstract version here we follow Moschovakis [1970] (which was written after Barwise–Gandy–Moschovakis [1971]).

8A.1. Abstract Kleene Theorem. *Every Π^1_1 second order relation on a countable acceptable structure is inductive.*

Proof. We give the argument for first order relations since the version for second order relations follows by a trivial relativization.

Suppose then that $\mathfrak{A} = \langle A, R_1, \ldots, R_l\rangle$ is countable, acceptable and R satisfies an equivalence

$$R(\bar{x}) \Leftrightarrow (\exists Y_1) \ldots (\exists Y_m)\varphi(Y_1, \ldots, Y_m, \bar{x}),$$

with φ elementary. If we bring φ into prenex normal form and then apply repeatedly the trivial equivalence

$$(\forall \bar{u})(\exists \bar{v})P(\bar{u}, \bar{v}) \Leftrightarrow (\exists X)\{(\forall \bar{u})(\exists \bar{v})X(\bar{u}, \bar{v}) \;\&\; (\forall \bar{u})(\forall \bar{v})[X(\bar{u}, \bar{v}) \Rightarrow P(\bar{u}, \bar{v})]\},$$

we obtain an equivalence of the form

(1) $$R(\bar{x}) \Leftrightarrow (\exists Y_1) \ldots (\exists Y_m)(\forall \bar{u})(\exists \bar{v})\psi(Y_1, \ldots, Y_m, \bar{u}, \bar{v}, \bar{x}),$$

where ψ now is quantifier free. Let

$$\bar{a} = a_1, \ldots, a_k$$

be the individual constants that occur in ψ. Notice that we can evaluate the truth or falsity of $\psi(Y_1, \ldots, Y_m, \bar{u}, \bar{v}, \bar{x})$ if we know $\bar{u}, \bar{v}, \bar{x}$ and the truth value of each $Y_j(t_1, \ldots, t_{n_j})$ when $t_1, \ldots, t_{n_j}$ are chosen from the finite sequence $\bar{a}, \bar{u}, \bar{v}, \bar{x}$.

For each $\bar{x} = x_1, \ldots, x_n$, consider all *finite* structures of the form

$$\mathfrak{B} = \langle B, Z_1, \ldots, Z_m \rangle$$

such that

$$a_1, \ldots, a_k, \qquad x_1, \ldots, x_n \in B$$

and such that each relation Z_j has the same number of arguments as the relation variable Y_j in (1). We can think of such structures as providing *approximations* $Z_1, \ldots, Z_m$ to relations $Y_1, \ldots, Y_m$ such that $(\forall \bar{u})(\exists \bar{v})\psi(Y_1, \ldots, Y_m, \bar{u}, \bar{v}, \bar{x})$. Of course in general we cannot expect that any finite structure will make $(\forall \bar{u})(\exists \bar{v})\psi(Y_1, \ldots, Y_m, \bar{u}, \bar{v}, \bar{x})$ true, since this sentence may be satisfiable only by infinite $Y_1, \ldots, Y_m$.

As usual, $\mathfrak{B}_1 \subseteq \mathfrak{B}_2$ means that $\mathfrak{B}_1$ is a *substructure* of $\mathfrak{B}_2$, i.e. the domain of $\mathfrak{B}_1$ is a subset of the domain of $\mathfrak{B}_2$ and the relations of $\mathfrak{B}_2$ agree with those of $\mathfrak{B}_1$ on the domain of $\mathfrak{B}_1$.

We will use the much stronger relation of extension for structures, where $\mathfrak{B}_1 < \mathfrak{B}_2$ implies that $\mathfrak{B}_2$ has *witnesses* $\bar{v}$ for each $\bar{u}$ in $\mathfrak{B}_1$ verifying $\psi(Z_1, \ldots, Z_m, \bar{u}, \bar{v})$. Let

$$\mathfrak{B} \models \psi(\bar{u}, \bar{v})$$

abbreviate $\psi(Z_1, \ldots, Z_m, \bar{u}, \bar{v}, \bar{x})$, if $\mathfrak{B} = \langle B, Z_1, \ldots, Z_m \rangle$ and the sequences $\bar{u}, \bar{v}$ lie in $\mathfrak{B}$, and put

$$\mathfrak{B}_1 < \mathfrak{B}_2 \Leftrightarrow \mathfrak{B}_1 \subseteq \mathfrak{B}_2 \;\&\; (\forall \bar{u} \in \mathfrak{B}_1)(\exists \bar{v} \in \mathfrak{B}_2)[\mathfrak{B}_2 \models \psi\,(\bar{u}, \bar{v})].$$

If $\mathfrak{B}_0 < \mathfrak{B}_1 < \mathfrak{B}_1 < \ldots$ is an infinite sequence of finite structures each containing $\bar{a}, \bar{x}$ and each extending the preceding, it is immediate that the structure

$$\mathfrak{B} = \bigcup_i \mathfrak{B}_i$$

satisfies

$$(\forall \bar{u} \in \mathfrak{B})(\exists \bar{v} \in \mathfrak{B})[\mathfrak{B} \models \psi(\bar{u}, \bar{v})].$$

If the domain of $\mathfrak{B}$ is all of A, this easily implies $R(\bar{x})$. We wish to verify that whether such a sequence of finite structures exists for a given $\bar{x}$ is a coinductive relation of $\bar{x}$.

For each $\bar{x}$ consider the game $G_{\bar{x}}$, where I plays members of A and II plays finite structures $\mathfrak{B}$ containing $\bar{a}$, $\bar{x}$ as follows:

I	II
z_0	$\mathfrak{B}_0$
z_1	$\mathfrak{B}_1$
.	.
.	.
.	.

At the end of the run, II wins if

$$\bigwedge_i [z_i \in \mathfrak{B}_i \;\&\; \mathfrak{B}_i < \mathfrak{B}_{i+1}].$$

The game is obviously open for I and closed for II. We prove

(*) $\quad R(\bar{x}) \Leftrightarrow$ II *has a winning strategy in* $G_{\bar{x}}$.

Proof of direction ($\Rightarrow$) *of* (*). If $R(\bar{x})$, then there exist $Y_1, \ldots, Y_m$ such that $(\forall \bar{u})(\exists \bar{v})\psi(Y_1, \ldots, Y_m, \bar{u}, \bar{v}, \bar{x})$. We instruct II to play on finite substructures of the structure

$$\langle A, Y_1, \ldots, Y_m \rangle.$$

It is a simple matter to verify that given any such $\mathfrak{B}_i$ and any z_{i+1} that I plays, II can find another such $\mathfrak{B}_{i+1}$, with $z_{i+1} \in \mathfrak{B}_{i+1}$ and $\mathfrak{B}_i < \mathfrak{B}_{i+1}$: he simply adds to $\mathfrak{B}_i$ the element z_{i+1} and he throws in for each $\bar{u}$ in $\mathfrak{B}_i$ some $\bar{v}$ such that $\psi(Y_1, \ldots, Y_m, \bar{u}, \bar{v}, \bar{x})$.

Proof of direction ($\Leftarrow$) *of* (*). Have I enumerate A against II's winning strategy. The resulting sequence $\mathfrak{B}_0 < \mathfrak{B}_1 < \ldots$ of finite structures has union $\bigcup_i \mathfrak{B}_i$ whose domain is precisely A and it satisfies $(\forall \bar{u})(\exists \bar{v})\psi(\bar{u}, \bar{v})$. Thus if the relations of $\bigcup_i \mathfrak{B}_i$ are $Y_1, \ldots, Y_m$, we have $(\forall \bar{u})(\exists \bar{v})\psi(Y_1, \ldots, Y_m, \bar{u}, \bar{v}, \bar{x})$, which implies $R(\bar{x})$.

To finish the proof using (*), we code all finite structures so that the following relation is elementary:

$Q(b_1, b_2, z, \bar{x}) \Leftrightarrow b_1, b_2$ *code finite structures* $\mathfrak{B}_1 = \langle B, Z_1, \ldots, Z_m \rangle$ *and* $\mathfrak{B}_2 = \langle C, W_1, \ldots, W_m \rangle$ *such that* $\bar{a}, \bar{x} \in B$, $z \in B$ *and* $\mathfrak{B}_1 < \mathfrak{B}_2$.

This is a bit messy but obviously possible. Now from (*),

$$R(\bar{x}) \Leftrightarrow \{(\forall z_0)(\exists b_0)(\forall z_1)(\exists b_1) \ldots\}(\forall i) Q(b_i, b_{i+1}, z_i, \bar{x}),$$

which implies easily that Q is coinductive. ⊣

The Kleene Theorem establishes for countable acceptable structures several results which fail for arbitrary acceptable structures. We list two of these here and leave the rest for the exercises.

8A.2. SECOND ORDER CLOSURE THEOREM. *Let $\varphi(\bar{x}, \bar{Y}, S)$ be S-positive in the language of a countable acceptable structure $\mathfrak{A}$. Then the fixed point $\mathcal{I}_\varphi$ determined by φ is hyperelementary on $\mathfrak{A}$ if and only if the closure ordinal $\|\varphi\|$ of φ is smaller than $\kappa^{\mathfrak{A}}$.*

PROOF. If $\|\varphi\| < \kappa$, then $\mathcal{I}_\varphi = \mathcal{I}_\varphi^{\|\varphi\|}$ is hyperelementary by Theorem 6C.3. Towards proving the converse by contradiction, assume that $\mathcal{I}_\varphi$ is hyperelementary but $\|\varphi\| \geqslant \kappa^{\mathfrak{A}}$, let $\psi(\bar{y}, T)$ be any T-positive formula in the language of $\mathfrak{A}$. Since $\|\psi\| \leqslant \kappa^{\mathfrak{A}}$, we obviously have

$$\begin{aligned} \bar{y} \in I_\psi &\Leftrightarrow (\exists \bar{x})(\exists \bar{Y})[(\bar{x}, \bar{Y}) \in \mathcal{I}_\varphi \;\&\; |\bar{y}|_\psi \leqslant |\bar{x}, \bar{Y}|_\varphi] \\ &\Leftrightarrow (\exists \bar{x})(\exists \bar{Y})[(\bar{x}, \bar{Y}) \in \mathcal{I}_\varphi \;\&\; \neg((\bar{x}, \bar{Y}) <^*_{\varphi,\psi} \bar{y})]. \end{aligned}$$

Now $<^*_{\varphi,\psi}$ is inductive by the Stage Comparison Theorem 6C.2, so that we have an equivalence

$$\bar{y} \in I_\psi \Leftrightarrow (\exists \bar{x})(\exists \bar{Y})\mathcal{R}(\bar{x}, \bar{Y}, \bar{y})$$

with a coinductive $\mathcal{R}$. Since coinductive relations are Σ^1_1 by 6B.5, this implies that I_ψ is Σ^1_1, hence coinductive by the Kleene Theorem. Thus every inductive relation on $\mathfrak{A}$ is coinductive, contradicting 5D.3. ⊣

8A.3. THEOREM. *If $\mathfrak{A}$ is countable, acceptable, then the class $\mathcal{HE}^1(\mathfrak{A})$ of hyperelementary subsets of A is inductive but not coinductive.*

PROOF. In the proof of the abstract Spector–Gandy Theorem 7D.2, we showed that if $\mathcal{HE}^1(\mathfrak{A})$ is hyperelementary, then every inductive relation on $\mathfrak{A}$ is Σ^1_1-definable with basis $\mathcal{HE}^1$—in particular every inductive relation on $\mathfrak{A}$ is Σ^1_1. By the Abstract Kleene Theorem every inductive relation is then coinductive, once more contradicting 5D.3. ⊣

8B. The Perfect Set Theorem

In addition to the Kleene Theorem, there is one more hard fact about countable acceptable structures which is interesting in itself and rich in its consequences. The strong version we need is due to Mansfield [1970] for the structure $\mathbb{N}$ of arithmetic, but weaker versions for $\mathbb{N}$ can be traced to classical work in Descriptive Set Theory. Our proof is an elaboration of the Mansfield proof.

8B.1. Perfect Set Theorem. *Let* $\mathfrak{A} = \langle A, R_1, \ldots, R_l \rangle$ *be a countable acceptable structure and let* $\mathscr{P} \subseteq Power(A)$ *be a* Σ_1^1 *class of sets. If* $\mathscr{P}$ *contains a nonhyperelementary set, then* $\mathscr{P}$ *has* $2^{\aleph_0}$ *members.*

Proof. If $\mathscr{P}$ is Σ_1^1, so is

$$\{X: X \in \mathscr{P} \ \& \ X \notin \mathscr{HE}\}$$

by 6D.2. Hence it will be enough to prove that if $\mathscr{P}$ is Σ_1^1, nonempty and without hyperelementary members, then $\mathscr{P}$ has $2^{\aleph_0}$ members.

As in the proof of 8A.1, we may assume that there is a quantifier free formula $\psi(Y_1, \ldots, Y_m, X, \bar{u}, \bar{v})$ in the language of $\mathfrak{A}$ such that

$$X \in \mathscr{P} \Leftrightarrow (\exists Y_1) \ldots (\exists Y_m)(\forall \bar{u})(\exists \bar{v})\psi(Y_1, \ldots, Y_m, X, \bar{u}, \bar{v}).$$

We now consider all finite structures of the form

$$\mathfrak{B} = \langle B, Z_1, \ldots, Z_m, W \rangle$$

which contain all the constants $a_1, \ldots, a_k$ in ψ and which give approximations to some $X \in \mathscr{P}$ and to $Y_1, \ldots, Y_m$ which verify that $X \in \mathscr{P}$ by satisfying $(\forall \bar{u})(\exists \bar{v})\psi(Y_1, \ldots, Y_m, X, \bar{u}, \bar{v})$. Recall that

$$\mathfrak{B}_1 < \mathfrak{B}_2 \Leftrightarrow \mathfrak{B}_1 \subseteq \mathfrak{B}_2 \ \& \ (\forall \bar{u} \in \mathfrak{B}_1)(\exists \bar{v} \in \mathfrak{B}_2)[\mathfrak{B}_2 \models \psi(\bar{u}, \bar{v})].$$

Call $\mathfrak{B}$ *good* if it does give a desired approximation, i.e.

$$\begin{aligned} \mathfrak{B} = \langle B, Z_1, \ldots, Z_m, W \rangle &\text{ is good} \\ \Leftrightarrow (\exists Y_1) \ldots (\exists Y_m)(\exists X)\{&(\forall \bar{u})(\exists \bar{v})\psi(Y_1, \ldots, Y_m, X, \bar{u}, \bar{v}) \\ &\& \ Y_1 \restriction B = Z_1 \ \& \ldots \& \ Y_m \restriction B = Z_m \ \& \ X \restriction B = W\}. \end{aligned}$$

The following is immediate from the definition:

(i) *If* $\mathfrak{B}$ *is good, then for every* z *there is some good* $\mathfrak{B}'$ *such that*

$$z \in \mathfrak{B}' \text{ and } \mathfrak{B} < \mathfrak{B}'.$$

Call $\mathfrak{B}_1$ and $\mathfrak{B}_2$ *incompatible* if their respective approximations to some member of $\mathscr{P}$ differ at some point:

$$\begin{aligned} \langle B^1, Z_1^1, \ldots, Z_m^1, W^1 \rangle, \langle B^2, Z_1^2, \ldots, Z_m^2, W^2 \rangle &\text{ are incompatible} \\ \Leftrightarrow \text{for some } t \in B^1 \cap B^2, \neg(t \in W^1 &\Leftrightarrow t \in W^2). \end{aligned}$$

The key to the proof is the following observation:

(ii) *If* $\mathfrak{B}$ *is good, then there exist incompatible good* $\mathfrak{B}_1, \mathfrak{B}_2$ *such that*

$$\mathfrak{B} < \mathfrak{B}_1, \qquad \mathfrak{B} < \mathfrak{B}_2.$$

Proof of (ii). Let $\mathfrak{B} = \langle B, Z_1, \ldots, Z_m, W\rangle$ and assume that $\mathfrak{B}$ is good but any two good extensions of $\mathfrak{B}$ are compatible. This implies immediately that

$$(\forall Y_1)\ldots(\forall Y_m)(\forall X)(\forall Y_1')\ldots(\forall Y_m')(\forall X')\{[\mathfrak{B} \subseteq \langle A, Y_1, \ldots, Y_m, X\rangle \\ \& \ \mathfrak{B} \subseteq \langle A, Y_1', \ldots, Y_m', X'\rangle \ \& \ (\forall \bar{u})(\exists \bar{v})\psi(Y_1, \ldots, Y_m, X, \bar{u}, \bar{v}) \\ \& \ (\forall \bar{u})(\exists \bar{v})\psi(Y_1', \ldots, Y_m', X', \bar{u}, \bar{v})] \Rightarrow X = X'\},$$

i.e. $\mathfrak{B}$ gives an approximation to exactly one $X^* \in \mathscr{P}$. But then we have a Δ_1^1 definition of this X^*,

$$t \in X^* \Leftrightarrow (\exists Y_1)\ldots(\exists Y_m)(\exists X)\{\mathfrak{B} \subseteq \langle A, Y_1, \ldots, Y_m, X\rangle \\ \& \ (\forall \bar{u})(\exists \bar{v})\psi(Y_1, \ldots, Y_m, X, \bar{u}, \bar{v}) \ \& \ t \in X\} \\ \Leftrightarrow (\forall Y_1)\ldots(\forall Y_m)(\forall X)\{[\mathfrak{B} \subseteq \langle A, Y_1, \ldots, Y_m, X\rangle \\ \& \ (\forall \bar{u})(\exists \bar{v})\psi(Y_1, \ldots, Y_m, X, \bar{u}, \bar{v})] \Rightarrow t \in X\},$$

so that X^* is a hyperelementary member of $\mathscr{P}$ contrary to hypothesis.

By combining (i) and (ii) we immediately get

(iii) *If* $\mathfrak{B}$ *is good, then for every* z *there exist incompatible good* $\mathfrak{B}_1, \mathfrak{B}_2$ *such that*

$$z \in \mathfrak{B}_1, \ z \in \mathfrak{B}_2, \ \mathfrak{B} < \mathfrak{B}_1, \ \mathfrak{B} < \mathfrak{B}_2.$$

We simply first get $\mathfrak{B}'$ such that $z \in \mathfrak{B}'$, $\mathfrak{B} < \mathfrak{B}'$ by (i) and then we get incompatible good $\mathfrak{B}_1, \mathfrak{B}_2$ extending $\mathfrak{B}'$ by (ii).

Let $z_0, z_1, \ldots$ be a fixed enumeration of A and construct a binary tree of finite good structures by repeated applications of (iii). This is shown in Fig. 8.1.

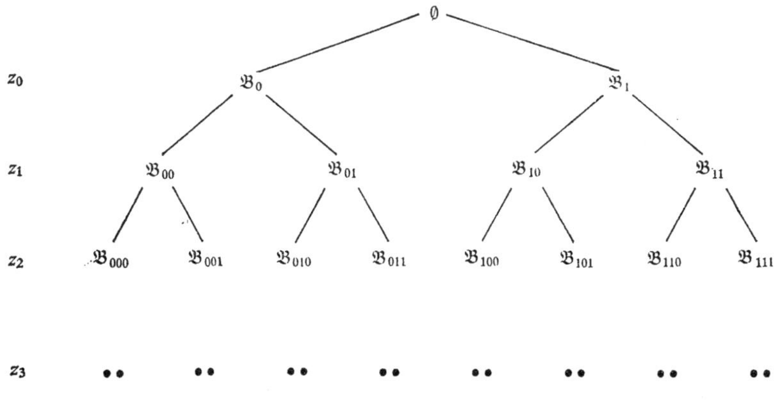

Fig. 8.1.

All structures at the 0th level of the tree contain z_0, all those at the 1st level contain z_1, etc. Moreover, if $\mathfrak{B}'$ is below $\mathfrak{B}$ in the tree, we have $\mathfrak{B} < \mathfrak{B}'$, and if $\mathfrak{B}, \mathfrak{B}'$ do not lie on the same branch coming down, then they are incompatible.

Every infinite branch $\mathfrak{B}_0 < \mathfrak{B}^1 < \mathfrak{B}^2 < \ldots$ through the tree determines relations $Y_1, \ldots, Y_m, X$ such that $(\forall \bar{u})(\exists \bar{v})\psi(Y_1, \ldots, Y_m, X, \bar{u}, \bar{v}, \bar{x})$, so in particular $X \in \mathscr{P}$. Since distinct branches determine distinct X's by the incompatibility condition and since there are $2^{\aleph_0}$ branches, there are at least $2^{\aleph_0}$ members of $\mathscr{P}$. ⊣

The theorem gets its name from the fact that we have actually constructed a *perfect* subset of $\mathscr{P}$, i.e. a subset of $\mathscr{P}$ which is closed and with no isolated points in the topology on *Power*(A) generated by the neighbourhoods

$$N(a_1, \ldots, a_k; b_1, \ldots, b_m) = \{X: a_1, \ldots, a_k \in X \;\&\; b_1, \ldots, b_m \notin X\}.$$

We do not pursue here the topological or descriptive set theoretic aspects of this result.

Most of the applications of the Perfect Set Theorem use the following trivial consequence of its contrapositive:

8B.2. COROLLARY. *If* $\mathfrak{A} = \langle A, R_1, \ldots, R_l\rangle$ *is countable, acceptable and* $\mathscr{P}$ *is a countable,* Σ_1^1 *collection of n-ary relations on A, then* $\mathscr{P}$ *contains only hyperelementary relations.* ⊣

8C. The intersection of all $\mathfrak{A}$-models of second order comprehension

The main result of Gandy–Kreisel–Tait [1960] is that the intersection of all ω-models of second order number theory consists precisely of (ω and) the hyperarithmetical relations. Barwise and Grilliot have extended this theorem to all countable acceptable structures using a version of the Omitting Types Theorem. We prove here this result by a very different method which is more in the spirit of our approach to inductive definability in this book. The key idea is the next lemma which is due to Kechris [1972] for the case $\mathfrak{A} = \mathbb{N}$.

8C.1. LEMMA. *Let* $\mathfrak{A}$ *be countable, acceptable, let* $\mathscr{P}$ *be a countable, inductive collection of n-ary relations such that*

$$\text{if } Y \in \mathscr{P} \text{ and } X \text{ is hyperelementary on } (\mathfrak{A}, Y), \text{ then } X \in \mathscr{P}.$$

Then either $\mathscr{P} \subseteq \mathscr{H}\mathscr{E}(\mathfrak{A})$ *or* $\mathscr{P}$ *contains all inductive n-ary relations.*

PROOF. We may assume that $\mathscr{P} \neq \emptyset$ and hence that $\mathscr{P}$ contains all hyperelementary relations.

Let Y be a fixed inductive, nonhyperelementary relation. If there is some $X \in \mathscr{P}$ such that Y is hyperelementary on $(\mathfrak{A}, X)$, then $Y \in \mathscr{P}$ by hypothesis. The other possibility is that for every $X \in \mathscr{P}$, Y is not hyperelementary on $(\mathfrak{A}, X)$. By Theorem 3D.2, we know then that

$$(*) \qquad X \in \mathscr{P} \Rightarrow \kappa^{(\mathfrak{A},X)} \leqslant \kappa = \kappa^{\mathfrak{A}}.$$

We complete the proof by showing that (*) implies $\mathscr{P} \subseteq \mathscr{H}\mathscr{E}(\mathfrak{A})$.

Since $\mathscr{P}$ is inductive, there is some $\psi \equiv \psi(\bar{u}, X, S)$ and parameters $\bar{a}$ such that

$$x \in \mathscr{P} \Leftrightarrow (\bar{a}, X) \in \mathscr{I}_\psi.$$

Now for each $X \in \mathscr{P}$, $|\bar{a}, X|_\psi < \kappa^{(\mathfrak{A},X)} \leqslant \kappa$; hence

$$\mathscr{P} = \bigcup_{\lambda<\kappa} \{X\colon |\bar{a}, X|_\psi \leqslant \lambda\},$$

and since each $\{X\colon |\bar{a}, X|_\psi \leqslant \lambda\}$ is hyperelementary by 6C.3 and countable by hypothesis, it contains only hyperelementary relations by 8B.2, i.e. $\mathscr{P} \subseteq \mathscr{H}\mathscr{E}(\mathfrak{A})$. ⊣

Lemma 8C.1 gives an obvious method for attempting to prove that certain countable inductive sets have only hyperelementary members. For the case $\mathfrak{A} = \mathbb{N}$, Kechris [1972] actually shows under the same hypotheses that $\mathscr{P} \subseteq \mathscr{H}\mathscr{E}$. It is not known whether this stronger result holds for arbitrary countable acceptable $\mathfrak{A}$, so that applying the lemma is a bit more complicated in the general case.

We have already considered some theories in the second order language $\mathscr{L}_2^{\mathfrak{A}}$ over a structure $\mathfrak{A}$, e.g. the theory of Σ_1^1-Collection or the theory of Δ_1^1-Comprehension. By a *theory* in $\mathscr{L}_2^{\mathfrak{A}}$ we simply mean a collection of sentences in $\mathscr{L}_2^{\mathfrak{A}}$. Another such interesting theory is that of *full second order comprehension* over $\mathfrak{A}$, i.e. the collection of (universal closures of the) formulas

$$(\Delta_\infty^1\textit{-Comp}) \qquad (\exists Z)(\forall \bar{x})[\bar{x} \in Z \Leftrightarrow \varphi(\bar{x}, \bar{Y})],$$

where $\varphi(\bar{x}, \bar{Y})$ is in $\mathscr{L}_2^{\mathfrak{A}}$.

By an $\mathfrak{A}$-*model* of a theory $\mathscr{T}$ in $\mathscr{L}_2^{\mathfrak{A}}$ we always understand a collection $\mathscr{F}$ of relations on $\mathfrak{A}$, such that all sentences in $\mathscr{T}$ are true when we interpret them on $\mathfrak{A}$ with *range* $\mathscr{F}$, as in section 7E. A theory $\mathscr{T}$ is *inductive, coinductive*, Π_1^1, etc., if the set

$$Cod(\mathscr{T}) = \{a\colon a \textit{ codes a sentence of } \mathscr{T}\}$$

is inductive, coinductive, Π_1^1, etc., where the coding is assumed reasonable so that Lemma 7E.2 holds.

It is trivial to check that the theory of full second order comprehension over $\mathfrak{A}$ is hyperelementary when $\mathfrak{A}$ is acceptable.

The next result is due to Gandy–Kreisel–Tait [1960] for $\mathbb{N}$, to Barwise (unpublished) and Grilliot [1972] for arbitrary $\mathfrak{A}$.

8C.2. Barwise–Grilliot Theorem. *Let $\mathfrak{A}$ be a countable acceptable structure, let $\mathcal{T}$ be an inductive theory in $\mathcal{L}_2^{\mathfrak{A}}$ which contains all instances of Δ_1^1-Comprehension and has an $\mathfrak{A}$-model. Then the intersection of all $\mathfrak{A}$-models of $\mathcal{T}$ is precisely $\mathcal{HE}(\mathfrak{A})$.*

Proof. Since every $\mathfrak{A}$-model of $\prec$ is closed under elementary definability, the model is completely determined by the unary relations in it. Thus it will be enough to show that the set

$$\mathcal{P} = \{X \subseteq A: X \textit{ belongs to every } \mathfrak{A}\textit{-model of } \mathcal{T}\}$$

is precisely the set of hyperelementary sets.

First we verify that $\mathcal{P}$ satisfies the hypotheses of Lemma 8C.1.

(i) *$\mathcal{P}$ is countable.* This is because by the Skolem–Löwenheim Theorem $\mathcal{T}$ has countable $\mathfrak{A}$-models.

(ii) *$\mathcal{P}$ is inductive.* The relation

$$Mod^{\mathcal{T}}(Z) \Leftrightarrow \{Z_a^{(n)}: n \geqslant 1, a \in A\} \textit{ is an } \mathfrak{A}\textit{-model of } \mathcal{T}$$

is Σ_1^1, since

$$Mod^{\mathcal{T}}(Z) \Leftrightarrow (\forall b)[b \in Cod(\mathcal{T}) \Rightarrow Sat_2(b, Z, \emptyset, \langle\emptyset\rangle)].$$

(We are using the notation of Lemma 7E.2.) Now

$$X \in \mathcal{P} \Leftrightarrow (\forall Z)[Mod^{\mathcal{T}}(Z) \Rightarrow (\exists a)[X = Z_a^{(1)}]],$$

so that $\mathcal{P}$ is Π_1^1.

(iii) *If $Y \in \mathcal{P}$ and X is hyperelementary in Y, then $X \in \mathcal{P}$.* This is because $\mathcal{T}$ extends the theory of Δ_1^1-Comprehension. Since we allow relation parameters in that schema, every $\mathfrak{A}$-model of $\mathcal{T}$ satisfies Δ_1^1-Comprehension on the expanded structure $(\mathfrak{A}, Y)$, for $Y \in \mathcal{P}$, and it must contain all sets hyperelementary on $(\mathfrak{A}, Y)$ by Theorem 7F.1.

Now Lemma 8C.1 applies, and to complete the proof it is sufficient to derive a contradiction from the hypothesis that $\mathcal{P}$ contains every inductive set.

Suppose R is an arbitrary coinductive relation. By 7B.4 we know that there is an elementary $\mathcal{Q}(Y, \bar{x})$ such that

$$\begin{aligned} R(\bar{x}) &\Leftrightarrow (\exists Y)\mathcal{Q}(Y, \bar{x}) \\ &\Leftrightarrow (\exists Y)\{Y \textit{ is coinductive } \& \ \mathcal{Q}(Y, \bar{x})\}. \end{aligned}$$

If $\mathcal{P}$ contains all inductive sets, then $\mathcal{P}$ contains all coinductive sets. Hence

$$R(\bar{x}) \Leftrightarrow (\exists Y)[Y \in \mathcal{P} \ \& \ R(Y, \bar{x})].$$

Substituting the definition of $\mathscr{P}$ we have

$$R(\bar{x}) \Leftrightarrow (\forall Z)[Mod^{\mathscr{T}}(Z) \Rightarrow (\exists a)\mathscr{R}(Z_a^{(1)}, \bar{x})]$$

which implies that R is Π_1^1. This is a contradiction, since R was an arbitrary coinductive relation. ⊣

8D. Counterexamples to special properties of arithmetic; the language $\mathscr{L}_{\omega_1,G}$

There are some very special properties of induction on the structure $\mathbb{N}$ of arithmetic which cannot be extended to all countable, acceptable structures. We discuss two of them here and we also give a general method of obtaining *countable* structures which reflect many of the features of induction on given *uncountable* structures. In Section 8E we will look at the Suslin–Kleene Theorem, the most important special property of $\mathbb{N}$.

Consider first the fact that the closure ordinal of $\mathbb{N}$ is the supremum of all arithmetical wellorderings on ω. We prove something a bit stronger.

8D.1. THEOREM. *Let $\omega_1 = \kappa^{\mathbb{N}}$ be the ordinal of the structure of arithmetic $\mathbb{N}$. Then there is a fixed arithmetical relation $P(u, v, x)$ such that for all x, the relation*

$$<_x = \{(u, v) : P(x, u, v)\}$$

is a (strict) linear ordering and

(1) $\omega_1 = supremum\{rank(<_x) : <_x \text{ is a wellordering}\}$.

In fact, P can be chosen recursive, so that ω_1 is the supremum of order types of recursive wellorderings on ω. (Spector [1961].)

PROOF. We outline how to obtain an arithmetical P with the properties in the theorem. Those who are familiar with the simple properties of recursive relations will recognize that our argument actually yields a recursive P.

Let $R(x)$ be some Π_1^1 relation which is not Σ_1^1. By Exercise 3.1 there is an arithmetical relation $P(u, v, x)$ such that for all x the relation

$$<_x = \{(u, v) : P(u, v, x)\}$$

is a linear ordering and

$$R(x) \Leftrightarrow <_x \textit{ is a wellordering}.$$

If there were a $\xi < \omega_1$ such that

$$supremum\{rank(<_x) : R(x)\} \leqslant \xi,$$

then

$$R(x) \Leftrightarrow rank(<_x) \leqslant \xi$$

and by Exercise 3.5, R would be hyperarithmetical; hence (1) holds.

It is easy to check that P can be chosen recursive by following up the hints given for the exercises leading up to 1.12 and 3.1. ⊣

In contrast to this, we have the following general result for structures on which wellfoundedness is elementary.

8D.2. THEOREM. *Let $\mathfrak{A}$ be an acceptable structure on which the class $\mathscr{WF}^1$ of wellfounded binary relations is elementary. For every hyperelementary relation P,*

$$supremum\{rank(\prec): \prec \text{ is wellfounded and elementary on } (\mathfrak{A}, P)\} < \kappa^{\mathfrak{A}}.$$

PROOF. Choose a hyperelementary $E^2 \subseteq A^3$ which parametrizes the binary relations elementary on $(\mathfrak{A}, P)$ – this exists by 5D.1. For each $a \in A$, put

(2) $$\prec_a = \{(u, v): E^2(a, u, v)\}$$

and assume towards a contradiction that

(3) $$supremum\{rank(\prec_a): \prec_a \text{ is wellfounded}\} = \kappa^{\mathfrak{A}}.$$

Now the relation

$$(a, u) \prec (b, v) \Leftrightarrow a = b \;\&\; \prec_a \text{ is wellfounded } \&\; u \prec_a v$$

is hyperelementary, wellfounded and has rank $\kappa^{\mathfrak{A}}$, which contradicts 2B.5. ⊣

Theorem 6C.3 gives us examples of structures in which $\mathscr{WF}^1$ is elementary, but they are all uncountable, as they must be by 7A.2 and 8A.1. The natural way to define countable structures on which

(4) $$\kappa^{\mathfrak{A}} > \text{supremum of ranks of elementary wellfounded relations}$$

is to find a language which is rich enough so that we can express (4) in it but which is nice enough so that it has the Skolem–Löwenheim property. We can then use countable, elementary substructures of the counterexamples given by 8D.2. From the several available languages which are known to have these properties, we prefer to use one with *game quantifiers*, in which we find natural explicit definitions for inductive relations. The same language has been studied by Barwise [1972].

For each set A, the language $\mathscr{L}^A_{\omega_1,G}$ is defined by adding to the symbols and formation rules for the first order language $\mathscr{L}^A$ the following four *infinitary* formation rules. Recall that we allow individual and relation constants and variables in the formation rules of $\mathscr{L}^A$.

If Φ is a countable collection of formulas all of whose free individual and relation variables are contained in some fixed finite list, then

$$\bigwedge\Phi, \qquad \bigvee\Phi$$

are also formulas.

We often have *indexed sequences* of formulas,

$$\Phi = \{\varphi_j : j \in J\},$$

in which case it is natural to write the infinite conjunctions and disjunctions using the indexing,

$$\bigwedge\Phi \equiv \bigwedge_j \varphi_j,$$
$$\bigvee\Phi \equiv \bigvee_j \varphi_j.$$

In many cases the indexing itself is by formulas. For example we may put down the conjunction

$$\bigwedge_\varphi (\exists\bar{u})\varphi(\bar{u}, \bar{x}),$$

where φ varies over all formulas of the form

$$\varphi \equiv \varphi(u_1, \ldots, u_k, x_1, \ldots, x_n)$$

(k may vary with φ) which have no individual constants from A and whose relation constants are included in some fixed finite list. Some such restrictions are necessary to ensure that the conjunction is over a countable set.

If $\varphi_0(x_0), \varphi_1(x_0, x_1), \varphi_2(x_0, x_1, x_2), \ldots$ is a sequence of formulas such that the free variables of each $\varphi_i(x_0, \ldots, x_i)$ are among $x_0, \ldots, x_i$ and the variables in some fixed finite list, and if $Q_0, Q_1, Q_2, \ldots$ is an infinite sequence of quantifiers, then

$$\{(Q_0x_0)(Q_1x_1)\ldots\} \bigwedge_i \varphi_i(x_0, \ldots, x_i), \tag{5}$$

$$\{(Q_0x_0)(Q_1x_1)\ldots\} \bigvee_i \varphi_i(x_0, \ldots, x_i) \tag{6}$$

are also formulas.

It is understood that the free variables of $\bigwedge\Phi$ or $\bigvee\Phi$ are the variables which are free in some $\varphi \in \Phi$. Similarly, the free variables of the *game formulas* in (5) or (6) are the variables which are free in some $\varphi_i(x_0, \ldots, x_i)$, exclusive of $x_0, x_1, \ldots$ which are all bound by the infinite quantifier string. The restrictions in the formation rules ensure that every formula has finitely many free variables of either type.

As usual, *sentences* are formulas with no free variables. We define *truth* for sentences by extending the truth definition of $\mathscr{L}^A$ in the obvious way. Thus $\bigwedge\Phi$ is true if every φ in Φ is true and $\bigvee\Phi$ is true if some φ in Φ is true. For the game sentences we use the game interpretation of infinite quantifier strings that we discussed in Chapter 4: The sentence in (5) is true if ($\exists$) has a winning strategy in the closed game determined by (5) and similarly for (6).

If $\mathfrak{A} = \langle A, R_1, \ldots, R_l\rangle$ is a structure with domain A, then the formulas of $\mathscr{L}^{\mathfrak{A}}_{\omega_1,G}$ are those formulas of $\mathscr{L}^{A}_{\omega_1,G}$ whose relation constants are among $=, R_1, \ldots, R_l$.

A second order relation $\mathscr{P}(\bar{x}, \bar{Y})$ is $\mathscr{L}^{\mathfrak{A}}_{\omega_1,G}$*-definable* (on $\mathfrak{A}$) if there is a formula $\varphi(\bar{x}, \bar{Y})$ of $\mathscr{L}^{\mathfrak{A}}_{\omega_1,G}$ with the appropriate free individual and relation variables such that

$$\mathscr{P}(\bar{x}, \bar{Y}) \Leftrightarrow \varphi(\bar{x}, \bar{Y}).$$

The Fixed Point Normal Form Theorem 6C.6 implies immediately that all inductive relations are $\mathscr{L}^{\mathfrak{A}}_{\omega_1,G}$-definable. Hence coinductive relations, conjunctions of inductive and coinductive relations, etc., are all $\mathscr{L}^{\mathfrak{A}}_{\omega_1,G}$-definable. Thus $\mathscr{L}^{\mathfrak{A}}_{\omega_1,G}$ is quite powerful and we can express in it a good part of the theory of inductive definability on $\mathfrak{A}$.

To be precise, let us assign to each $\varphi(\bar{x}, \bar{Y}, S)$ in the language of a structure $\mathfrak{A}$ which is S-positive in canonical form a fixed formula "$(\bar{x}, \bar{Y}) \in \mathscr{I}_\varphi$" of $\mathscr{L}^{\mathfrak{A}}_{\omega_1,G}$ which defines the fixed point $\mathscr{I}_\varphi$. We will naturally write "$\bar{x} \in I_\varphi$" if φ has no relation variables. It is also useful to notice that every inductive relation $\mathscr{R}(x_1, \ldots, x_n, Y_1, \ldots, Y_m)$ on $\mathfrak{A}$ satisfies

$$\mathscr{R}(x_1, \ldots, x_n, Y_1, \ldots, Y_m) \Leftrightarrow (a_1, \ldots, a_k, x_1, \ldots, x_n, Y_1, \ldots, Y_m) \in \mathscr{I}_\varphi,$$

with suitable constants $a_1, \ldots, a_k$ and some $\varphi(\bar{u}, \bar{x}, \bar{Y}, S)$ *which has no individual constants*; we do this by counting any individual constants that may occur in φ among the parameters of the induction. This is useful because there are only *countably many* formulas of the first order language $\mathscr{L}^{\mathfrak{A}}$ with no constants from $\mathfrak{A}$.

Notice that

$$X \textit{ is elementary} \Leftrightarrow \bigvee_\chi (\exists\bar{u})(\forall\bar{x})[\bar{x} \in X \Leftrightarrow \chi(\bar{u}, \bar{x})],$$

where X is an n-ary relation variable and χ varies over the countably many formulas of $\mathscr{L}^{\mathfrak{A}}$ which have no individual constants and whose free variables $u_1, \ldots, u_k, x_1, \ldots, x_n$ include $x_1, \ldots, x_n$. It is only slightly more complicated to define the class of hyperelementary n-ary relations,

$$X \textit{ is hyperelementary} \Leftrightarrow \bigvee_\varphi \bigvee_\psi (\exists\bar{u})(\exists\bar{v})(\forall\bar{x})[\bar{x} \in X \Leftrightarrow (\bar{u}, \bar{x}) \in I_\varphi \;\&\; \bar{x} \in X \Leftrightarrow (\bar{v}, \bar{x}) \notin I_\psi];$$

here φ varies over the constant-free S-positive formulas of the form $\varphi(u_1, \ldots, u_k, x_1, \ldots, x_n, S)$ and similarly for ψ.

We can say that "every inductive relation is elementary" by

$$\bigwedge_{\varphi} (\forall \bar{u}) \bigvee_{\chi} (\exists \bar{v})(\forall \bar{x})[(\bar{u}, \bar{x}) \in I_{\varphi} \Leftrightarrow \chi(\bar{v}, \bar{x})],$$

where the ranges of φ and ψ are as above and countable. It is only a bit more complicated to say in $\mathscr{L}^{\mathfrak{A}}_{\omega_1,G}$ that "every hyperelementary relation is elementary",

$$\bigwedge_{\varphi} \bigwedge_{\psi}(\forall \bar{u})(\forall \bar{v})\{(\forall \bar{x})[(\bar{u}, \bar{x}) \in I_{\varphi} \Leftrightarrow (\bar{v}, \bar{x}) \notin I_{\psi}]$$
$$\Rightarrow \bigvee_{\chi} (\exists \bar{t})(\forall \bar{x})[(\bar{u}, \bar{x}) \in I_{\varphi} \Leftrightarrow \chi(\bar{t}, \bar{x})]\}.$$

These simple tricks which allow us to quantify over all elementary, hyperelementary or inductive relations in $\mathscr{L}^{\mathfrak{A}}_{\omega_1,G}$ are very useful. They make it conceptually easy if a bit messy to write down explicit sentences of this language which assert that

"$\mathfrak{A}$ *is acceptable*"

or

"$\mathfrak{A}$ *admits a hyperelementary coding scheme*".

For example, the first of these can be expressed by

"there is an elementary relation $\leqslant$ which is an ordering such that every initial segment is finite and for each $n \geqslant 0$ there are elementary maps of ${}^nA$ into A, one-to-one with disjoint ranges and such that there are elementary relations $Seq(x)$, $lh(x) = t$, $q(x, i) = y$ having the appropriate properties".

In fact the formal version of this sentence does not have any infinite alternating quantifier strings.

Using Exercise 6.1 and the explicit definitions of inductive relations in $\mathscr{L}^{\mathfrak{A}}_{\omega_1,G}$, it is easy to find a formula $X \leqslant Y$ of this language such that

$$X \leqslant Y \Leftrightarrow X, Y \textit{ are wellfounded binary relations and } rank(X) \leqslant rank(Y).$$

Now the sentence

$$\bigvee_{\varphi} \bigvee_{\psi} (\exists \bar{u})(\exists \bar{v})\{(\forall s)(\forall t)[(\bar{u}, s, t) \in I_{\varphi} \Leftrightarrow (\bar{v}, s, t) \notin I_{\psi}]$$
$$\& \{(s, t) \colon (\bar{u}, s, t) \in I_{\varphi}\} \leqslant \{(s, t) \colon (\bar{u}, s, t) \in I_{\varphi}\}$$
$$\& \bigwedge_{\chi} (\forall \bar{z})[\{(s, t) \colon \chi(\bar{z}, s, t)\} \leqslant \{(s, t) \colon \chi(\bar{z}, s, t)\}$$
$$\Rightarrow \{(s, t) \colon \chi(\bar{z}, s, t)\} \leqslant \{(s, t) \colon (\bar{u}, s, t) \in I_{\varphi}\}]\}$$

asserts that "some wellfounded hyperelementary binary relation has rank greater than or equal to the rank of every elementary wellfounded relation." Of course we must interpret correctly the ranges of φ, ψ, χ as above. This assertion is equivalent to (4) above for acceptable structures. It is only a bit messier to put down a single sentence ψ of $\mathscr{L}^{\mathfrak{A}}_{\omega_1,G}$ with no individual constants such that

(7) *ψ is true in $\mathfrak{A} \Leftrightarrow \mathfrak{A}$ is acceptable & for every hyperelementary P,*

$$\kappa^{\mathfrak{A}} > \textit{supremum}\{\textit{rank}(\prec): \prec \textit{ is wellfounded, elementary on } (\mathfrak{A}, P)\}.$$

Consider all structures $\mathfrak{A} = \langle A, R_1, \ldots, R_l \rangle$ of a given *signature* $(n_1, \ldots, n_l)$, i.e. where each R_i is n_i-ary. If $\mathfrak{A} = \langle A, R_1, \ldots, R_l \rangle$, $\mathfrak{B} = \langle B, P_1, \ldots, P_l \rangle$ are of the same signature and φ is a formula of $\mathscr{L}^{\mathfrak{A}}_{\omega_1,G}$ *all of whose individual constants are in the intersection $A \cap B$*, it is natural to associate with φ the formula $\varphi^{\mathfrak{A}}$ of $\mathscr{L}^{\mathfrak{A}}_{\omega_1,G}$ obtained by replacing each P_i by R_i in φ. In this notation $\varphi^{\mathfrak{B}} \equiv \varphi$. If φ is a sentence, it is customary to write

"$\mathfrak{A} \models \varphi$" *for* "$\varphi^{\mathfrak{A}}$ *is true.*"

A common situation where we apply this notation is when $\mathfrak{B}$ is a *substructure* of $\mathfrak{A}$, $\mathfrak{B} \subseteq \mathfrak{A}$, in which case every φ in $\mathscr{L}^{\mathfrak{B}}_{\omega_1,G}$ has some $\varphi^{\mathfrak{A}}$ associated with it.

If $\mathfrak{B} \subseteq \mathfrak{A}$ and $\varphi(x_1, \ldots, x_m)$ is a formula of $\mathscr{L}^{\mathfrak{A}}_{\omega_1,G}$ with the indicated free variables, we call φ *absolute for* $\mathfrak{B}$ if the individual constants of φ are in $\mathfrak{B}$ and

$$(\forall b_1, \ldots, b_n \in \mathfrak{B})[\mathfrak{B} \models \varphi(b_1, \ldots, b_n) \Leftrightarrow \mathfrak{A} \models \varphi(b_1, \ldots, b_n)].$$

In the special case that φ is a sentence, this means that φ holds in $\mathfrak{B}$ if and only if it holds in $\mathfrak{A}$.

The next result was independently noticed by Barwise. It is a special case of very general theorems of Barwise [1972].

8D.3. Skolem–Löwenheim Theorem for $\mathscr{L}_{\omega_1,G}$. *Let $\mathfrak{A} = \langle A, R_1, \ldots, R_l \rangle$ be an infinite structure, let Φ be a countable set of formulas of $\mathscr{L}^{\mathfrak{A}}_{\omega_1,G}$. There exists a countable substructure $\mathfrak{B}$ of $\mathfrak{A}$ which contains all the individual constants occurring in formulas of Φ and such that every φ in Φ is absolute for $\mathfrak{B}$.*

Proof. If φ is a formula of $\mathscr{L}^{\mathfrak{A}}_{\omega_1,G}$, a *set of Skolem functions* for φ is any set $\mathscr{S}$ of functions of any number of arguments such that φ is absolute for every substructure of $\mathfrak{A}$ which is closed under all the functions in $\mathscr{S}$.

Here we allow *individual constants* in $\mathscr{S}$, call them 0-ary functions, and a substructure $\mathfrak{B} = \langle B, P_1, \ldots, P_l \rangle$ is *closed under* a k-ary f if

$$b_1, \ldots, b_k \in B \Rightarrow f(b_1, \ldots, b_k) \in B.$$

To prove the theorem, we assign to each formula φ of $\mathscr{L}^{\mathfrak{A}}_{\omega_1,G}$ a *countable* set of Skolem functions $\mathscr{S}(\varphi)$. This will surely do it, because given a countable set Φ of formulas we can take B to be the smallest set closed under $\bigcup_{\varphi \in \Phi} \mathscr{S}(\varphi)$

and let $\mathfrak{B}$ be the restriction of $\mathfrak{A}$ to B. Surely $\mathfrak{B}$ is countable and it satisfies the conclusion of the theorem.

The assignment of $\mathscr{S}(\varphi)$ to φ is by induction on the construction of φ and is completely trivial, except perhaps for the sentences introduced by the game quantifiers. For example, if

$$\varphi(x_1, \ldots, x_n) \equiv (\exists y)\psi(y, x_1, \ldots, x_n),$$

we can take

$$\mathscr{S}(\varphi) = \mathscr{S}(\psi) \cup \{f\},$$

where f is n-ary and chosen so that in $\mathfrak{A}$,

$$(\exists y)\psi(y, x_1, \ldots, x_n) \Rightarrow \psi(f(x_1, \ldots, x_n), x_1, \ldots, x_n).$$

To consider one of the formation rules that involves game quantifiers, suppose

$$\varphi(y) \equiv \{(Q_0 x_0)(Q_1 x_1) \ldots\} \bigwedge_i \varphi_i(y, x_0, \ldots, x_i),$$

where we have exhibited all the free variables and for simplicity we have assumed that φ has only y free. For each $y \in A$, either $(\exists)$ or $(\forall)$ wins the game determined by $\varphi(y)$; let (Q^y) be the winning player and choose a winning strategy for him, a collection of functions

$$\mathscr{T}^y = \{f_i^y : Q_i = Q^y\},$$

where as in Chapter 4 each f_i^y is i-ary. For each i, put

$$f_i(y, x_0, \ldots, x_{i-1}) = \begin{cases} f_i^y(x_0, \ldots, x_{i-1}) & \textit{if } Q^y = Q_i, \\ a & \textit{if } Q^y \neq Q_i, \end{cases}$$

where a is some fixed element of A, and set

$$\mathscr{S}(\varphi) = \{f_i : i = 0, 1, 2, \ldots\} \cup \bigcup_i \mathscr{S}(\varphi_i).$$

If B is closed under $\mathscr{S}(\varphi)$ and $\mathfrak{A} \models \varphi(y)$ for some $y \in B$, then by the definition B is closed under some winning strategy for $(\exists)$ in the associated game. If we follow this strategy, we can ensure that for all plays of $(\forall)$, $\mathfrak{A} \models \bigwedge_i \varphi_i(y, x_0, \ldots, x_i)$, and since B is also closed under Skolem functions for all the φ_i, we have by induction hypothesis $\mathfrak{B} \models \bigwedge_i \varphi_i(y, x_0, \ldots, x_i)$, so $\varphi(y)$ holds in $\mathfrak{B}$. The argument is similar if $\mathfrak{A} \models \neg\varphi(y)$ and both the construction and the proof are similar for the open game quantifier. ⊣

If $\mathfrak{A}$, $\mathfrak{B}$ are related as in the theorem, we call $\mathfrak{B}$ an *$\mathscr{L}_{\omega_1,G}$-elementary substructure of $\mathfrak{A}$ with respect to Φ.*

This result yields immediately counterexamples to the possibility of extending 8D.1 to all countable acceptable structures.

8D.4. THEOREM. *There is a countable acceptable structure such that for every hyperelementary P,*

(8) $\kappa^{\mathfrak{A}} > supremum\{rank(\prec): \prec$ *is wellfounded, elementary on* $(\mathfrak{B}, P)\}$.

PROOF. Choose an $\mathfrak{A}$ such that (8) holds by 8D.2, then choose a countable structure $\mathfrak{B}$ for which the sentence ψ satisfying (7) above is absolute. ⊣

It is clear that the same method can be used to obtain countable structures that reflect many complicated properties of induction on uncountable structures. Some caution is needed since no truly second order property is expressible in $\mathscr{L}^{\mathfrak{A}}_{\omega_1,G}$—this is why we cannot construct a countable acceptable structure on which every inductive relation is Σ^1_1. However, we can often get very close to reflecting second order properties by a bit of trickery, as in the next example.

The following important *basis property* for Σ^1_1 sets of sets of integers is due to Kleene.

8D.5. BASIS THEOREM FOR Σ^1_1 SETS ON $\mathbb{N}$. *If* $\mathscr{P} \subseteq \text{Power}(\omega)$ *is nonempty and* Σ^1_1 *on the structure* $\mathbb{N}$, *then* $\mathscr{P}$ *contains a set X which is arithmetical in some* Σ^1_1 *set* $R \subseteq \omega$.

In fact, under the same hypothesis, $\mathscr{P}$ *contains a set X which is recursive in some* Σ^1_1 *set* $R \subseteq \omega$.

PROOF. As in the proof of 8B.1, we have

$$X \in \mathscr{P} \Leftrightarrow (\exists Y_1) \ldots (\exists Y_m)(\forall \bar{u})(\exists \bar{v})\psi(Y_1, \ldots, Y_m, X, \bar{u}, \bar{v}),$$

where ψ is quantifier free in the language of $\mathbb{N}$. Again we consider finite structures of the form

$$\mathfrak{B} = \langle B, Z_1, \ldots, Z_m, W\rangle$$

which give approximations to $Y_1, \ldots, Y_m, X$ such that $(\forall \bar{u})(\exists \bar{v})\psi(Y_1, \ldots, Y_m, X, \bar{u}, \bar{v})$. The relation

$$\mathfrak{B}_1 < \mathfrak{B}_2$$

is that defined in the proof of 8B.1, and again as in that proof

$$\begin{aligned}\mathfrak{B} = \langle B, Z_1, \ldots, Z_m, W\rangle \text{ is good} \\ \Leftrightarrow (\exists Y_1) \ldots (\exists Y_m)(\exists X)\{(\forall \bar{u})(\exists \bar{v})\psi(Y_1, \ldots, Y_m, X, \bar{u}, \bar{v}) \\ \& \; Y_1 \upharpoonright B = Z_1 \; \& \ldots \& \; Y_m \upharpoonright B = Z_m \; \& \; X \upharpoonright B = W\}.\end{aligned}$$

It is easy to *code* all finite structures of this signature using integers, so that as in the proof of 8A.1 the relation

$$Q(b_1, b_2, z) \Leftrightarrow b_1 \textit{ codes some } \mathfrak{B}_1 \textit{ and } b_2 \textit{ codes some } \mathfrak{B}_2 \textit{ and } \mathfrak{B}_1 < \mathfrak{B}_2 \textit{ and } z \in \mathfrak{B}_2$$

is arithmetical. For any reasonable coding, the relation

$$G(b) \Leftrightarrow b \textit{ codes some good } \mathfrak{B}$$

will be Σ^1_1.

Let b_0 be the code of the empty (good) finite structure and define by induction

(9) $\quad b_{n+1} = (\textit{least } b)[G(b) \ \& \ Q(b_n, b, n)].$

The function

$$f(n) = b_n$$

is arithmetical in the relations Q, G, in fact it is recursive in Q, G. It determines a sequence

$$\mathfrak{B}_0 < \mathfrak{B}_1 < \dots$$

of good structures whose union is of the form $\langle \omega, Y_1, \dots, Y_m, X \rangle$ and satisfies $(\forall \bar{u})(\exists \bar{v})\psi(Y_1, \dots, Y_m, X, \bar{u}, \bar{v})$. Moreover this X is arithmetical (in fact recursive) in Q, G and it satisfies $(\exists Y_1) \dots (\exists Y_m)(\forall \bar{u})(\exists \bar{v})\psi(Y_1, \dots, Y_m, X, \bar{u}, \bar{v})$, so that $X \in \mathscr{P}$. ⊣

We have put down a version of this classical argument which uses as few of the special properties of $\mathbb{N}$ as possible, so that at first glance it looks as if the theorem may hold for more general structures. However, the definition of b_{n+1} in (9) uses the basic property of $\mathbb{N}$ that we can arithmetically *enumerate* ω, so that in fact the argument works only for structures of the form $(\mathbb{N}, R_1, \dots, R_l)$.

8D.6. THEOREM. *There is a countable acceptable structure $\mathfrak{B}$ and an elementary set $\mathscr{P} \subseteq Power(B)$ which is nonempty but contains no set which is elementary in Σ^1_1 sets. In fact we can choose $\mathfrak{B}$ so that it admits an elementary wellordering of its domain.*

PROOF. Let λ be any uncountable cardinal, for example $\lambda = \aleph_1$, take

$$\mathfrak{A} = \langle \lambda, \leqslant, R_1, \dots, R_l \rangle,$$

where the R_i are hyperelementary on $\langle \lambda, \leqslant \rangle$ and chosen so that $\mathfrak{A}$ is acceptable. Since λ is uncountable, there is no function with domain ω enumerating

λ. This is a second order property, but it is simple to find a sentence φ of $\mathscr{L}^{\mathfrak{A}}_{\omega_1,G}$ which asserts that

> *there is no function from ω onto the structure which is elementary in coinductive relations.*

Let ψ assert that $\mathfrak{A}$ is acceptable and take a countable substructure $\mathfrak{B}$ of $\mathfrak{A}$ for which both φ and ψ are absolute.

Clearly $\mathfrak{B}$ is acceptable, it admits an elementary wellordering of its domain and it admits no enumeration of its domain by a function elementary in coinductive (i.e. Σ^1_1) relations. Thus the elementary set $\mathscr{P}$ defined by

$$X \in \mathscr{P} \Leftrightarrow \{(u, v): \langle u, v\rangle \in X\} \textit{ maps } \omega \textit{ onto } B$$

contains no such set, although it is nonempty since $\mathfrak{B}$ is countable. ⊣

8E. The Suslin–Kleene Theorem

Perhaps the most significant single result in the theory of inductive relations on $\mathbb{N}$ is the theorem of Kleene [1955b], [1955c] which identifies the classes of Δ^1_1 and "hyperarithmetical" relations on ω. It has been recognized as a *construction principle*, since it allows us to "construct" the Δ^1_1 sets by iterating the elementary operations on $\mathbb{N}$. It is known today as the *Suslin–Kleene Theorem*, because it is the effective analog of Suslin's classical theorem identifying the Borel with the analytic–coanalytic sets of reals. In fact it is not too hard to show that both the Kleene and the Suslin theorems are simple corollaries of one unifying principle.

We put the word "hyperarithmetical" in quotation marks above because Kleene's definition of this term was quite involved. Similar definitions of the hyperarithmetical relations had been given by Davis [1950] and Mostowski [1951], and eventually Spector [1961] pointed out explicitly that these are precisely the inductive–coinductive relations on $\mathbb{N}$. Because of this, it is tempting to understand the Suslin–Kleene Theorem as saying simply that

(1) $\quad \Pi^1_1 = \textit{inductive},$

what we called the "Kleene Theorem" in Section 8A. This was my own approach in Moschovakis [1970], and of course it is satisfying that (1) holds for all countable acceptable structures.

Nevertheless, when we examine the proof in Kleene [1955c] or any other known proof of this result expressed in terms of Kleene's original definition of "hyperarithmetical", it becomes obvious that something much stronger than (1) is established. This was obvious to Kleene who tried very hard to prove this theorem although he had already established (1) in Kleene [1955a].

We shall formulate and prove here a version of the Suslin–Kleene Theorem which is very close to Kleene's version, although free of the notational complications of the theory of constructive ordinals. It is essentially the version given in Shoenfield [1967]. It turns out that this is indeed much harder to establish than (1) and that it cannot be extended to all countable acceptable structures.

Since we will use recursive functions both in the statement and in the proof of the Suslin–Kleene Theorem, we assume just for this section the very basic facts of recursion theory.

For each $n \geqslant 1$, there is an enumeration of all n-ary recursive partial functions

$$\varphi_0(\bar{x}),\ \varphi_1(\bar{x}),\ \varphi_2(x), \ldots .$$

so that the $(n+1)$-ary partial function

$$\varphi(e, \bar{x}) = \varphi_e(\bar{x})$$

is recursive. (*Enumeration Theorem.*)

Moreover, there are total recursive functions $S^m_n(e, y_1, \ldots, y_m)$ such that

$$\textit{if } z = S^m_n(e, y_1, \ldots, y_m), \textit{ then for every } x_1, \ldots, x_m,$$

$$\varphi_z(x_1, \ldots, x_n) = \varphi_e(y_1, \ldots, y_m, x_1, \ldots, x_n).$$

(*Iteration Theorem.*)

If $f(e, \bar{x})$ is a recursive partial function of $n+1$ arguments, then there is some fixed number e^* such that for all $\bar{x}$,

$$f(e^*, \bar{x}) = \varphi_{e^*}(\bar{x}).$$

(*Second Recursion Theorem.*)

Except for these basic facts which can be found easily in Kleene [1952] (or any other elementary text in recursion theory), we will only need to assume that certain functions whose definitions are obviously effective are in fact recursive.

The motivation for this formulation of the Suslin–Kleene Theorem lies in taking seriously the contention that it should be the effective analog of the classical Suslin Theorem. "Analytic–coanalytic" clearly corresponds to "Δ^1_1" and the Borel sets are the smallest σ-ring of sets of reals which contains all the intervals. Assuming that the intervals in this definition simply give us a starting point of very simple sets, we aim to show that the Δ^1_1 subsets of ω form the smallest *effective* σ-ring of sets of integers which contains all singletons.

Let $\mathscr{B}$ be a family of subsets of ω. A *coding* for $\mathscr{B}$ is any mapping

$$\pi : \mathscr{B} \to Power(\omega)$$

which assigns to every $X \in \mathscr{B}$ a *nonempty* set of integers $\pi(X)$ such that

$$X \neq Y \Rightarrow \pi(X) \cap \pi(Y) = \emptyset.$$

We call the members of $\pi(X)$ *the codes of* X (in the coding π) and if $n \in \pi(X)$ we say that X *is in* $\mathscr{B}$ *with code* n. Thus a coding of $\mathscr{B}$ assigns at least one integer code to each $X \in \mathscr{B}$ and does not assign the same code to two distinct members of $\mathscr{B}$.

A family $\mathscr{B}$ of subsets of ω is *an effective σ-ring* (containing the singletons) if there is some coding π of $\mathscr{B}$ so that the following three conditions hold:

(i) There is a recursive function $\tau_1(n)$ so that for each n the singleton $\{n\}$ is in $\mathscr{B}$ with code $\tau_1(n)$.

(ii) There is a recursive function $\tau_2(x)$ so that if X is in $\mathscr{B}$ with code x, then $\omega - X$ is in $\mathscr{B}$ with code $\tau_2(x)$.

(iii) There is a recursive function $\tau_3(e)$ such that whenever the recursive partial function $\varphi_e(n)$ is defined for each n and codes for each n a member X_n of $\mathscr{B}$, then $\bigcup_{n \in \omega} X_n$ is in $\mathscr{B}$ with code $\tau_3(e)$.

Speaking loosely, $\mathscr{B}$ is an effective σ-ring if it admits a coding relative to which $\mathscr{B}$ contains all singletons (effectively) and $\mathscr{B}$ is effectively closed under complementation and recursive union.

8E.1. SUSLIN–KLEENE THEOREM. *The collection of Δ^1_1 subsets of ω is the smallest effective σ-ring.*

At this point it is not even obvious that there is an effective σ-ring, much less that there is a smallest one and that it contains precisely the Δ^1_1 sets. Before going on to the proof of 8E.1 we define a particular effective σ-ring and a coding for it.

The definition is by a simultaneous induction which determines a set of integers G and for each $x \in G$, a set of integers G^x. There are three clauses to the induction.

(a) For each n, $\langle 1, n\rangle \in G$ and $G^{\langle 1,n\rangle} = \{n\}$.

(b) If $x \in G$, then $\langle 2, x\rangle \in G$ and $G^{\langle 2,x\rangle} = \omega - G^x$.

(c) If the recursive partial function $\varphi_e(n)$ is defined for each n and if for each n,

$$\varphi_e(n) = e_n \in G,$$

then $\langle 3, e\rangle \in G$ and

$$G^{\langle 3,e\rangle} = \bigcup_n G^{e_n}.$$

By $\langle x, y\rangle$ we understand here the pair associated with some recursive coding scheme on ω, e.g.

$$\langle x_1, \ldots, x_n\rangle = 2^{x_1+1} \cdot 3^{x_2+1} \cdot \ldots \cdot (\textit{the } n^{th} \textit{ prime})^{x_n+1}.$$

To cast this definition as a generalized induction of the type we introduced in Section 1A, let A be the set of all pairs (x, X) with $x \in \omega$, $X \subseteq \omega$ and define Γ: $Power(A) \to Power(A)$ by

$$\text{(2)} \qquad \Gamma(S) = \{(\langle 1, n\rangle, \{n\}) : n \in \omega\} \cup \{(\langle 2, x\rangle, \omega - X) : (x, X) \in S\} \cup \{(\langle 3, e\rangle, \textstyle\bigcup_n X_n) : \varphi_e \textit{ is total and } (\forall n)(\varphi_e(n), X_n) \in S\}.$$

Clearly Γ is a monotone operator. We interpret the definition of G, G^x above to mean that we take

$$\text{(3)} \qquad G = \{x : \textit{for some } X, (x, X) \in I_\Gamma\}$$

and for $x \in G$,

$$\text{(4)} \qquad G^x = \textit{the unique } X \textit{ such that } (x, X) \in I_\Gamma.$$

This is justified since we can easily verify that

$$(x, X), (x, Y) \in I_\Gamma \Rightarrow X = Y.$$

Notice that

$$\text{(5)} \qquad I_\Gamma = \{(x, G^x) : x \in G\}$$

and put

$$\text{(6)} \qquad \mathscr{G} = \{G^x : x \in G\}.$$

Proof of 8E.1 is in three lemmas.

Lemma 1. *The collection of sets $\mathscr{G}$ is an effective σ-ring and every set in $\mathscr{G}$ is Δ^1_1.*

Proof. The first assertion is trivial from the definitions, taking

$$\pi(X) = \{x \in G : X = G^x\}$$

for the coding and putting

$$\tau_1(n) = \langle 1, n\rangle, \qquad \tau_2(x) = \langle 2, x\rangle, \qquad \tau_3(e) = \langle 3, e\rangle.$$

To prove the second assertion, let us associate with each $Y \subseteq \omega$ the subset Y^* of the basic set A defined by

$$Y^* = \{(x, \{t : \langle x, t+1\rangle \in Y\}) : \langle x, 0\rangle \in Y\}.$$

Easily, if

$$\text{(7)} \qquad Z = \{\langle x, 0\rangle : x \in G\} \cup \{\langle x, t+1\rangle : x \in G \;\&\; t \in G^x\},$$

then

$$Z^* = I_\Gamma.$$

We claim that for each $x \in G$,

$$t \in G^x \Leftrightarrow (\forall Y \subseteq \omega)[\Gamma(Y^*) = Y^* \Rightarrow \langle x, t+1 \rangle \in Y]. \tag{8}$$

Proof of direction ($\Leftarrow$) of (8) is immediate by choosing $Y = Z$. To prove direction ($\Rightarrow$), notice that if $\Gamma(Y^*) = Y^*$, then $I_\Gamma \subseteq Y^*$, and this easily implies $t \in G^x \Rightarrow \langle x, t+1 \rangle \in Y$.

Now (8) implies via a simple computation that each G^x is Π^1_1, and since (for $x \in G$) $G^x = \omega - G^{\langle 2, x \rangle}$, each G^x is Δ^1_1.

Lemma 2. *If $\mathscr{B}$ is an effective σ-ring, then for each $x \in G$, $G^x \in \mathscr{B}$.*

Proof. Let τ_1, τ_2, τ_3 be fixed recursive functions which satisfy conditions (i), (ii) and (iii) of the definition of an effective σ-ring relative to some coding of $\mathscr{B}$. Choose m so that for every z, e, n,

$$\varphi_m(z, e, n) = \varphi_z(\varphi_e(n))$$

and using the Second Recursion Theorem choose z^* such that if

$$f = \varphi_{z^*},$$

then the following equations hold:

$$f(\langle 1, n \rangle) = \tau_1(n),$$

$$f(\langle 2, x \rangle) = \tau_2(f(x)),$$

$$f(\langle 3, e \rangle) = \tau_3(S^2_1(m, z^*, e)),$$

$$f(x) = 0 \textit{ if } x \textit{ is not of the form } \langle i, y \rangle \textit{ with } 1 \leqslant i \leqslant 3.$$

It is now easy to verify by induction on the definition of G, G^x that

$$x \in G \Rightarrow G^x \textit{ is in } \mathscr{B} \textit{ with code } f(x).$$

For example, if $x = \langle 3, e \rangle$, then $\varphi_e(n)$ is total and by induction hypothesis, for each n, G^{e_n} is in $\mathscr{B}$ with code $f(\varphi_e(n))$. Letting

$$w = S^2_1(m, z^*, e),$$

we have for each n,

$$\varphi_w(n) = \varphi_m(z^*, e, n) = f(\varphi_e(n));$$

hence $\tau_3(w)$ is a code in $\mathscr{B}$ of $\bigcup_n G^{e_n}$ so that $f(\langle 3, e \rangle)$ has the required property.

To complete the proof of the theorem there remains the hard part of showing that every Δ^1_1 set of integers is G^x for some $x \in G$. We do this by an adaptation of a very simple classical proof of the Suslin Theorem, see Kuratowski [1966], §39, III. (The observation that this proof has an effective version is independently due to Donald A. Martin.)

Lemma 3. *If X is a Δ^1_1 set of integers, then $X \in \mathscr{G}$.*

Proof. By the basic representation of Exercise 1.12, suppose

$$x \in X \Leftrightarrow (\exists \alpha)(\forall t)P(\bar{\alpha}(t), x),$$
$$x \notin X \Leftrightarrow (\exists \beta)(\forall s)Q(\bar{\beta}(s), x),$$

where P and Q are recursive. The idea of the proof is to define a recursive function f such that

(9) *for every* x, $f(x) \in G$,

(10) $x \in X \Rightarrow G^{f(x)} = \{x\}$,

(11) $x \notin X \Rightarrow G^{f(x)} = \emptyset$.

From this the lemma will follow immediately, since if $f = \varphi_e$, then

$$X = G^{\langle 3, e \rangle}.$$

In point of fact we will set

(12) $$f(x) = h(x, \langle \emptyset \rangle, \langle \emptyset \rangle),$$

where $h(x, u, v)$ will be a recursive function whose values will be significant for the proof only when u, v are sequence codes of the same length.

For each fixed x, consider the set of pairs of sequence codes

(13) $T^x = \{(u, v) \colon Seq(u) \;\&\; Seq(v) \;\&\; lh(u) = lh(v) \;\&\; P(u, x) \;\&\; Q(v, x)\}$.

The relation

(14) $(u, v) \succ^x (u', v') \Leftrightarrow (u, v) \in T^x$

& u' *codes a one-point extension of* u

& v' *codes a one-point extension of* v

is obviously wellfounded, since an infinite descending chain in it would prove that both $x \in X$ and $x \notin X$. We assign to each pair (u, v) of sequence codes of the same length a set of integers $C(x, u, v)$ by recursion on $\prec^x$ as follows:

(15) $$C(x, u, v) = \begin{cases} \emptyset & \text{if } P(u, x), \\ \{x\} & \text{if } P(u, x) \;\&\; \neg Q(v, x), \\ \bigcup_s \bigcap_t C(x, u^\frown\langle s \rangle, v^\frown\langle t \rangle) & \text{if } P(u, x) \;\&\; Q(v, x). \end{cases}$$

Clearly, for all x, u, v,

$$C(x, u, v) = \emptyset \text{ or } C(x, u, v) = \{x\}.$$

We claim that

(16) *if* $x \in X$, *then* $C(x, \langle \emptyset \rangle, \langle \emptyset \rangle) = \{x\}$,

(17) *if* $x \notin X$, *then* $C(x, \langle \emptyset \rangle, \langle \emptyset \rangle) = \emptyset$.

To prove (16) by contradiction, assume $x \in X$ and $C(x, \langle\emptyset\rangle, \langle\emptyset\rangle) = \emptyset$, choose

$$\alpha = s_0, s_1, \ldots$$

such that for every n,

(18) $$P(\langle s_0, s_1, \ldots, s_n\rangle, x).$$

By definition (15),

$$C(x, \langle\emptyset\rangle, \langle\emptyset\rangle) = \bigcup_s \bigcap_t C(x, \langle s\rangle, \langle t\rangle),$$

so there must be some t_0 such that

$$C(x, \langle s_0\rangle, \langle t_0\rangle) = \emptyset.$$

Again,

$$C(x, \langle s_0\rangle, \langle t_0\rangle) = \bigcup_s \bigcap_t C(x, \langle s_0, s\rangle, \langle t_0, t\rangle),$$

so there must exist some t_1 such that

$$C(x, \langle s_0, s_1\rangle, \langle t_0, t_1\rangle) = \emptyset.$$

Proceeding in the same way we find

$$\beta = t_0, t_1, \ldots$$

such that for every n,

$$C(x, \langle s_0, \ldots, s_n\rangle, \langle t_0, \ldots, t_n\rangle) = \emptyset$$

which by the definition implies that for all n,

$$Q(\langle t_0, t_1, \ldots, t_n\rangle, x),$$

so that $x \notin X$, contradicting $x \in X$.

The proof of (17) is similar.

The definition of the sets $C(x, u, v)$ is by an "effective" recursion and an application of the Second Recursion Theorem like that in the proof of Lemma 2 yields a recursive function h such that for all x, u, v with u, v sequence codes of the same length,

(19) $$h(x, u, v) \in G, \qquad G^{h(x,u,v)} = C(x, u, v).$$

To be precise, choose n such that

$$n \in G, \qquad G^n = \emptyset,$$

choose m such that

$$\varphi_m(z, x, u, v, s, t) = \langle 2, \varphi_z(x, u^\frown\langle s\rangle, v^\frown\langle t\rangle)\rangle,$$

choose e such that

$$\varphi_e(z, x, u, v, s) = \langle 2, \langle 3, S^5_1(m, z, x, u, v, s)\rangle\rangle$$

and by the Second Recursion Theorem choose z^* such that if

$$h = \varphi_{z^*},$$

then the following equations hold:

$$(20)\quad h(x, u, v) = \begin{cases} n \textit{ if } Seq(u) \ \& \ Seq(v) \ \& \ lh(u) = lh(v) \ \& \ \neg P(u, x), \\ \langle 1, x\rangle \textit{ if similarly and } P(u, x) \ \& \ \neg Q(v, x), \\ \langle 3, S^4_1(e, z^*, x, u, v)\rangle \textit{ if similarly and } P(u, x) \ \& \ Q(v, x), \\ 0, \textit{ otherwise.} \end{cases}$$

We now prove (19) with this h by induction on the wellfounded relation $\prec^x$ defined by (14). To treat one of the cases, if the third clause in the definition of $h(x, u, v)$ applies, then by induction hypothesis, for all s, t,

$$a(s, t) = h(x, u^\frown\langle s\rangle, v^\frown\langle t\rangle) \in G,$$

$$G^{a(s,t)} = C(x, u^\frown\langle s\rangle, v^\frown\langle t\rangle).$$

Hence, by the choice of m, for each s,

$$b(s) = \langle 3, S^5_1(m, z^*, x, u, v, s)\rangle \in G,$$

$$G^{b(s)} = \bigcup_t (\omega - C(x, u^\frown\langle s\rangle, v^\frown\langle t\rangle)).$$

Hence, for each s,

$$c(s) = \langle 2, b(s)\rangle \in G,$$

$$G^{c(s)} = \bigcap_t C(x, u^\frown\langle s\rangle, v^\frown\langle t\rangle)$$

and thus

$$d = h(x, u, v) \in G,$$

$$G^d = \bigcup_s G^{c(s)} = \bigcup_s \bigcap_t C(x, u^\frown\langle s\rangle, v^\frown\langle t\rangle) = C(x, u, v).$$

Now Lemma 3 and hence the theorem follow by defining f from this h by (12). ⊣

Much of the significance attributed to the Suslin–Kleene Theorem is because of its proof rather than just its statement. It seems clear that each set G^x ($x \in G$) can be "constructed" starting with the singletons and then iterating "effectively" the operations of complementation and recursive union.

There are many ways to formulate the Suslin–Kleene Theorem so that it makes sense for arbitrary acceptable structures. We will see in the Exercises that none of these versions holds generally, in fact they all fail for some countable acceptable structures. Thus the hierarchy of Theorem 7E.1 is the only construction of the class of hyperelementary sets "from below" which is known now to hold in any generality.

Exercises for Chapter 8

8.1. Prove that if $\mathfrak{A} = \langle A, R_1, \ldots, R_l \rangle$ is countable and acceptable and if

$$\prec \subseteq Power(A) \times Power(A)$$

is a Σ_1^1 wellfounded second order relation on A, then $rank(\prec) < \kappa^{\mathfrak{A}}$.

HINT: Assume towards a contradiction that $rank(\prec) \geqslant \kappa^{\mathfrak{A}}$ and put

$$\prec^X = \{(s, t) : X_s^{(1)} \prec X_t^{(1)}\},$$

where $X_a^{(1)} = \{x : \langle a, x \rangle \in X\}$ as in (15) of 7E. Let $\rho(X, s)$ be the rank of s in $\prec^X$ and prove that

$$supremum\{\rho(X, s) : X \subseteq A, X_s^{(1)} \in Field(\prec)\} \geqslant \kappa^{\mathfrak{A}}.$$

Show then that if $\varphi(\bar{x}, S)$ is any S-positive formula in the language of $\mathfrak{A}$, the relation

$$Q(\bar{x}, X, s) \Leftrightarrow X_s^{(1)} \in Field(\prec) \;\&\; \bar{x} \in I_\varphi^{\rho(X,s)}$$

s Σ_1^1. From this a contradiction follows easily. ⊣

8.2. Let $\mathscr{P}$ be a Π_1^1 second order relation on a countable acceptable structure $\mathfrak{A}$ with ordinal κ, let $\sigma : \mathscr{P} \twoheadrightarrow \lambda$ be an inductive norm on $\mathscr{P}$, let $\mathscr{Q}$ be a Σ_1^1 second order relation and assume that f is a second order function of the appropriate type of arguments and values whose graph is Σ_1^1 and such that

$$f(\mathscr{Q}) \subseteq \mathscr{P}.$$

Prove that there is some $\xi < \kappa$ such that

$$\mathscr{Q}(\bar{x}, \bar{Y}) \Rightarrow \sigma(f(\bar{x}, Y)) \leqslant \xi.$$

(*Second Order Covering Theorem.*)

8.3. Let $\mathfrak{A}$ be countable and acceptable. Prove that for each signature $\nu = (n, r_1, \ldots, r_k)$ there is a Π_1^1 relation $\mathscr{H}^\nu(a, \bar{x}, \bar{Y})$ of signature $(n+1, r_1, \ldots, r_k)$ which parametrizes the ν-ary Δ_1^1 relations on $\mathfrak{A}$. Moreover, there is a Π_1^1 non Σ_1^1 set $I^\nu \subseteq A$ and a Σ_1^1 relation $\breve{\mathscr{H}}^\nu(a, \bar{x}, \bar{Y})$ such that whenever $\mathscr{R}$ is of signature ν,

$$\mathscr{R} \textit{ is } \Delta_1^1 \Leftrightarrow \textit{for some } a \in I^\nu, \mathscr{R} = \mathscr{H}_a^\nu,$$

and

$$a \in I^\nu \Rightarrow \mathscr{H}_a^\nu = \breve{\mathscr{H}}_a^\nu.$$

(*Parametrization Theorem* for Δ_1^1 second order relations.)

HINT: Use 6C.11. ⊣

8.4. Prove that if λ is a cardinal of cofinality ω, then every Π_1^1 relation on the structure $\langle V_\lambda, \in \restriction V_\lambda \rangle$ is inductive. (Chang–Moschovakis [1970].) $\dashv$

8.5. Prove that there is a countable structure on which not every Δ_1^1 relation is inductive. (K. Kunen.)

HINT: Take $\mathfrak{A} = \langle A, \sim \rangle$, where $\sim$ is an equivalence relation on A with the equivalence classes $C_1, C_2, \ldots$, where C_n has exactly n elements. For each $x \in A$ let

$$f(x) = \textit{the number of elements equivalent to } x,$$

and put

$$P(x, y) \Leftrightarrow f(x) < f(y).$$

Easily P is Δ_1^1, so it is enough to show that it is not inductive.

For each integer j define the equivalence relation $\approx_j$ on tuples from A by

$$(x_1, \ldots, x_n) \approx_j (y_1, \ldots, y_n)$$
$$\Leftrightarrow \textit{the mapping } x_i \mapsto y_i \textit{ is an isomorphism of the finite substructures of } \mathfrak{A} \textit{ with domains } \{x_1, \ldots, x_n\}, \{y_1, \ldots, y_n\} \textit{ and for } i = 1, \ldots, n, \textit{ if } f(x_i) \leqslant j \textit{ or } f(y_i) \leqslant j, \textit{ then } x_i = y_i.$$

Call a set S of n-tuples *j-closed* if

$$(x_1, \ldots, x_n) \in S \;\&\; (x_1, \ldots, x_n) \approx_j (y_1, \ldots, y_n) \Rightarrow (y_1, \ldots, y_n) \in S.$$

Now show by induction on the construction of formulas $\varphi(x_1, \ldots, x_m, S)$ (not necessarily positive) that for each φ there is an integer k such that for all $j \geqslant k$ and all sets of n-tuples S,

$$\textit{if } S \textit{ is } j\textit{-closed, then } \{(x_1, \ldots, x_m) : \varphi(x_1, \ldots, x_m, S\} \textit{ is } j\textit{-closed.}$$

From this the result follows easily. $\dashv$

8.6. Let $\mathfrak{A}$ be a countable acceptable structure, let $\mathcal{T}$ be an inductive theory in $\mathcal{L}_2^{\mathfrak{A}}$ which has an $\mathfrak{A}$-model. Prove that the intersection of all $\mathfrak{A}$-models of $\mathcal{T}$ is a subset of $\mathcal{HE}(\mathfrak{A})$. (Barwise, Grilliot [1972].)

HINT: Put

$$X \in \mathcal{P} \Leftrightarrow (\forall Z)[Mod^{\mathcal{T}}(Z) \Rightarrow (\exists a)[X \in \mathcal{HE}(\mathfrak{A}, Z_a^{(1)})]]$$

and apply 8C.1. $\dashv$

For each enumeration without repetitions of a countable set A,

$$\pi : \omega \rightarrowtail\!\!\!\!\twoheadrightarrow A$$

and each $R \subseteq A^n$, let R^π be the pullback of R to ω,

$$R^\pi(x_1, \ldots, x_n) \Leftrightarrow R(\pi(x_1), \ldots, \pi(x_n)).$$

If $\mathfrak{A} = \langle A, R_1, \ldots, R_l \rangle$ is a countable structure and R is a relation on A, we call R *$\forall$-hyperarithmetical* on $\mathfrak{A}$ if for every $\pi: \omega \rightarrowtail\!\!\!\twoheadrightarrow A$, R^π is hyperelementary on $(\mathbb{N}, R_1, \ldots, R_l)$. This notion was introduced by Grilliot [1972], inspired by a similar notion of Lacombe's in abstract recursion theory.

8.7. Prove that if $\mathfrak{A}$ is countable and acceptable, then a relation R is $\forall$-hyperarithmetical on $\mathfrak{A}$ if and only if R is hyperelementary on $\mathfrak{A}$. (Grilliot [1972].) ⊣

The next problem shows that for countable acceptable structures we may use monotone operators (as defined in Section 1A) rather than positive operators to determine the inductive relations. The result is essentially due to Spector [1961], see Exercise 1.14.

8.8. Let $\mathfrak{A} = \langle A, R_1, \ldots, R_l \rangle$ be countable and acceptable, let Γ: $Power(A^n) \to Power(A^n)$ be a monotone operator which is elementary on $\mathfrak{A}$, i.e. for some (not necessarily positive) formula $\varphi(\bar{x}, S)$,

$$x \in \Gamma(S) \Leftrightarrow \varphi(\bar{x}, S).$$

Prove that I_Γ is inductive. ⊣

We now consider some plausible abstract formulations of the Suslin–Kleene Theorem. Of course the most direct abstractions of the definition of an effective σ-ring involve abstract recursion theory, but we can see why the result fails to extend by using elementary rather than recursive functions.

8.9. Let $\mathfrak{A} = \langle A, R_1, \ldots, R_l \rangle$ be acceptable, let $E^1(a, x)$, $E^2(a, x, y)$ be fixed hyperelementary relations which parametrize the unary and binary elementary relations by Theorem 5D.1. Relative to a fixed coding scheme define a set $G \subseteq A$ and for each $x \in G$ a set G^x by the following inductive clauses:

(a′) For each a, $\langle 1, a \rangle \in G$ and $G^{\langle 1,a \rangle} = \{x: E^1(a, x)\}$.

(b′) If $x \in G$, then $\langle 2, x \rangle \in G$ and $G^{\langle 2,x \rangle} = A - G^x$.

(c′) If $(\forall t)(\exists! s)E^2(a, t, s)$ & $(\forall t)(\forall s)[E^2(a, t, s) \Rightarrow s \in G]$, then $\langle 3, a \rangle \in G$ and $G^{\langle 3,a \rangle} = \cup \{G^s: (\exists t)E^2(a, t, s)\}$.

Prove that each G^x is hyperelementary. Prove also that if the relation $\mathscr{WF}^1(Y)$ of wellfoundedness is hyperelementary on $\mathfrak{A}$, then there is a hyperelementary set which is not G^x for any $x \in G$.

HINT: For the second part, find an elementary formula $\theta(x)$ and an elementary function $f(x, t)$ such that if

$$\varphi(x, S) \Leftrightarrow \theta(x) \vee (\forall t)[f(x, t) \in S],$$

then

$$G = I_\Gamma.$$

From this follows that if $\mathscr{W}\mathscr{F}^1(Y)$ is hyperelementary, then G is hyperelementary and then the result follows easily. ⊣

8.10. Prove that there is a countable acceptable structure $\mathfrak{A}$ such that if G, G^x are defined as in Exercise 8.9, then some Δ_1^1 set in $\mathfrak{A}$ is not G^x for any $x \in G$.

HINT: Use Exercise 8.9 and the technique of 8D. ⊣

In the definition of G, G^x of Exercise 8.9 we stayed as close as possible to the definition of the smallest effective σ-ring of Section 8E. It is clear, however, that the proofs in Exercise 8.9, 8.10 will extend to cover any reasonable modifications of that definition.

A nontrivial modification in this type of construction is to allow unions which are elementary (in the coding) *relative to some set which we have already constructed.* This idea was introduced by Kleene and has been used successfully to obtain hierarchies of some interesting classes of sets which arise in recursion theory of higher types over ω. It does not give a generalization of the Suslin–Kleene Theorem, but the proof that it fails is not entirely trivial. We outline this argument in the hint for the next exercise, which is meant for those familiar with Moschovakis [1967].

8.11. Let $\mathfrak{A} = \langle A, R_1, \ldots, R_l \rangle$ be acceptable, let $E^1(a, x)$ and $\mathscr{E}^{(2,1)}(a, x, y, Z)$ be hyperelementary relations which parametrize the unary and (2, 1)-ary elementary relations on $\mathfrak{A}$ by 5D.1 and 6C.8. Relative to a fixed elementary coding scheme, define $G \subseteq A$ and for each $x \in G$, $G^x \subseteq A$ by the following inductive clauses:

(a″) For each a, $\langle 1, a \rangle \in G$ and $G^{\langle 1,a \rangle} = \{x\colon E^1(a, x)\}$.

(b″) If $x \in G$, then $\langle 2, x \rangle \in G$ and $G^{\langle 2,x \rangle} = A - G^x$.

(c″) If $b \in G$ and $(\forall x)(\exists! y)\mathscr{E}^{(2,1)}(a, x, y, G^b)$ and $(\forall x)(\forall y)[\mathscr{E}^{(2,1)}(a, x, y, G^b) \Rightarrow y \in G]$, then $\langle 3, a, b \rangle \in G$ and $G^{\langle 3,a,b \rangle} = \bigcup \{G^y\colon (\exists x)\mathscr{E}^{(2,1)}(a, x, y, G^b)\}$.

Prove that every G^x is hyperelementary on $\mathfrak{A}$. Prove also that if $\mathfrak{A} = \mathbb{R}$, the structure of analysis defined in 1D, then there is some hyperelementary set which is not G^x for any $x \in G$.

HINT: For the second part, use Corollary 3.1 of Moschovakis [1967] to prove that every G^x is *hyperanalytic* in some $\alpha \in {}^\omega\omega$, with the obvious definition of "hyperanalytic" for subsets of $\omega \cup {}^\omega\omega$. Use then Theorem 10 of the same paper to argue that not every hyperprojective set is hyperanalytic in some $\alpha \in {}^\omega\omega$. ⊣

8.12. Prove that there is a countable acceptable structure $\mathfrak{A}$ such that if G, G^x are defined as in Exercise 8.11, then some Δ_1^1 set in $\mathfrak{A}$ is not G^x for any $x \in G$. ⊣

The use of the parameters of induction was illustrated already in the proofs of the very basic facts 1C.1, 1C.2. Up till now, however, we have not proved that these parameters are *necessary*, i.e. that there are inductive sets which are not fixed points.

8.13.* If $\mathfrak{A}$ is countable and acceptable, then there exists a hyperelementary set in $\mathfrak{A}$ which is not a fixed point. (Essentially Feferman [1965].)

HINT: Feferman's proof depends on a simple version of Cohen's forcing method.

Consider all finite structures of the form

$$\mathfrak{B} = \langle B, X_1, \ldots, X_l, Z \rangle$$

such that

$$\langle B, X_1, \ldots, X_l \rangle \subseteq \mathfrak{A},$$

and Z is a unary relation on B. We define the relation of *forcing* between such a finite structure $\mathfrak{B}$ and a sentence φ in the language of $\mathfrak{B}$ by induction on the construction of φ:

$$\begin{aligned} &\mathfrak{B} \Vdash \varphi && \Leftrightarrow \mathfrak{B} \models \varphi, \textit{ for prime } \varphi, \\ &\mathfrak{B} \Vdash \varphi \;\&\; \psi && \Leftrightarrow \mathfrak{B} \Vdash \varphi \;\&\; \mathfrak{B} \Vdash \psi, \\ &\mathfrak{B} \Vdash \neg\varphi && \Leftrightarrow (\forall \mathfrak{B}' \supseteq \mathfrak{B}) \textit{ not } \mathfrak{B}' \Vdash \varphi, \\ &\mathfrak{B} \Vdash (\exists x)\varphi(x) && \Leftrightarrow \textit{for some } x \in \mathfrak{B},\ \mathfrak{B} \Vdash \varphi(x). \end{aligned}$$

Call a set $Z \subseteq A$ *generic* if for every sentence φ in the language of the structure $(\mathfrak{A}, Z)$ there is a finite structure $\mathfrak{B} \subseteq (\mathfrak{A}, Z)$ such that $\mathfrak{B} \Vdash \varphi$ or $\mathfrak{B} \Vdash \neg\varphi$.

Prove by induction on φ that

$$\mathfrak{B} \Vdash \varphi \;\&\; \mathfrak{B} \subseteq \mathfrak{B}' \Rightarrow \mathfrak{B}' \Vdash \varphi,$$

and then show by enumerating all sentences in the language of every $\mathfrak{B}$ that for every finite $\mathfrak{B}$ there is a generic Z such that $\mathfrak{B} \subseteq (\mathfrak{A}, Z)$. Then prove that for generic Z and for every φ in the language of $(\mathfrak{A}, Z)$,

$$(\mathfrak{A}, Z) \models \varphi \Leftrightarrow \textit{for some finite } \mathfrak{B} \subseteq (\mathfrak{A}, Z),\ \mathfrak{B} \Vdash \varphi.$$

*Note: The hint to Exercise 8.13 works for the structure of arithmetic, but not for arbitrary countable, acceptable structures, for which the claim in the exercise is still open. Exercise 8.14 is equivalent to 8.13, and so it, too, is open.

Use these two properties to show that no generic set can be a fixed point. Finally, prove that there are hyperelementary generic sets, by arguing that in some appropriate codings of all the finite structures and all the sentences, the relation

$$Q(b, a) \Leftrightarrow \textit{b codes a finite structure } \mathfrak{B}$$
$$\textit{and a codes a formula } \varphi$$
$$\textit{and } \mathfrak{B} \Vdash \varphi$$

is hyperelementary. ⊣

8.14. Prove that every acceptable structure admits hyperelementary relations which are not fixed points. ⊣

CHAPTER 9

THE NEXT ADMISSIBLE SET

Our main aim in this chapter is to connect the theory of inductive relations as we have developed it here with recursion theory on admissible sets. We will give a very brief summary of the basic facts about admissible sets and we will establish the principal result of Barwise–Gandy–Moschovakis [1971]: *If A is a transitive set closed under pairing, then the inductive relations on the structure $\langle A, \in \restriction A\rangle$ are precisely those relations on A which are Σ_1 on the next admissible set.*

Actually we will work in the axiomatic context of *Spector classes of relations* and a good part of the chapter will be devoted to developing this axiomatic framework. It is the key to extending the ideas and results of this book to many notions of definability, including various nonmonotone inductive definabilities. The main result of the chapter in 9E is substantially stronger than the Barwise–Gandy–Moschovakis [1971] theorem and some of its applications are described in 9F and in the Exercises.

9A. Spector classes of relations

The reader with a taste for axiomatics must have noticed that the development of the theory of inductive relations was based on very few, key facts. At this point it pays to collect some of these basic properties into a definition and formulate axiomatic versions of the most significant results.

Let Γ be a collection of relations (of all numbers of arguments) on some infinite set A. We call Γ *closed under* & if whenever P, Q are n-ary relations in Γ, then their intersection $P \cap Q$ is also in Γ. Closure under the other propositional connectives $\vee$, $\neg$ is defined similarly.

If $P \subseteq A^{n+1}$ is an $(n+1)$-ary relation on A, let $\exists^A P$ be the n-ary relation obtained by quantifying P in its first variable,

$$\bar{x} \in \exists^A P \Leftrightarrow (\exists y)P(y, \bar{x}).$$

Similarly,

$$\bar{x} \in \forall^A P \Leftrightarrow (\forall y)P(y, \bar{x}).$$

We call Γ *closed under* $\exists^A$ if whenever P is in Γ so is $\exists^A P$, and similarly for closure under $\forall^A$.

A function

$$f: A^n \to A$$

is *trivial combinatorial* if it is definable by a quantifier-free formula of the trivial structure $\langle A \rangle$, i.e. if

$$f(\bar{x}) = y \Leftrightarrow \varphi(\bar{x}, y),$$

where φ is built up by &, $\vee$, $\neg$ from prime formulas of the form

$$t_1 = t_2,$$

where each t_i is either a variable in the list $\bar{x}, y$ or a constant from A. We call Γ *closed under trivial combinatorial substitutions* if whenever $P(y_1, \ldots, y_m)$ is in Γ and $f_1(\bar{x}), \ldots, f_m(\bar{x})$ are trivial combinatorial functions, then the relation Q defined by

$$Q(\bar{x}) \Leftrightarrow P(f_1(\bar{x}), \ldots, f_m(\bar{x}))$$

is also in Γ. This implies for example that if $R(x, y, z)$ is in Γ and $c \in A$, then

$$Q(x) \Leftrightarrow R(x, x, c)$$

is also in Γ, since

$$Q(x) \Leftrightarrow R(f_1(x), f_2(x), f_3(x))$$

with $f_1(x) = f_2(x) = x$ and $f_3(x) = c$. Also if Γ is closed under trivial combinatorial substitutions and $\exists^A$, then easily Γ is closed under existential quantification on variables other than the first.

A coding scheme

$$\mathscr{C} = \langle N^{\mathscr{C}}, \leqslant^{\mathscr{C}}, \langle\,\rangle^{\mathscr{C}} \rangle$$

is in Γ if all the associated relations

$$N^{\mathscr{C}}(x), \qquad x \leqslant^{\mathscr{C}} y, \qquad Seq^{\mathscr{C}}(x), \qquad lh^{\mathscr{C}}(x) = y, \qquad q^{\mathscr{C}}(x, i) = y$$

and their negations are in Γ.

Finally, we call a class of relations Γ *positive, elementary, rich* if it contains equality $x = y$, inequality $x \neq y$ and some coding scheme $\mathscr{C}$ and if it is closed under the positive elementary operations &, $\vee$, $\exists^A$, $\forall^A$ and trivial combinatorial substitutions.

This is the trivial base of our axiomatization. We are interested in classes of relations which have all these natural closure properties and which also satisfy the *parametrization* and the *prewellordering* theorems.

A class Γ is *parametrized* if for every $n \geqslant 1$ there is some $U^n \subseteq A^{n+1}$ in Γ which parametrizes the n-ary relations in Γ, i.e. for $P \subseteq A^n$,

$$P \in \Gamma \Leftrightarrow \textit{for some } a \in A,\ P = U^n_a = \{\bar{x}: (a, \bar{x}) \in U^n\}.$$

The *dual class* $\neg\Gamma$ of a class Γ is the class of all *complements* or *negations* of relations in Γ,

$$\neg\Gamma = \{R \subseteq A^n: \neg R = A^n - R \in \Gamma\}.$$

A *norm* $\sigma: P \twoheadrightarrow \lambda$ is a Γ-*norm* if there are relations $J_\sigma, \check{J}_\sigma$ in Γ and $\neg\Gamma$, respectively, such that

$$\textit{if } \bar{y} \in P, \textit{ then } (\forall \bar{x})\{[\bar{x} \in P \ \&\ \sigma(\bar{x}) \leqslant \sigma(\bar{y})] \Leftrightarrow J_\sigma(\bar{x}, \bar{y}) \Leftrightarrow \check{J}_\sigma(\bar{x}, \bar{y})\}.$$

It follows as in 3A.1 that if Γ is positive, elementary rich and $\sigma: P \twoheadrightarrow \lambda$ is a norm on some $P \in \Gamma$, then σ is a Γ-norm if and only if both $\leqslant^*_\sigma$, $<^*_\sigma$ are in Γ.

A class Γ is *normed* if every relation in Γ admits a Γ-norm, i.e. if Γ satisfies the Prewellordering Theorem.

Finally a class Γ of relations on some infinite set A is a *Spector class* if it is positive, elementary rich, parametrized and normed.

It should be fairly obvious that the whole theory of the class of inductive relations on some acceptable structure $\mathfrak{A}$ can be extended to an arbitrary Spector class Γ. The only plausible source of difficulty is the fact that from time to time we defined specific relations and proved them inductive—need they belong to an arbitrary Spector class? What makes this axiomatization work is the next theorem which shows that Spector classes are closed under relative inductive definability. The key to the proof is Exercise 3.6 and we first establish here a strong version of it.

9A.1. Lemma. *Let $\varphi(\bar{x}, S)$ be S-positive in the language over a set A. Then the fixed point I_φ is the unique relation P on A which admits a norm $\sigma: P \twoheadrightarrow \lambda$ such that for every $\bar{x}$,*

(1) $$\bar{x} \in P \Leftrightarrow \varphi(\bar{x}, \{\bar{y}: \bar{y} <^*_\sigma \bar{x}\}).$$

Moreover, if σ is any norm for which (1) *holds, then for every $\bar{x} \in I_\varphi$,*

(2) $$|\bar{x}|_\varphi \leqslant \sigma(\bar{x}).$$

Proof. Clearly I_φ admits a norm with which (1) holds, namely $\sigma(\bar{x}) = |\bar{x}|_\varphi$. Assume then that P is any relation with a norm $\sigma: P \twoheadrightarrow \lambda$ such that (1) holds.

Step 1: $\bar{x} \in P \Rightarrow \bar{x} \in I_\varphi \ \&\ |\bar{x}|_\varphi \leqslant \sigma(\bar{x})$.

Proof of Step 1 is by induction on $\sigma(\bar{x})$. If $\bar{x} \in P$ and $\sigma(\bar{x}) = \xi$, then by induction hypothesis

$$\bar{y} <^*_\sigma \bar{x} \Rightarrow [\bar{y} \in I_\varphi \ \&\ |\bar{y}|_\varphi < \xi];$$

hence by (1) and the monotonicity of φ in S we have

$$\varphi(\bar{x}, \{\bar{y}: \bar{y} \in I_\varphi \;\&\; |\bar{y}|_\varphi < \xi\})$$

which implies immediately

$$\bar{x} \in I_\varphi \;\&\; |\bar{x}|_\varphi \leqslant \xi.$$

Step 2: $\bar{x} \in I_\varphi \Rightarrow \bar{x} \in P$.

Proof of Step 2 is by induction on $|\bar{x}|_\varphi$. Suppose $\bar{x} \in I_\varphi$ and assume towards a contradiction that $\bar{x} \notin P$. Then $\{\bar{y}: \bar{y} <^*_\sigma \bar{x}\} = P$ and since $\bar{x} \notin P$, (1) yields

$$\neg\varphi(\bar{x}, P). \tag{3}$$

On the other hand, if $\xi = |\bar{x}|_\varphi$, the induction hypothesis yields $I_\varphi^{<\xi} \subseteq P$ and the fact that $\bar{x} \in I_\varphi$ yields $\varphi(\bar{x}, I_\varphi^{<\xi})$. Now the monotonicity of φ in S implies $\varphi(\bar{x}, P)$ which contradicts (3).

Clearly (1) and (2) follow immediately from these two steps. ⊣

9A.2. THEOREM. *Let Γ be a Spector class, let $Q_1, \ldots, Q_m$ be in Γ. If R is inductive in $Q_1, \ldots, Q_m$, then R is in Γ.*

PROOF. It is enough to show that if

$$\varphi \equiv \varphi(\bar{x}, S) \equiv \varphi(\bar{x}, Q_1, \ldots, Q_m, S)$$

is positive in $Q_1, \ldots, Q_m, S$ and $Q_1, \ldots, Q_m$ are in Γ, then so is I_φ. By the lemma, it will be sufficient to find some P in Γ which admits some norm σ so that (1) of the lemma holds.

Let $U^{n+1} \subseteq A^{n+2}$ be in Γ and parametrize the $(n+1)$-ary relations in Γ, let

$$\tau: U^{n+1} \twoheadrightarrow \kappa$$

be some Γ-norm on U^{n+1} and put

$$(t, \bar{x}) \in Q \Leftrightarrow \varphi(\bar{x}, \{\bar{y}: (t, t, \bar{y}) <^*_\tau (t, t, \bar{x})\}). \tag{4}$$

Clearly Q is in Γ and since it is $(n+1)$-ary, there is a fixed $a \in A$ such that

$$(t, \bar{x}) \in Q \Leftrightarrow (a, t, \bar{x}) \in U^{n+1}.$$

Now put

$$P(\bar{x}) \Leftrightarrow (a, \bar{x}) \in Q \Leftrightarrow (a, a, \bar{x}) \in U^{n+1}.$$

For this relation P, we have

$$\begin{aligned} P(\bar{x}) &\Leftrightarrow (a, \bar{x}) \in Q \\ &\Leftrightarrow \varphi(\bar{x}, \{\bar{y}: (a, a, \bar{y}) <^*_\tau (a, a, \bar{x})\}) && \text{(by (4))} \\ &\Leftrightarrow \varphi(\bar{x}, \{\bar{y}: \bar{y} <^*_\sigma \bar{x}\}), \end{aligned}$$

where $\sigma: P \twoheadrightarrow \lambda$ is the obvious norm on P chosen so that

$$\sigma(\bar{x}) \leqslant \sigma(\bar{y}) \Leftrightarrow \tau(a, a, \bar{x}) \leqslant \tau(a, a, \bar{y}).$$

This establishes (1) for P with σ and completes the proof. ⊣

The "tricky" part of this argument is very similar to that we used in proving 6C.10.

One of the by-products of this theorem is an elegant structural characterization of the inductive relations on an "almost acceptable" structure.

9A.3. COROLLARY. *Let* $\mathfrak{A} = \langle A, R_1, \ldots, R_l \rangle$ *be a structure which admits a hyperelementary coding scheme. Then the collection of inductive relations on* $\mathfrak{A}$ *is the smallest Spector class on* A *which contains* $R_1, \neg R_1, \ldots, R_l, \neg R_l$. ⊣

We do not know any structural characterization of this type for the class of inductive relations on an arbitrary infinite structure $\mathfrak{A}$.

The term "Spector class" derives from the relation of these ideas to the axiomatization of abstract computation theory in Moschovakis [1971b]. It is easy to verify that a class Γ of relations on a set A is a Spector class if and only if there is a *Spector computation theory* Θ on A such that Γ consists of the Θ-semicomputable relations.

9B. Examples of Spector classes

The last remark in 9A suggests that many natural examples of Spector classes arise in abstract recursion theory. Some of these are important and we will list them in the exercises for those who are familiar with the *theory of recursion in higher types* introduced by Kleene [1959b]. Still more examples arise in the study of nonmonotone inductive definability. Here we restrict ourselves to two interesting Spector classes which are much closer to the theory of inductive relations.

The first theorem is due (in its full generality) to P. Aczel who has obtained recently many interesting results about inductive definability in the language $\mathscr{L}^{\mathfrak{A}}(\mathrm{Q})$. (See also his earlier Aczel [1970].)

9B.1. THEOREM. *Let* $\mathfrak{A} = \langle A, R_1, \ldots, R_l \rangle$ *be an acceptable structure, let* Q *be a nontrivial, monotone unary quantifier on* A *and let*

$$\Gamma = \textit{all relations on } A \textit{ which are positive } \mathscr{L}^{\mathfrak{A}}(\mathrm{Q})\textit{-inductive},$$

as we defined this notion in (2) *of Section* 3D. *Then* Γ *is a Spector class.*

Proof. According to the discussion in Section 3D, we only need verify that Γ is parametrized. We will outline this proof omitting all details.

It is easy to check that the methods of Section 5B extend easily to the language $\mathscr{L}^{\mathfrak{A}}(\mathrm{Q})$, so that for example the satisfaction relation for $\mathscr{L}^{\mathfrak{A}}(\mathrm{Q})$ is hyper-$\mathscr{L}^{\mathfrak{A}}(\mathrm{Q})$-definable. We show then, as in 5D.1, that for every $n \geqslant 1$ there is a hyper-$\mathscr{L}^{\mathfrak{A}}(\mathrm{Q})$-definable relation $E^n \subseteq A^{n+1}$ which parametrizes the n-ary $\mathscr{L}^{\mathfrak{A}}(\mathrm{Q})$-definable relations.

Lemma 4B.1 also extends easily to $\mathscr{L}^{\mathfrak{A}}(\mathrm{Q})$, so that for each S-positive formula $\varphi(\bar{x}, S)$ in $\mathscr{L}^{\mathfrak{A}}(\mathrm{Q})$ there is a quantifier free $\theta(\bar{x}, z_1, \ldots, z_m, \bar{y})$ and

$$\varphi(\bar{x}, S) \Leftrightarrow (\mathrm{Q}_1 z_1)(\mathrm{Q}_2 z_2) \ldots (\mathrm{Q}_m z_m)(\forall \bar{y})[\theta(\bar{x}, \bar{z}, \bar{y}) \vee S(\bar{y})],$$

where each Q_i is $\forall$, $\exists$, Q or $\mathrm{Q}^{\smile}$. Introducing vacuous quantifiers if necessary, we get a canonical form

$$\varphi(\bar{x}, S) \Leftrightarrow (\forall s_1)(\exists t_1)(\mathrm{Q}u_1)(\mathrm{Q}^{\smile}v_1) \ldots (\forall s_m)(\exists t_m)(\mathrm{Q}u_m)(\mathrm{Q}^{\smile}v_m)(\forall \bar{y}) \\ [\theta(\bar{x}, \bar{s}, \bar{t}, \bar{u}, \bar{v}, \bar{y}) \vee S(\bar{y})]$$

and then using the hyper-$\mathscr{L}^{\mathfrak{A}}(\mathrm{Q})$-definable relation E that parametrizes the $(n+1+n)$-ary $\mathscr{L}^{\mathfrak{A}}(\mathrm{Q})$-definable relations on A, we finally have that each n-ary fixed point in this language is determined by some $\varphi(\bar{x}, S)$ which satisfies

$$\text{(1)} \quad \varphi(\bar{x}, S) \Leftrightarrow (\forall s_1)(\exists t_1)(\mathrm{Q}u_1)(\mathrm{Q}^{\smile}v_1) \ldots (\forall s_m)(\exists t_m)(\mathrm{Q}u_m)(\mathrm{Q}^{\smile}v_m)(\forall \bar{y}) \\ [E(b, \bar{x}, \langle s_1, t_1, \ldots, u_m, v_m\rangle, \bar{y}) \vee S(\bar{y})]$$

with some fixed b and m.

Put

$$\text{(2)} \quad \psi(m, j, b, w, \bar{x}, T) \Leftrightarrow Seq(w) \\ \& \{[1 \leqslant j < m \;\&\; (\forall s)(\exists t)(\mathrm{Q}u)(\mathrm{Q}^{\smile}v)T(m, j+1, b, w^\frown\langle s, t, u, v\rangle, \bar{x})] \\ \vee [j = m \;\&\; (\forall \bar{y})[E(b, \bar{x}, w, \bar{y}) \vee T(m, 1, b, \langle\emptyset\rangle, \bar{y})]]\}.$$

We now prove that if φ satisfies (1) with fixed b, m, then

$$\text{(3)} \quad \bar{x} \in I^{\xi}_{\varphi} \Rightarrow (m, 1, b, \langle\emptyset\rangle, \bar{x}) \in I_{\psi}.$$

Proof of (3) is by induction on ξ. Assuming $\bar{x} \in I^{\xi}_{\varphi}$ and towards a contradiction that $(m, 1, \langle\emptyset\rangle, \bar{x}) \notin I_{\psi}$ and using repeatedly the easily verified rule that

$$(\mathrm{Q}z)\chi_1(z) \;\&\; (\mathrm{Q}^{\smile}z)\chi_2(z) \Rightarrow (\exists z)[\chi_1(z) \;\&\; \chi_2(z)],$$

we obtain $s_1, t_1, u_1, v_1, \ldots, s_m, t_m, u_m, v_m$ such that on the one hand

$$(\forall \bar{y})[E(b, \bar{x}, \langle s_1, t_1, \ldots, u_m, v_m\rangle, \bar{y}) \vee \bar{y} \in I^{<\xi}_{\varphi}]$$

and on the other

$$(\exists \bar{y})[\neg E(b, \bar{x}, \langle s_1, t_1, \ldots, u_m, v_m\rangle, \bar{y}) \;\&\; (m, 1, b, \langle\emptyset\rangle, \bar{y}) \notin I_\psi],$$

which immediately contradicts the induction hypothesis.

A symmetric argument shows

(4) $$(m, 1, b, \langle\emptyset\rangle, \bar{x}) \in I_\psi^\xi \Rightarrow \bar{x} \in I_\varphi,$$

so that we have

(5) $$\bar{x} \in I_\varphi \Leftrightarrow (m, 1, b, \langle\emptyset\rangle, \bar{x}) \in I_\psi.$$

Now (5) implies that the n-ary fixed points are parametrized by the inductive relation

$$R(a, \bar{x}) \Leftrightarrow ((a)_1, 1, (a)_2, \langle\emptyset\rangle, \bar{x}) \in I_\psi,$$

and then it is easy to parametrize all the n-ary inductive relations. ⊣

An interesting special case of the above is when we take Q to be the game quantifier G corresponding to a fixed elementary coding scheme on $\mathfrak{A}$.

The next example is quite different. Recall that a relation $R(\bar{x})$ on A is Σ_2^1 on the structure $\mathfrak{A}$ if there is a formula $\varphi(Z_1, \ldots, Z_k, Y_1, \ldots, Y_m, \bar{x})$ in the language of $\mathfrak{A}$ such that

$$R(\bar{x}) \Leftrightarrow (\exists Z_1) \ldots (\exists Z_k)(\forall Y_1) \ldots (\forall Y_m)\varphi(Z_1, \ldots, Z_k, Y_1, \ldots, Y_m, \bar{x}).$$

9B.2. THEOREM. *If $\mathfrak{A}$ is a countable acceptable structure, then the collection of all Σ_2^1 relations on $\mathfrak{A}$ is a Spector class.*

PROOF. Using the coding scheme on $\mathfrak{A}$ to contract the relation variables and replace them by set variables, as in the proof of 7D.2, we easily show that every Σ_2^1 relation on $\mathfrak{A}$ satisfies an equivalence

$$R(\bar{x}) \Leftrightarrow (\exists Z)(\forall Y)\mathscr{R}(Z, Y, \bar{x}),$$

where $\mathscr{R}(Z, Y, \bar{x})$ is elementary and Z, Y vary over subsets of A. Equivalently,

(6) $$R(\bar{x}) \Leftrightarrow (\exists Z)\mathscr{P}(Z, \bar{x}),$$

where $\mathscr{P}(Z, \bar{x})$ is Π_1^1 on $\mathfrak{A}$. Now the Parametrization Theorem 6C.8 for second order Π_1^1 relations on $\mathfrak{A}$ implies immediately that the class of Σ_2^1 relations is parametrized.

Since the closure properties of Σ_2^1 relations are easily verified, it remains to check that every Σ_2^1 relation admits a Σ_2^1-norm.

Suppose then that R satisfies (6) with some Π_1^1 relation $\mathscr{P}$. By the Kleene Theorem 8A.1, $\mathscr{P}$ is inductive and by the Prewellordering Theorem 6C.4, $\mathscr{P}$ admits an inductive norm

$$\sigma : \mathscr{P} \twoheadrightarrow \lambda.$$

To begin with, put

$$\rho(\bar{x}) = \mathit{infimum}\{\sigma(Y, \bar{x}) \colon \mathscr{P}(Y, \bar{x})\}.$$

Clearly ρ maps R into the ordinals but it need not be a norm, since it need not be *onto* some ordinal. On the other hand, if we set

$$\tau(\bar{x}) = \textit{order type of } \{\rho(\bar{y}) \colon \rho(\bar{y}) < \rho(\bar{x})\},$$

then τ is a norm and $\leqslant_\tau^* = \leqslant_\rho^*$, $<_\tau^* = <_\rho^*$. Hence it is enough to check that both $\leqslant_\rho^*$, $<_\rho^*$ are Σ_2^1, and this follows immediately from the equivalences

$$\bar{x} \leqslant_\rho^* \bar{y} \Leftrightarrow (\exists Y)\{\mathscr{P}(Y, \bar{x}) \mathbin{\&} (\forall Z)[(Y, \bar{x}) \leqslant_\sigma^* (Z, \bar{y})]\},$$

$$\bar{x} <_\rho^* \bar{y} \Leftrightarrow (\exists Y)\{\mathscr{P}(Y, \bar{x}) \mathbin{\&} (\forall Z)[(Y, \bar{x}) <_\sigma^* (Z, \bar{y})]\}. \qquad \dashv$$

9C. Structure theory for Spector classes

We now give a list of the most important structure properties of Spector classes. In almost all cases proofs can be obtained by trivial modifications of the proofs for the class of inductive relations.

Throughout this section we let Γ be a Spector class on some set A. As usual,

$$\neg\Gamma = \{A^n - P \colon P \in \Gamma\}$$

is the *dual* class. It is also convenient to let

$$\Delta = \Gamma \cap \neg\Gamma$$

be the *self-dual* class of relations both in Γ and $\neg\Gamma$.

9C.1. Theorem. *There is some relation P in Γ which is not in Δ* (Hierarchy Property for Γ).

If P, Q are in Γ, then there are relations $P_1 \subseteq P$, $Q_1 \subseteq Q$, both P_1, Q_1 in Γ such that $P_1 \cap Q_1 = \emptyset$, $P_1 \cup Q_1 = P \cup Q$ (Reduction Property for Γ.)

If P, Q are in $\neg\Gamma$ and $P \cap Q = \emptyset$, then there is some R in Δ such that $P \subseteq R$ and $Q \cap R = \emptyset$ (Separation Property for $\neg\Gamma$.)

There are P, Q in Γ such that $P \cap Q = \emptyset$ and there is no R in Δ such that $P \subseteq R$, $Q \cap R = \emptyset$ (Inseparability Property for Γ.)

Proofs are identical to those of 5D.3, 3A.4, 3A.5 and 5.2. $\dashv$

9C.2. Δ Selection Theorem. *Let $P(\bar{x}, \bar{y})$ be in Γ. Then there exist relations P^* in Γ and P^{**} in $\neg\Gamma$ such that*

$$P^* \subseteq P,$$

$$(\exists \bar{y})P(\bar{x}, \bar{y}) \Rightarrow (\exists \bar{y})P^*(\bar{x}, \bar{y}),$$

$$(\exists \bar{y})P(\bar{x}, \bar{y}) \Rightarrow (\forall \bar{y})[P^*(\bar{x}, \bar{y}) \Leftrightarrow P^{**}(\bar{x}, \bar{y})].$$

In particular, if $P(\bar{x}, \bar{y})$ is in Γ, if B is in Δ and if $(\forall \bar{x} \in B)(\exists \bar{y})P(\bar{x}, \bar{y})$, then there exists some $P^* \subseteq P$, P^* in Δ, so that $(\forall \bar{x} \in B)(\exists \bar{y})P^*(\bar{x}, \bar{y})$.

PROOF is that of 3B.1. ⊣

9C.3. RANK COMPARISON THEOREM. *If $\prec_1$, $\prec_2$ are wellfounded relations, $\prec_1$ in $\neg\Gamma$ and $\prec_2$ in Γ, then there exists some P in Γ such that*

$$\bar{x} \in Field(\prec_1) \Rightarrow (\forall \bar{y})\{P(\bar{x}, \bar{y}) \Leftrightarrow [\bar{y} \in Field(\prec_2) \;\&\; \rho^{\prec_1}(\bar{x}) \leqslant \rho^{\prec_2}(\bar{y})]\}.$$

PROOF. We can choose P inductive in $\neg\prec_1$, $\prec_2$ by the proof of 3B.4, hence we can choose P in Γ by 9A.2. ⊣

With each class of relations Λ there is naturally associated the ordinal

$$o(\Lambda) = supremum\{rank(\prec) \colon \prec \text{ is a prewellordering in } \Lambda\}.$$

In the case of a Spector class Γ, the ordinal $o(\Delta)$ of the self-dual class yields an important measure of the complexity of Γ. To begin with, the Boundedness Theorem holds with $o(\Delta)$.

9C.4. BOUNDEDNESS THEOREM. *Let P be a relation in Γ, let $\sigma \colon P \twoheadrightarrow \lambda$ be a Γ-norm on P. Then*

(1) $$\lambda \leqslant o(\Delta),$$

(2) $$\lambda < o(\Delta) \Leftrightarrow P \text{ is in } \Delta.$$

PROOF is identical to that of 3C.1. ⊣

This says in particular that $o(\Delta)$ is the supremum of the ranks of Γ-norms on relations in Γ. Here are two more interesting and less trivial characterizations of $o(\Delta)$.

9C.5. THEOREM. *If $\varphi(\bar{x}, Q_1, \ldots, Q_m, S)$ is positive in $Q_1, \ldots, Q_m$, S and $Q_1, \ldots, Q_m$ are in Γ, then the closure ordinal $\|\varphi\|$ is $\leqslant o(\Delta)$.*

If $\prec$ is wellfounded in $\neg\Gamma$, then $rank(\prec) < o(\Delta)$.

PROOF. For the first assertion we must look into the proof of the basic Theorem 9A.2. We showed there that

(3) $$\bar{x} \in I_\varphi \Leftrightarrow (a, a, \bar{x}) \in U^{n+1},$$

where U^{n+1} is in Γ, universal for the $(n+1)$-ary relations in Γ and a is some constant. Moreover there was a norm σ on I_φ chosen so that on the one hand

(4) $$\sigma(\bar{x}) \leqslant \sigma(\bar{y}) \Leftrightarrow \tau(a, a, \bar{x}) \leqslant \tau(a, a, \bar{y}),$$

with τ some Γ-norm on U^{n+1} and on the other hand

(5) $$\bar{x} \in I_\varphi \Leftrightarrow \varphi(\bar{x}, \{\bar{y} : \bar{y} <^*_\sigma \bar{x}\}).$$

Now (3) and (4) imply that σ is a Γ-norm on I_φ, hence if $\sigma : I_\varphi \twoheadrightarrow \lambda$, we have $\lambda \leqslant o(\Delta)$ by 9B.3. But (5) together with Lemma 9A.1 implies that for every $\bar{x} \in I_\varphi$,

$$|\bar{x}|_\varphi \leqslant \sigma(\bar{x}) < o(\Delta),$$

so that

$$\|\varphi\| = \textit{supremum}\{|\bar{x}|_\varphi + 1 : \bar{x} \in I_\varphi\} \leqslant o(\Delta).$$

The second assertion follows easily from the first and the construction in the proof of 2B.5. Put

$$\varphi(\bar{x}, S) \Leftrightarrow (\forall \bar{y})[\bar{y} \prec \bar{x} \Rightarrow \bar{y} \in S],$$

so that by the proof of 2B.5,

$$\|\varphi\| \geqslant \textit{rank}(\prec).$$

If $\prec$ is in $\neg\Gamma$, then $\neg\prec$ is in Γ and it occurs positively in φ, so by the first part

$$\|\varphi\| \leqslant o(\Delta).$$

To prove that $\|\varphi\| < o(\Delta)$, it is enough to notice that there cannot be a wellfounded relation in $\neg\Gamma$ of maximum rank. ⊣

The Covering Theorem can be proved with the same trivial argument we gave for 3C.2, but it is useful and worth putting down.

9C.6. Covering Theorem. *Let P be in Γ, let $\sigma : P \twoheadrightarrow \lambda$ be a Γ-norm, let Q be in $\neg\Gamma$ and assume that f is a function in Δ such that $f[Q] \subseteq P$. Then there is some $\xi < o(\Delta)$ such that*

$$\bar{y} \in Q \Rightarrow \sigma(f(\bar{y})) \leqslant \xi. \qquad ⊣$$

One of the most useful tools in the theory of Spector classes is the axiomatic version of the parametrization theorem for the hyperelementary relations on an acceptable structure. Let us first notice a useful fact about parametrizations of the sets in Γ.

9C.7. Good Parametrization Theorem for Γ. *There is a sequence $\{U^n\}_{n\in\omega}$ of relations in Γ such that $U^n \subseteq A^{n+1}$, each U^n parametrizes the n-ary relations in Γ and for each m, n there is a function*

$$S^m_n(a, \bar{y}) = S^m_n(a, y_1, \ldots, y_m)$$

in Δ such that for all $a, \bar{y}, \bar{x} = a, y_1, \ldots, y_m, x_1, \ldots, x_n$,

$$(a, \bar{y}, \bar{x}) \in U^{m+n} \Leftrightarrow (S^m_n(a, \bar{y}), \bar{x}) \in U^n.$$

PROOF. Let $V \subseteq A^3$ be a relation in Γ which parametrizes the binary relations in Γ and for each k put

$$(a, u_1, \ldots, u_k) \in U^k \Leftrightarrow ((a)_1, (a)_2, \langle u_1, \ldots, u_k \rangle) \in V.$$

Each U^k is in Γ and it is a trivial exercise to verify that U^k parametrizes the k-ary relations in Γ.

To construct the functions S^m_n, fix m, n and put

$$Q(s, t) \Leftrightarrow U^{m+n}((s)_1, (s)_2, \ldots, (s)_{m+1}, (t)_1, \ldots, (t)_n).$$

Since Q is in Γ, there is some fixed b so that

$$Q(s, t) \Leftrightarrow V(b, s, t),$$

so that for all $a, \bar{y}, \bar{x}$,

$$\begin{aligned}(a, \bar{y}, \bar{x}) \in U^{m+n} &\Leftrightarrow (b, \langle a, \bar{y}\rangle, \langle \bar{x}\rangle) \in V \\ &\Leftrightarrow (\langle b, \langle a, \bar{y}\rangle\rangle, \bar{x}) \in U^n.\end{aligned}$$

Hence we can take

$$S^m_n(a, \bar{y}) = \langle b, \langle a, \bar{y}\rangle\rangle. \qquad \dashv$$

Using this we can prove a very strong axiomatic version of the Parametrization Theorem for Hyperelementary Relations 5D.4.

A set $J \subseteq A$ is *Γ-complete* if J is in Γ and for every n-ary relation R in Γ there is a function $f(\bar{x})$ in Δ such that

$$R(\bar{x}) \Leftrightarrow f(\bar{x}) \in J.$$

Clearly a Γ-complete set cannot be in Δ. The second assertion in the theorem below yields several Γ-complete sets. Much stronger statements can be proved along those lines.

9C.8. GOOD PARAMETRIZATION THEOREM FOR Δ. *For each $n \leqslant 1$ there is a set I^n in $\Gamma - \Delta$ and $(n+1)$-ary relations H^n, $\check{H}^n$ in Γ and $\neg\Gamma$ respectively such that:*

(6) *If $R \subseteq A^n$, then R is in Δ if and only if there is some $a \in I^n$ such that $R = H^n_a$.*

(7) *If $a \in I^n$, then $H^n_a = \check{H}^n_a$.*

Moreover, I^1 is Γ-complete and for every $J \subseteq I^1$, if J is in Γ and if there exists some t_0 such that

$$(\forall a)[H^1_a = \{t_0\} \Rightarrow a \in J],$$

then J is also Γ-complete.

Moreover, for each pair $P(\bar{y}, \bar{x})$, $Q(\bar{y}, \bar{x})$ of relations, P in Γ and Q in $\neg\Gamma$, there is a function $j(\bar{y})$ in Δ such that for every $\bar{y}$, if

$$(\forall\bar{x})[P(\bar{y}, \bar{x}) \Leftrightarrow Q(\bar{y}, \bar{x})],$$

then

$$j(\bar{y}) \in I^n, \quad H^n_{j(\bar{y})} = \{\bar{x}: P(\bar{y}, \bar{x})\}.$$

Proof. Following the proof of 5D.4, we set

$$I^n(a) \Leftrightarrow U^n((a)_1, (a)_2, \ldots, (a)_{n+1}),$$
$$H^n(a, \bar{x}) \Leftrightarrow I^n(a) \;\&\; ((a)_{n+2}, \bar{x}) \leqslant^*_\sigma ((a)_1, (a)_2, \ldots, (a)_{n+1}),$$
$$\breve{H}^n(a, \bar{x}) \Leftrightarrow \neg[((a)_1, \ldots, (a)_{n+1}) <^*_\sigma ((a)_{n+2}, \bar{x})],$$

where the parametrization scheme $\{U^n\}_{n\in\omega}$ is good in the sense of 9C.7 and for each fixed n, $\sigma: U^n \twoheadrightarrow \kappa$ is a Γ-norm on U^n. Proof of (6) and (7) is exactly as in 5D.4.

For the second assertion, choose a so that

$$R(\bar{x}) \;\&\; t = t_0 \Leftrightarrow (a, \bar{x}, t) \in U^{n+1}$$
$$\Leftrightarrow (S^n_1(a, \bar{x}), t) \in U^1$$

and let

$$f(\bar{x}) = \langle S^n_1(a, \bar{x}), t_0, S^n_1(a, \bar{x})\rangle.$$

It is immediate from the definition of I^1 that

$$R(\bar{x}) \Leftrightarrow R(\bar{x}) \;\&\; t_0 = t_0 \Leftrightarrow f(\bar{x}) \in I^1.$$

Moreover, if $R(\bar{x})$, then

$$t \in H^1_{f(\bar{x})} \Leftrightarrow (S^n_1(a, \bar{x}), t) \leqslant^*_\sigma (S^n_1(a, \bar{x}), t_0),$$

so that $t_0 \in H^1_{f(\bar{x})}$ is immediate and

$$t \in H^1_{f(\bar{x})} \Rightarrow (S^n_1(a, \bar{x}), t) \in U^1$$
$$\Rightarrow (a, \bar{x}, t) \in U^1$$
$$\Rightarrow t = t_0,$$

i.e. $H^1_{f(\bar{x})} = \{t_0\}$. Thus if J contains all the codes of $\{t_0\}$, then

$$R(\bar{x}) \Leftrightarrow f(\bar{x}) \in J.$$

Assume now that $P(\bar{y}, \bar{x})$, $Q(\bar{y}, \bar{x})$ are given, P in Γ and Q in $\neg\Gamma$, P and Q $(m+n)$-ary, where we may assume $m, n \geqslant 1$. Choose a so that

$$P(\bar{y}, \bar{x}) \Leftrightarrow (a, \bar{y}, \bar{x}) \in U^{m+n}$$

and put

$$R(\bar{y}, \bar{x}) \Leftrightarrow (\exists \bar{y}')(\exists \bar{x}')\{Q(\bar{y}', \bar{x}') \;\&\; \neg[(S^m_n(a, \bar{y}'), \bar{x}') \leqslant^*_\sigma (S^m_n(x_1, \bar{y}), \bar{x})]\},$$

where of course x_1 is the first term of the tuple $\bar{x}$, S^m_n the function of 9C.7 and $\sigma: U^{m+n} \twoheadrightarrow \kappa$ is a Γ-norm on U^{m+n}. Since R is in $\neg\Gamma$, there is a fixed b such that

$$R(\bar{y}, \bar{x}) \Leftrightarrow (b, \bar{y}, \bar{x}) \notin U^{m+n};$$

notice that this b depends only on the relations P, Q, not on any particular values of $\bar{x}$ or $\bar{y}$.

Let

$$\bar{b} = b, b, \ldots, b \text{ (n times)},$$

and fix $\bar{y}$ so that

$$(\forall \bar{x})[P(\bar{y}, \bar{x}) \Leftrightarrow Q(\bar{y}, \bar{x})].$$

For such $\bar{y}$ we can repeat the argument of 6C.10 and show that

$$(b, \bar{y}, \bar{b}) \in U^{m+n},$$

so that

$$(S^m_n(b, \bar{y}), \bar{b}) \in U^n,$$

and

$$(\forall \bar{x})[P(\bar{y}, \bar{x}) \Leftrightarrow (S^m_n(a, \bar{y}), \bar{x}) \leqslant^*_\sigma (S^m_n(b, \bar{y}), \bar{b})],$$

so by the way we chose the coding we can set

$$j(\bar{y}) = \langle S^m_n(b, \bar{y}), \bar{b}, S^m_n(a, \bar{y})\rangle. \qquad \dashv$$

The axiomatic setup deals only with first order relations, but we can use this parametrization theorem to introduce second order relations *with arguments in Δ*.

Fix relations $\{I^n, H^n, \breve{H}^n\}_{n\in\omega}$ which satisfy (6) and (7) of Theorem 9C.8. We say that a second order relation

$$\mathscr{P}(\bar{x}, \bar{Y}) \Leftrightarrow \mathscr{P}(x_1, \ldots, x_n, Y_1, \ldots, Y_k)$$

of signature $(n, r_1, \ldots, r_k)$ is Γ *on Δ* if the first order relation

(8) $\quad \mathscr{P}\#(\bar{x}, y_1, \ldots, y_k) \Leftrightarrow y_1 \in I^{r_1} \;\&\; \ldots \;\&\; y_k \in I^{r_k} \;\&\; \mathscr{P}(\bar{x}, H^{r_1}_{y_1}, \ldots, H^{r_k}_{y_k})$

is in Γ. This holds intuitively if for arguments in Δ the relation $\mathscr{P}$ is Γ *in the coding*.

We say that $\mathscr{P}$ *is Δ on Δ* if both $\mathscr{P}$ and $\neg\mathscr{P}$ are Γ on Δ. Notice that this does not imply that $\mathscr{P}\#$ is a relation in Δ.

It is important to notice that this definition is independent of the particular parametrization $\{I^n, H^n, \breve{H}^n\}_{n=0,1,\ldots}$.

9C.9. THEOREM. *If* $\{I^n, H^n, \breve{H}^n\}_{n=0,1,\ldots}$ *and* $\{J^n, G^n, \breve{G}^n\}_{n=0,1,\ldots}$ *are both parametrizations of the* Δ *relations satisfying conditions* (6), (7) *of Theorem* 9C.8 *and if* $\mathscr{P}(\bar{x}, \bar{Y})$ *is* Γ *on* Δ *relative to* $\{I^n, H^n, \breve{H}^n\}_{n=0,1,\ldots}$, *then* $\mathscr{P}(\bar{x}, \bar{Y})$ *is* Γ *on* Δ *relative to* $\{J^n, G^n, \breve{G}^n\}_{n=0,1,\ldots}$.

PROOF. For each fixed r put

$$Q^r(z, y) \Leftrightarrow z \in J^r \ \&\ y \in I^r \ \&\ G^r_z = H^r_y.$$

These relations Q^r are all in Γ, since

$$Q^r(z, y) \Leftrightarrow z \in J^r \ \&\ y \in I^r \ \&\ (\forall \bar{u})[\breve{G}^r(z, \bar{u}) \Rightarrow H^r(y, \bar{u})] \ \&\ (\forall \bar{u})[\breve{H}^r(y, \bar{u}) \Rightarrow G^r(z, \bar{u})].$$

From this the theorem follows by a trivial computation. ⊣

The closure properties of this class of relations are trivial but useful.

9C.10. THEOREM. *The class of second order relations which are* Γ *on* Δ *is closed under* &, $\vee$, $\exists^A$, $\forall^A$ *and trivial combinatorial substitutions.*

If $\mathscr{P}(Z, \bar{x}, \bar{Y})$ *is* Γ *on* Δ *and* $\mathscr{Q}(\bar{x}, \bar{Y})$ *is defined by*

$$\mathscr{Q}(\bar{x}, \bar{Y}) \Leftrightarrow (\exists Z \in \Delta)\mathscr{P}(Z, \bar{x}, \bar{Y}),$$

then $\mathscr{Q}$ *is* Γ *on* Δ.

If $\mathscr{R}(\bar{x}, Z, \bar{Y})$ *is* Γ *on* Δ *and* $\mathscr{Q}(\bar{u}, \bar{x}, \bar{Y})$ *is* Δ *on* Δ, *then the relation* $\mathscr{P}(\bar{x}, \bar{Y})$ *defined by*

$$\mathscr{P}(\bar{x}, \bar{Y}) \Leftrightarrow \mathscr{R}(\bar{x}, \{\bar{u} : \mathscr{Q}(\bar{u}, \bar{x}, \bar{Y})\}, \bar{Y})$$

is Γ *on* Δ.

If $\mathscr{P}(\bar{x}, \bar{Y})$ *is inductive in* $Q_1, \ldots, Q_m$ *and* $Q_1, \ldots, Q_m$ *are in* Γ, *then* $\mathscr{P}$ *is* Γ *on* Δ.

PROOF. The first two assertions are trivial, e.g. the second follows from the equivalence

$$\mathscr{Q}^{\#}(\bar{x}, \bar{y}) \Leftrightarrow (\exists z)\mathscr{P}^{\#}(z, \bar{x}, \bar{y}).$$

For the third assertion, notice first that if $\mathscr{Q}(\bar{u}, \bar{x}, \bar{Y})$ is Δ on Δ and $Y_1, \ldots, Y_k$ are all in Δ, then $\{\bar{u} : \mathscr{Q}(\bar{u}, \bar{x}, \bar{Y})\} \in \Delta$. This is because clearly

$$\mathscr{Q}(\bar{u}, \bar{x}, \bar{Y}) \Leftrightarrow \mathscr{Q}^{\#}(\bar{u}, \bar{x}, \bar{y}),$$

$$\neg\mathscr{Q}(\bar{u}, \bar{x}, \bar{Y}) \Leftrightarrow (\neg\mathscr{Q})^{\#}(\bar{u}, \bar{x}, \bar{y}),$$

with $\bar{y} = y_1, \ldots, y_k$ any set of codes for $Y_1, \ldots, Y_k$ in the parametrization. Hence

$$\mathscr{R}(\bar{x}, \{\bar{u} : \mathscr{Q}(\bar{u}, \bar{x}, \bar{Y})\}, \bar{Y}) \Leftrightarrow (\exists Z \in \Delta)\{(\forall \bar{u})[\bar{u} \in Z \Leftrightarrow \mathscr{Q}(\bar{u}, \bar{x}, \bar{Y})] \ \&\ \mathscr{R}(\bar{x}, Z, \bar{Y})\},$$

which implies that the substitution yields a relation Γ on Δ by the first two assertions.

For the last assertion, notice that by the definition of $\mathscr{P}^{\#}$ and the Substitution Theorem 6B.3, if $\mathscr{P}(\bar{x}, \bar{Y})$ is inductive in $Q_1, \ldots, Q_m$, then $\mathscr{P}^{\#}(\bar{x}, \bar{y})$ is inductive in $Q_1, \ldots, Q_m, H^{r_1}, \breve{H}^{r_1}, \ldots, H^{r_k}, \breve{H}^{r_k}$; hence $\mathscr{P}^{\#}$ is in Γ by 9A.2. ⊣

The last assertion of the theorem often allows us to avoid lengthy computations by appealing to known results about inductive second order relations. For example, the relations

$$\bar{x} \in Y, \qquad X = Y$$

are Δ on Δ—this is quite easy to verify directly. But also

$$\mathscr{W}\mathscr{F}^n(X) \Leftrightarrow X \text{ is wellfounded},$$

$$\mathscr{W}\mathscr{F}^n(X) \;\&\; \mathscr{W}\mathscr{F}^m(Y) \;\&\; rank(X) \leqslant rank(Y)$$

are Γ on Δ and direct proofs for these are not trivial.

9D. Admissible sets

Admissible sets were introduced by Platek [1966] as the natural domains on which to develop abstract recursion theory with a view towards applying it to set theory. Kripke [1964] defined the related concept of admissible ordinal. Both notions have proved to be interesting and useful and a substantial body of theory has evolved about them.

Unfortunately, there is no exposition of the elementary theory of admissible sets in the literature. Here we must confine ourselves to the definitions and the few basic facts which we will need in the remainder of this chapter. I am grateful to K. J. Barwise for patiently teaching these facts to me—the general approach to the subject that we follow in this section is due to him.

Let $\mathscr{M}$ be a nonempty transitive set, let $R_1, \ldots, R_l$ be relations on $\mathscr{M}$. We enlarge the language of the structure $\langle \mathscr{M}, \in \restriction \mathscr{M}, R_1, \ldots, R_l \rangle$ by adding the *restricted quantifiers* $(\exists x \in y)$, $(\forall x \in y)$, i.e. if φ is a formula, so are

$$(\exists x \in y)\varphi, \qquad (\forall x \in y)\varphi.$$

The interpretation of these is via the obvious equivalences

$$(\exists x \in y)\varphi \Leftrightarrow (\exists x)[x \in y \;\&\; \varphi],$$

$$(\forall x \in y)\varphi \Leftrightarrow (\forall x)[x \in y \Rightarrow \varphi].$$

Following Levy [1965], we say that a formula in this enlarged language is $\Delta_0(R_1, \ldots, R_l)$ if it can be built up from the prime formulas

$$t = s, \qquad t \in s, \qquad R_i(t_1, \ldots, t_{n_i}) \quad (i = 1, \ldots, l)$$

using the propositional connectives $\neg$, &, $\vee$, $\rightarrow$ and the restricted quantifiers.

The schema of Δ_0-*Separation* on $\langle \mathcal{M}, \in \restriction \mathcal{M}, R_1, \ldots, R_l \rangle$ is the class of all formulas

$$(\Delta_0\text{-}Sep) \qquad (\exists w)(\forall x)\{x \in w \Leftrightarrow [x \in z \;\&\; \varphi(x)]\},$$

where $\varphi(x)$ is $\Delta_0(R_1, \ldots, R_l)$. The schema of Δ_0-*Collection* on $\langle \mathcal{M}, \in \restriction \mathcal{M}, R_1, \ldots, R_l \rangle$ is the class of all formulas

$$(\Delta_0\text{-}Coll) \qquad (\forall x \in z)(\exists y)\varphi(x, y) \Rightarrow (\exists w)(\forall x \in z)(\exists y \in w)\varphi(x, y),$$

where again $\varphi(x, y)$ is $\Delta_0(R_1, \ldots, R_l)$.

The first of these schemas is a very weak form of the classical *Axiom of Separation* in Zermelo Set Theory. It only allows us to construct $\Delta_0(R_1, \ldots, R_l)$-definable subsets of a given set.

To appreciate Δ_0-Collection, take the special case where $\varphi(x, y)$ is the graph of a function f,

$$f(x) = y \Leftrightarrow \varphi(x, y),$$

whose domain contains the set z. By Δ_0-Collection there is some w which contains the image

$$f[z] = \{f(x) : x \in z\},$$

and then we can separate this image by Δ_0-Separation,

$$y \in f[z] \Leftrightarrow y \in w \;\&\; (\exists x \in z)\varphi(x, y).$$

Thus Δ_0-Collection implies a very weak form of the classical *Axiom of Replacement* of Zermelo–Fraenkel set theory; it allows us to construct the image of a given set by a function with $\Delta_0(R_1, \ldots, R_l)$-definable graph. Actually Δ_0-Collection is substantially stronger than this weak replacement property.

A nonempty, transitive set $\mathcal{M}$ is $(R_1, \ldots, R_l)$-*admissible* or *admissible relative to* $R_1, \ldots, R_l$, if

(1) $\mathcal{M}$ is closed under pairing and union,

(2) $\langle \mathcal{M}, \in \restriction \mathcal{M}, R_1, \ldots, R_l \rangle$ satisfies Δ_0-Separation,

(3) $\langle \mathcal{M}, \in \restriction \mathcal{M}, R_1, \ldots, R_l \rangle$ satisfies Δ_0-Collection.

Clearly every model of Zermelo–Fraenkel set theory is admissible (relative to the empty list of relations). Also, if κ is a regular cardinal, then *the set of sets of cardinality hereditarily less than* κ,

$$H_\kappa = \{x\text{: the transitive closure of } x \text{ has cardinality} < \kappa\},$$

is admissible—actually this is true even for singular κ, but it requires proof in that case. It is still harder to prove that for every cardinal κ the set L_κ (of sets constructible before κ) is admissible—this result is due to Kripke.

Many other admissible sets can be constructed starting from these and applying the Skolem–Löwenheim Theorem.

We say that ψ is $\Sigma_1(R_1, \ldots, R_l)$ if

$$\psi \equiv (\exists x)\varphi,$$

where φ is $\Delta_0(R_1, \ldots, R_l)$; similarly, ψ is $\Pi_1(R_1, \ldots, R_l)$ if

$$\psi \equiv (\forall x)\varphi$$

with some φ in $\Delta_0(R_1, \ldots, R_l)$.

A *relation R on $\mathcal{M}$ is* $\Delta_0(R_1, \ldots, R_l)$, $\Sigma_1(R_1, \ldots, R_l)$ or $\Pi_1(R_1, \ldots, R_l)$ accordingly as R is definable by a formula in the appropriate class. We say that R is $\Delta_1(R_1, \ldots, R_l)$ if R is both $\Sigma_1(R_1, \ldots, R_l)$ and $\Pi_1(R_1, \ldots, R_l)$.

The class of $\Sigma_1(R_1, \ldots, R_l)$ relations on a set $\mathcal{M}$ which is admissible relative to $R_1, \ldots, R_l$ has many nice properties. It is closed under &, $\vee$, restricted quantification of both kinds and existential quantification over $\mathcal{M}$. These are all easy to prove, e.g. closure under $\exists^{\mathcal{M}}$ follows from the equivalence

$$(\exists x)(\exists y)\varphi(x, y) \Leftrightarrow (\exists z)(\exists x \in z)(\exists y \in z)\varphi(x, y)$$

which is true because $\mathcal{M}$ is closed under pairing. Similarly, closure under $(\forall x \in y)$ follows from the equivalence

$$(\forall x \in y)(\exists z)\varphi(x, y, z) \Leftrightarrow (\exists w)(\forall x \in y)(\exists z \in w)\varphi(x, y, z)$$

which holds when φ is $\Delta_0(R_1, \ldots, R_l)$ because $\mathcal{M}$ satisfies Δ_0-Collection.

We illustrate the notion of admissible set by proving stronger versions of Δ_0-Separation and Δ_0-Collection.

9D.1. THEOREM. *Let $\mathcal{M}$ be admissible relative to $R_1, \ldots, R_l$.*

If $R(x)$ is $\Delta_1(R_1, \ldots, R_l)$ and $z \in \mathcal{M}$, then $\{x \in z: R(x)\} \in \mathcal{M}$ (Δ_1-Separation).

If $R(x, y)$ is $\Sigma_1(R_1, \ldots, R_l)$ and $(\forall x \in z)(\exists y)R(x, y)$, then there is some $w \in \mathcal{M}$ such that $(\forall x \in z)(\exists y \in w)R(x, y)$ (Σ_1-Collection).

PROOF. For the first assertion, by hypothesis

$$R(x) \Leftrightarrow (\exists y)\varphi(x, y)$$
$$\Leftrightarrow (\forall y)\psi(x, y)$$

where both $\varphi(x, y)$, $\psi(x, y)$ are $\Delta_0(R_1, \ldots, R_l)$. Hence

$$(\forall x \in z)(\exists y)[\varphi(x, y) \vee \neg\psi(x, y)],$$

and Δ_0-Collection yields some $w \in \mathcal{M}$ so that

$$(\forall x \in z)(\exists y \in w)[\varphi(x, y) \vee \neg\psi(x, y)].$$

Clearly then

$$\{x \in z: R(x)\} = \{x \in z: (\exists y \in w)\varphi(x, y)\}$$

and the set on the right is in $\mathscr{M}$ by Δ_0-Separation.

For the second assertion, by hypothesis

$$R(x, y) \Leftrightarrow (\exists t)\varphi(x, y, t)$$

with $\varphi(x, y, t)$ in $\Delta_0(R_1, \ldots, R_l)$, and

$$(\forall x \in z)(\exists y)(\exists t)\varphi(x, y, t).$$

Since $\mathscr{M}$ is closed under pairing, $y, t \in \mathscr{M} \Rightarrow \{y, t\} \in \mathscr{M}$, hence

$$(\forall x \in z)(\exists u)[(\exists y \in u)(\exists t \in u)\varphi(x, y, t)].$$

Now by Δ_0-Collection there is an $s \in \mathscr{M}$ so that

$$(\forall x \in z)(\exists u \in s)[(\exists y \in u)(\exists t \in u)\varphi(x, y, t)].$$

Since $\mathscr{M}$ is closed under union, take

$$w = \bigcup s = \{y: (\exists u \in s) y \in u\}.$$

Clearly for this w,

$$(\forall x \in z)(\exists y \in w)R(x, y),$$

which completes the proof. ⊣

Still one more collection of formulas will be useful. We say that ψ is $\Sigma(R_1, \ldots, R_l)$ if ψ can be built up from $\Delta_0(R_1, \ldots, R_l)$ formulas by the positive connectives &, $\vee$, the restricted quantifiers $(\exists x \in y)$, $(\forall x \in y)$ and the unrestricted existential quantifier $(\exists y)$. For example,

$$\psi \equiv (\exists y)(\forall x \in v)[(\exists z)\varphi(x, v, y, z) \vee \chi(x, v, y)]$$

is $\Sigma(R_1, \ldots, R_l)$, if φ, χ are $\Delta_0(R_1, \ldots, R_l)$. Clearly every $\Sigma_1(R_1, \ldots, R_l)$ formula is $\Sigma(R_1, \ldots, R_l)$ but not vice-versa.

The remarks preceding 9D.1 make it obvious that if $\mathscr{M}$ is admissible relative to $R_1, \ldots, R_l$, then every relation definable by a $\Sigma(R_1, \ldots, R_l)$ formula is $\Sigma_1(R_1, \ldots, R_l)$. We need these formulas for a reason other than definability.

If ψ is $\Sigma(R_1, \ldots, R_l)$ and w is a variable which does not occur at all in ψ, let $\psi^{(w)}$ be the $\Delta_0(R_1, \ldots, R_l)$ formula obtained by restricting all the unbounded existential quantifiers in ψ to w. For example, for the ψ above,

$$\psi^{(w)} \equiv (\exists y \in w)(\forall x \in v)[(\exists z \in w)\varphi(x, v, y, z) \vee \chi(x, v, y)].$$

In this transformation we do not interfere with the restricted quantifiers in ψ.

The next result is simple but very useful.

9D.2. Σ-Reflection Principle for admissible sets. *Let $\mathcal{M}$ be admissible relative to $R_1, \ldots, R_l$, let ψ be a $\Sigma(R_1, \ldots, R_l)$ formula, let w be a variable not occurring in ψ. Then*

$$(4) \qquad \psi \Leftrightarrow (\exists w)\psi^{(w)}.$$

Proof. A trivial induction on the construction of Σ formulas shows that

$$(5) \qquad w \subseteq z \;\&\; \psi^{(w)} \Rightarrow \psi^{(z)} \Rightarrow \psi;$$

in particular we get direction ($\Leftarrow$) of (4).

The other direction of (4) is proved also by induction on the construction of ψ. For example,

$$\varphi \;\&\; \chi \Rightarrow \varphi^{(u)} \;\&\; \chi^{(v)} \text{ for some } u, v \qquad \text{(by ind. hyp.)}$$

$$\Rightarrow (\exists w)[\varphi^{(w)} \;\&\; \chi^{(w)}]$$

taking $w = u \cup v$ and using (5). Also

$$(\forall x \in y)\varphi \Rightarrow (\forall x \in y)(\exists u)\varphi^{(u)} \qquad \text{(by ind. hyp.)}$$

$$\Rightarrow (\forall x \in y)(\exists u \in z)\varphi^{(u)} \text{ for some } z \qquad \text{(by } \Delta_0\text{-Collection)}$$

$$\Rightarrow (\exists w)(\forall x \in y)\varphi^{(w)},$$

taking $w = \bigcup z$ and using again (5). ⊣

In addition to the simple operations of pairing and union, an admissible set is closed under many operations defined by constructive transfinite recursions of various sorts. It is these closure properties that make admissible sets a natural domain for abstract recursion theory. We cite here three simple results of this type which are typical and which we will need.

9D.3. Theorem. *Let $\mathcal{M}$ be admissible relative to $R_1, \ldots, R_l$, let $\prec$ be a wellfounded relation which is a member of $\mathcal{M}$, let $G(f, x)$ be a $\Delta_1(R_1, \ldots, R_l)$ function mapping $\mathcal{M} \times \mathcal{M}$ into $\mathcal{M}$. Then there is a unique function f in $\mathcal{M}$ such that*

$$\mathit{Domain}(f) = \mathit{Field}(\prec)$$

and for every $x \in \mathit{Field}(\prec)$,

$$f(x) = G(\{\langle t, f(t)\rangle : t \prec x\}, x).$$

Proof. Put

$$R(f) \Leftrightarrow f \text{ is a function } \& \; \mathit{Domain}(f) \subseteq \mathit{Field}(\prec)$$

$$\& \; (\forall x \in \mathit{Domain}(f))[f(x) = G(\{\langle t, f(t)\rangle : t \prec x\}, x)].$$

Here all notions are understood in the usual settheoretic manner, i.e. a function is a set of ordered pairs, etc.

It takes a bit of checking to verify that R is $\Delta_1(R_1, \ldots, R_l)$ and we shall omit the computation.

A trivial induction on $\prec$ shows that

$$R(f) \mathbin{\&} R(g) \mathbin{\&} x \in Domain(f) \cap Domain(g) \Rightarrow f(x) = g(x).$$

We now prove by induction on $\prec$ that

$$(\forall x \in Field(\prec))(\exists f)[R(f) \mathbin{\&} x \in Domain(f)].$$

From the induction hypothesis,

$$(\forall t \prec x)(\exists f)[R(f) \mathbin{\&} t \in Domain(f)],$$

whence by Σ_1-Collection there is some $w \in \mathcal{M}$ such that

$$(\forall t \prec x)(\exists f \in w)[R(f) \mathbin{\&} t \in Domain(f)].$$

Put now

$$g = \{\langle t, f(t)\rangle : f \in w \mathbin{\&} R(f) \mathbin{\&} t \prec x\}$$

which is in $\mathcal{M}$ by Δ_1-Separation and closure under union. Finally, take

$$f = g \cup \{\langle x, G(g, x)\rangle\}$$

and verify immediately that $R(f)$ and $x \in Domain(f)$.

Using Σ_1-Collection once more, there is a $w \in \mathcal{M}$ such that

$$(\forall x \in Field(\prec))(\exists g \in w)[R(g) \mathbin{\&} x \in Domain(g)];$$

the required function f is given by

$$f = \{\langle t, g(t)\rangle : g \in w \mathbin{\&} R(g)\}. \qquad \dashv$$

The wellfounded relation $\prec$ occurred as a parameter in this proof and the definition of the required function f was uniform in $\prec$. An examination of the proof shows that we have established the following more complicated but also more useful result.

9D.4. THEOREM. *Let $\mathcal{M}$ be admissible relative to $R_1, \ldots, R_l$, let $G(w, f, x)$ be a $\Delta_1(R_1, \ldots, R_l)$ function mapping $\mathcal{M}^3$ into $\mathcal{M}$. There is a $\Delta_1(R_1, \ldots, R_l)$ function $F(w, x)$ such that whenever w is a wellfounded relation which is a member of $\mathcal{M}$ and $x \in Field(w)$, then*

$$F(w, x) = G(w, \{\langle t, F(w, t)\rangle : (t, x) \in w\}, x). \qquad \dashv$$

It is easy to verify that the relation

$$Ord(\xi) \Leftrightarrow \xi \text{ is an ordinal}$$

is Δ_0 on any admissible set $\mathcal{M}$. Using this and 9D.3 we can show that if w is a wellfounded relation which is a member of $\mathcal{M}$, then $rank(w)$ is an ordinal of $\mathcal{M}$. Hence

$$\begin{aligned} o(\mathcal{M}) &= supremum\{rank(w) \colon w \text{ is wellfounded}, w \in \mathcal{M}\} \\ &= supremum\{\xi \colon Ord(\xi), \xi \in \mathcal{M}\}. \end{aligned}$$

It is worth citing explicitly an easy corollary of 9D.4 which allows us to define functions by transfinite inductions on the ordinals.

9D.5. THEOREM. *Let $\mathcal{M}$ be admissible relative to $R_1, \ldots, R_l$, let $G(f, \xi)$ be a $\Delta_1(R_1, \ldots, R_l)$ function mapping $\mathcal{M} \times o(\mathcal{M})$ into $\mathcal{M}$. The unique function $F \colon o(\mathcal{M}) \times \mathcal{M} \to \mathcal{M}$ which satisfies*

(6)
$$\begin{aligned} F(\xi) &= G(F \upharpoonright \xi, \xi) \\ &= G(\{\langle \eta, F(\eta)\rangle \colon \eta < \xi\}, \xi) \end{aligned}$$

is $\Delta_1(R_1, \ldots, R_l)$.

PROOF. For each ξ, let

$$W(\xi) = \{(\eta, \zeta) \colon \eta < \zeta < \xi\}.$$

Clearly each $W(\xi)$ is a wellfounded relation in $\mathcal{M}$ and

$$\xi \geqslant 2 \Rightarrow rank(W(\xi)) = \xi.$$

Moreover, the mapping W is easily Δ_0.

Put

$$G^*(w, f, x) = G(f, x)$$

and let $F^*(w, x)$ be the $\Delta_1(R_1, \ldots, R_l)$ function given by 9D.3 such that when w is wellfounded and $x \in Field(w)$,

$$\begin{aligned} F^*(w, x) &= G^*(w, \{(t, F^*(w, t)) \colon (t, x) \in w\}, x) \\ &= G(\{(t, F^*(w, t)) \colon (t, x) \in w\}, x). \end{aligned}$$

An easy induction on ξ shows that if $\lambda, \lambda' > 2$ and $\xi < \lambda, \lambda'$ then

$$F^*(W(\lambda), \xi) = F^*(W(\lambda'), \xi),$$

and then another trivial induction on ξ establishes that the function

$$F(\xi) = F^*(W(\xi+2), \xi)$$

satisfies (6). This completes the proof, since the uniqueness of the function satisfying (1) is obvious. ⊣

One of the consequences of this result is that the function

$$\xi \to L_\xi$$

which defines the constructible hierarchy is Δ_1 on every admissible set, and in particular, if $\mathcal{M}$ is admissible,

$$\xi < o(\mathcal{M}) \Rightarrow L_\xi \in \mathcal{M}.$$

It is a little harder to prove that if $\mathcal{M}$ is admissible and $\kappa = o(\mathcal{M})$, then L_κ is admissible.

An ordinal κ is *admissible* if $\kappa = o(\mathcal{M})$ for some admissible set $\mathcal{M}$, i.e.

κ is admissible $\Leftrightarrow L_\kappa$ is admissible.

We will not use this fact here, but it is useful to keep in mind.

There is one very basic fact about the Σ_1 relations on an admissible set whose proof is quite complicated. We cite the result here and we include an outline for a proof in the hint for Exercise 9.5.

9D.6. Parametrization Theorem for Σ_1 Relations. *If $\mathcal{M}$ is admissible relative to $R_1, \ldots, R_l$, then for each n there is a $\Sigma_1(R_1, \ldots, R_l)$ $(n+1)$-ary relation S^n which parametrizes the n-ary $\Sigma_1(R_1, \ldots, R_l)$ relations on $\mathcal{M}$.* ⊣

We close this section with a brief discussion of two properties of admissible sets which are relevant to the construction in 9E.

If $\mathcal{M}$ is a transitive set and $z \in \mathcal{M}$, a *projection of $\mathcal{M}$ on z* is a function

$$\pi \colon D \twoheadrightarrow \mathcal{M}$$

with domain some $D \subseteq z$ which is *onto $\mathcal{M}$. Notice that D need not be* (and usually is not) *an element of $\mathcal{M}$*. We call an admissible set $\mathcal{M}$ *projectible on z relative to $R_1, \ldots, R_l$* if $\mathcal{M}$ admits a projection $\pi \colon D \twoheadrightarrow \mathcal{M}$, $D \subseteq z$, which is $\Delta_1(R_1, \ldots, R_l)$, i.e. whose graph is $\Delta_1(R_1, \ldots, R_l)$. The domain D of a projection π which is $\Delta_1(R_1, \ldots, R_l)$ is $\Sigma_1(R_1, \ldots, R_l)$, since

$$x \in D \Leftrightarrow (\exists y)[\pi(x) = y].$$

Also the inverse map

$$\pi^{-1}(y) = \{x \in z \colon \pi(x) = y\}$$

is $\Delta_1(R_1, \ldots, R_l)$ and maps $\mathcal{M}$ into $Power(z) \cap \mathcal{M}$, so that

$$y \neq z \Rightarrow \pi^{-1}(y) \cap \pi^{-1}(z) = \emptyset.$$

All the admissible sets which we will construct in the next section will be projectible. On the other hand, it is clear that H_κ (κ regular) is not projectible.

Finally, a *resolution* of an admissible set $\mathcal{M}$ is a function

$$\tau: o(\mathcal{M}) \to \mathcal{M}$$

such that

$$\mathcal{M} = \bigcup_\xi \tau(\xi).$$

We call $\mathcal{M}$ *resolvable relative to* $R_1, \ldots, R_l$ if $\mathcal{M}$ admits a resolution τ which is $\Delta_1(R_1, \ldots, R_l)$.

Again, all the admissible sets which we will construct will be resolvable. But there are nonresolvable admissible sets, in fact it is possible to define an uncountable admissible $\mathcal{M}$ such that $o(\mathcal{M})$ and every member of $\mathcal{M}$ are countable.

9E. The companion of a Spector class

We prove here the main result of this chapter, a representation theorem for Spector classes.

9E.1. THEOREM. *Let A be a transitive, infinite set, let Γ be a Spector class on A such that*

$$\in \restriction A \in \varDelta,$$

i.e. both the relation $\in$ and its negation are in Γ. Let $\mathcal{M}(\varDelta)$ be the intersection of all admissible sets which contain every $\varDelta$-subset of A,

(1) $\quad \mathcal{M}(\varDelta) = \bigcap \{\mathcal{M}: \mathcal{M}$ *is admissible and*
$$(\forall X)[(X \subseteq A \;\&\; X \in \varDelta) \Rightarrow X \in \mathcal{M}]\}.$$

Then:

(a) *$\mathcal{M}(\varDelta)$ is admissible, $o(\mathcal{M}(\varDelta)) = o(\varDelta)$ and for every $X \subseteq A$,*

$$X \in \mathcal{M}(\varDelta) \Leftrightarrow X \in \varDelta.$$

Moreover, there is a relation

$$R = R_\Gamma$$

on $\mathcal{M}(\varDelta)$ such that the following hold:

(b) *$\mathcal{M}(\varDelta)$ is admissible, resolvable and projectible on A relative to R.*

(c) *A relation $P \subseteq A^n$ is in Γ if and only if P is $\Sigma_1(R)$ on $\mathcal{M}(\varDelta)$.*

PROOF. If x is a set, let

$$\mathcal{T}(x) = \{(x_1, \ldots, x_n): x \ni x_1 \ni \ldots \ni x_n\}$$

and for $(x_1, \ldots, x_n), (y_1, \ldots, y_m)$ in $\mathcal{T}(x)$, put

$$(x_1, \ldots, x_n) \succ (y_1, \ldots, y_m) \Leftrightarrow n < m \;\&\; x_1 = y_1 \;\&\ldots\&\; x_n = y_n.$$

Clearly $\prec$ is a *wellfounded* relation, indeed a *tree*. It is not hard to verify that the set x is completely determined by the homomorphism type of $\prec$. This is the basic idea of the proof: we will take all *wellfounded trees* in Δ, we will show that the sets they determine are precisely those in $\mathcal{M}(\Delta)$ and then we will use this representation of $\mathcal{M}(\Delta)$ to help prove the other assertions of the theorem.

It will be convenient to define wellfounded trees as sets of *sequence codes* rather than sequences of elements of A. So fix a coding scheme in Δ and put

(2) *T is a wellfounded tree*

$$\begin{aligned} &\Leftrightarrow T \neq \emptyset \\ &\quad \& \ (\forall u)[u \in T \Rightarrow Seq(u)] \\ &\quad \& \ (\forall u)(\forall v)[(Seq(u) \ \& \ Seq(v) \ \& \ u^\frown v \in T) \Rightarrow u \in T] \\ &\quad \& \ \{(\forall x_0)(\forall x_1) \ldots\}(\exists n)[\langle x_0, \ldots, x_n\rangle \notin T]. \end{aligned}$$

The code of the empty sequence belongs to every wellfounded tree and the set $\{\langle\emptyset\rangle\}$ is a wellfounded tree.

We assign to each wellfounded tree T and to each $u \in T$ the set $m(T, u)$ by the following induction on the wellfounded relation

(3) $\prec_T = \{(u, v)\colon u \in T, v \in T \ \& \ u$ *codes a proper extension of the sequence coded by* $v\}$:

(4) $$m(T, u) = \{m(T, u^\frown\langle y\rangle)\colon u^\frown\langle y\rangle \in T\}.$$

The *set determined* by a wellfounded tree is defined by

(5) $$m(T) = m(T, \langle\emptyset\rangle).$$

The function $m(T, u)$ assigns sets to the *nodes* of the tree T, where the set assigned to u is the set of all sets assigned to the immediate extensions of u. Then $m(T)$ is the set that is assigned to the *root* $\langle\emptyset\rangle$ of T. For example, if $T = \{\langle\emptyset\rangle\}$, then

$$m(T) = m(T, \langle\emptyset\rangle) = \{m(T, \langle x\rangle)\colon \langle x\rangle \in T\} = \emptyset.$$

In Fig. 9.1 we picture a tree S such that $m(S) = 3$.

If T is a wellfounded tree, $u \in T$ and we put

(6) $$T_u = \{v\colon u^\frown v \in T\},$$

then T_u is a wellfounded tree and for each v,

$$m(T_u, v) = m(T, u^\frown v);$$

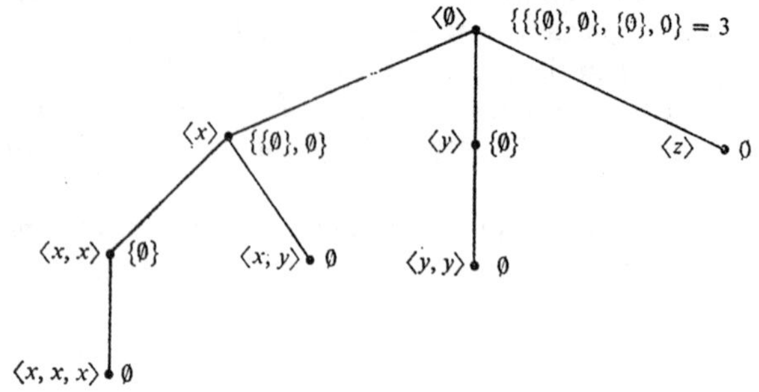

Fig. 9.1. $S = \{\langle\emptyset\rangle, \langle x\rangle, \langle y\rangle, \langle z\rangle, \langle x, x\rangle, \langle x, y\rangle, \langle y, y\rangle, \langle x, x, x\rangle\}, x \neq y, y \neq z, x \neq z.$

in particular, taking $v = \langle\emptyset\rangle$,

(7) $$m(T_u) = m(T, u).$$

Put

(8) $$\mathscr{M} = \{m(T)\colon T \textit{ is a wellfounded tree, } T \in \Delta\}.$$

The proof that $\mathscr{M} = \mathscr{M}(\Delta)$ and that (a)–(c) hold is by a sequence of lemmas.

Lemma 1. $\mathscr{M}$ *is transitive.*

Proof. If $z = m(T)$ is in $\mathscr{M}$ and $x \in z$, then $x = m(T, \langle y\rangle) = m(T_{\langle y\rangle})$ for some y, and $T_{\langle y\rangle}$ is obviously a wellfounded tree in Δ.

Lemma 2. $A^n \subseteq \mathscr{M}$.

Proof. For $n = 1$ we proceed as in the beginning of the proof. For each $x \in A$, put

(9) $$T(x) = \{\langle x_1, \ldots, x_n\rangle\colon x \ni x_1 \ni \ldots \ni x_n\}.$$

It is trivial to check that $T(x)$ is a wellfounded tree in Δ, and an easy $\in$-induction (using the fact that A is transitive) shows that for each $x \in A$,

$$m(T(x)) = x.$$

For higher n, take for simplicity the case $n = 2$ and put

$$\begin{aligned} T = T(x, y) = {} & \{\langle x, x, x_1, \ldots, x_n\rangle\colon x \ni x_1 \ni \ldots \ni x_n\} \\ & \cup \{\langle\langle x, y\rangle, x, x_1, \ldots, x_n\rangle\colon x \ni x_1 \ni \ldots \ni x_n\} \\ & \cup \{\langle\langle x, y\rangle, y, y_1, \ldots, y_m\rangle\colon y \ni y_1 \ni \ldots \ni y_m\}. \end{aligned}$$

Clearly T is a wellfounded tree in Δ and

$$\begin{aligned} m(T) &= \{m(T, \langle x\rangle), m(T, \langle\langle x, y\rangle\rangle)\} \\ &= \{\{m(T, \langle x, x\rangle)\}, \{m(T, \langle\langle x, y\rangle, x\rangle), m(T, \langle\langle x, y\rangle, y\rangle)\}\} \\ &= \{\{m(T(x))\}, \{m(T(x)), m(T(y))\}\} \\ &= \{\{x\}, \{x, y\}\} \\ &= (x, y) = \textit{the ordered pair of } x \textit{ and } y. \end{aligned}$$

Lemma 3. *If* $P \subseteq A^n$ *and* P *is in* Δ, *then* $P \in \mathcal{M}$.

Proof is similar to that of Lemma 2. For $n = 1$, put

$$T = T(P) = \{\langle x, x_1, \ldots, x_n\rangle : x \in P \;\&\; x \ni x_1 \ni \ldots \ni x_n\}$$

and notice that T is a wellfounded tree in Δ and

$$\begin{aligned} m(T) &= \{m(T, \langle x\rangle) : \langle x\rangle \in T\} \\ &= \{m(T(x)) : x \in P\} \\ &= \{x : x \in P\} = P. \end{aligned}$$

For higher n the tree in question is a bit more complicated, but the idea is simple. For example, if $P \subseteq A^2$, we can take

$$T = T(P) = \{\langle\langle x, y\rangle\rangle^\frown u : P(x, y) \;\&\; u \in T(x, y)\} \cup \{\langle\emptyset\rangle\},$$

where $T(x, y)$ is defined in the proof of Lemma 2 so that

$$m(T(x, y)) = (x, y).$$

Then T is clearly a wellfounded tree in Δ and

$$\begin{aligned} m(T) &= \{m(T, \langle\langle x, y\rangle\rangle) : P(x, y)\} \\ &= \{m(T(x, y)) : P(x, y)\} \\ &= \{(x, y) : P(x, y)\} = P^2. \end{aligned}$$

It should be pointed out that the slight complication in these proofs for the case $n > 1$ would not be necessary if we assumed that A is closed under the ordinary settheoretic pair so that we could take

$$\langle x, y\rangle = (x, y).$$

But it may be for example that $A = \lambda$ is an infinite ordinal, which admits hyperelementary pairs but is not closed under the ordinary settheoretic pair.

Lemma 4. *If* $\mathcal{N}$ *is an admissible set such that for each* $X \subseteq A$,

$$\textit{if } X \in \Delta, \textit{ then } X \in \mathcal{N},$$

then $\mathcal{M} \subseteq \mathcal{N}$.

Proof. If $\mathcal{N}$ contains all the sets in Δ, in particular it contains all wellfounded trees in Δ. Now an easy application of Theorem 9D.3 implies that $\mathcal{N}$ contains $m(T)$ for every wellfounded tree T in Δ, i.e. $\mathcal{M} \subseteq \mathcal{N}$.

To complete the proof of (a) it will be sufficient to show that $\mathcal{M}$ is admissible. For this and for the proofs of (b), (c) we introduce a coding of the sets in $\mathcal{M}$ by a set in Γ.

Fix a good parametrization scheme $\{I^n, H^n, \breve{H}^n\}_{n \in \omega}$ for the relations in Δ in the sense of Theorem 9C.8, and to avoid superscripts let

$$I = I^1, \qquad H = H^1, \qquad \breve{H} = \breve{H}^1.$$

Put

(10) $\quad a \in M \Leftrightarrow a \in I \;\&\; H_a$ *is a wellfounded tree.*

Lemma 5. M is in Γ, it is Γ-complete and the function

$$a \to m(H_a)$$

maps M onto $\mathcal{M}$.

Proof. The second assertion is immediate. The first follows from Theorem 9C.10 if we notice that the second order relation

$$\mathscr{R}(T) \Leftrightarrow T \textit{ is a wellfounded tree}$$

is inductive in relations in Γ (the coding scheme) and hence $\mathscr{R}$ is Γ on Δ by 9C.10. But

$$a \in M \Leftrightarrow \mathscr{R}^{\#}(a)$$

in the notation introduced by (8) of Section 9C, hence M is in Γ. That M is Γ-complete follows from the second assertion of 9C.8, since $M \subseteq I$ and $\{\langle\emptyset\rangle\}$ is a wellfounded tree, so that

$$a \in I \;\&\; H_a = \{\langle\emptyset\rangle\} \Rightarrow a \in M.$$

The key property of the coding of $\mathcal{M}$ given by M and m is embodied in the next lemma.

Lemma 6. If X is a set in Δ, $X \subseteq M$ and if $P(a, u)$ is a relation in Δ, then

$$\{m(H_a, u) : a \in X \;\&\; u \in H_a \;\&\; P(a, u)\}$$

is a set in $\mathcal{M}$.

In particular, if $X \subseteq M$ and X is in Δ, then

$$\{m(H_a) : a \in X\} \in \mathcal{M}.$$

Proof. Put

$$T = \{\langle\emptyset\rangle\} \cup \{\langle\langle a, u\rangle\rangle^\frown v : a \in X \;\&\; u^\frown v \in H_a \;\&\; P(a, u)\}.$$

Clearly T is a wellfounded tree in Δ. We compute:

$$m(T) = \{m(T, \langle\langle a, u\rangle\rangle): a \in X \;\&\; u \in H_a \;\&\; P(a, u)\}$$
$$= \{m(T_{\langle\langle a,u\rangle\rangle}): a \in X \;\&\; u \in H_a \;\&\; P(a, u)\} \qquad \text{(by (7))}.$$

But if $a \in X \;\&\; u \in H_a = S$, then directly from the definition of T,

$$T_{\langle\langle a,u\rangle\rangle} = S_u,$$

so that by (7) again,

$$m(T_{\langle\langle a,u\rangle\rangle}) = m(S_u) = m(H_a, u)$$

which completes the proof of the first assertion. The second follows immediately.

As an example in the use of this lemma, consider:

Lemma 7. *$\mathcal{M}$ is closed under pairing and union.*

Proof. If $z = m(H_a)$, $w = m(H_b)$ with $a, b \in M$, take $X = \{a, b\}$. Then $\{m(H_x): x \in X\} = \{z, w\} \in \mathcal{M}$.

For closure under union, if $z = m(H_a)$ with $a \in M$, take

$$X = \{a\},$$
$$P(x, u) \Leftrightarrow Seq(u) \;\&\; lh(u) = 2.$$

Then

$$\{m(H_a, u): P(a, u) \;\&\; u \in H_a\} = \{m(H_a, \langle x, y\rangle): \langle x, y\rangle \in H_a\}$$
$$= \bigcup z.$$

Another nice property of M is that we can imbed A in it via a Δ function

Lemma 8. *There is a function*

$$i: A \to M$$

which is in Δ and which assigns to each $x \in A$ a code of x, i.e.

$$m(H_{i(x)}) = x.$$

Proof. Put

$$Q(x, u) \Leftrightarrow Seq(u) \;\&\; (u)_1 \in x \;\&\; (\forall i < lh(u))[(u)_{i+1} \in (u)_i].$$

For each x, clearly

$$\{u: Q(x, u)\} = T(x)$$

in the notation introduced by (9), so that by Lemma 2,

$$m(\{u: Q(x, u)\}) = x.$$

Now by the good parametrization property, there is a function i in Δ such that for every x,

$$i(x) \in I, \qquad H_{i(x)} = \{u\colon Q(x, u)\} = T(x).$$

There are many natural candidates for the relation R which will satisfy (b) and (c) of the theorem. Here we choose one which makes the proof transparent.

Let

$$\sigma\colon M \twoheadrightarrow o(\Delta) \tag{11}$$

be a fixed Γ-norm on the set M and define on $\mathscr{M}$,

$$R(a, u, \xi) \Leftrightarrow a \in M \;\&\; u \in H_a \;\&\; \xi \textit{ is an ordinal } \&\; \xi > 1 \;\&\; \sigma(a) \leqslant \xi. \tag{12}$$

We now begin the proof that $\mathscr{M}$ is admissible relative to this R.

Lemma 9. *The relations*

$$P^{+}(a, u, b, v) \Leftrightarrow a, b \in M \;\&\; u \in H_a \;\&\; v \in H_b \;\&\; m(H_a, u) = m(H_b, v),$$

$$P^{-}(a, u, b, v) \Leftrightarrow a, b \in M \;\&\; u \in H_a \;\&\; v \in H_b \;\&\; m(H_a, u) \neq m(H_b, v)$$

are both in Γ.

Proof. Choose φ so that

$$\begin{aligned}\varphi(u, v, S, T, U) \Leftrightarrow{}& Seq(u) \;\&\; u \in S \;\&\; Seq(v) \;\&\; v \in T\\ &\& (\forall x)[u^\frown\langle x\rangle \in S \rightarrow (\exists y)[v^\frown\langle y\rangle \in T \;\&\; (u^\frown\langle x\rangle, v^\frown\langle y\rangle) \in U]]\\ &\& (\forall y)[v^\frown\langle y\rangle \in T \rightarrow (\exists x)[u^\frown\langle x\rangle \in S \;\&\; (u^\frown\langle x\rangle, v^\frown\langle y\rangle) \in U]].\end{aligned}$$

If both S and T are wellfounded trees, then we can easily verify that

$$(u, v, S, T) \in \mathscr{I}^{\xi}_{\varphi} \Rightarrow m(S, u) = m(T, v)$$

by induction on ξ and

$$m(S, u) = m(T, v) \Rightarrow (u, v, S, T) \in \mathscr{I}_{\varphi}$$

by induction on the wellfounded tree S. Hence the second order relation

$$\mathscr{R}(u, v, S, T) \Leftrightarrow S, T \textit{ are wellfounded trees and } m(S, u) = m(T, v)$$

is inductive in relations in Γ (the coding scheme) and hence $\mathscr{R}$ is Γ on Δ by 9C.10. Now we notice as in the proof of Lemma 5 that

$$P^{+}(a, u, b, v) \Leftrightarrow \mathscr{R}^{\#}(u, v, a, b),$$

so that P^{+} is in Γ.

The proof for P^{-} is similar and we omit it.

Lemma 10. *Let R be defined by* (11), *let $\varphi(x_1, \ldots, x_n)$ be any $\Delta_0(R)$ formula on $\mathcal{M}$ with the indicated free variables, put*

$$P_\varphi^+(a_1, \ldots, a_n) \Leftrightarrow a_1, \ldots, a_n \in M \;\&\; \varphi(m(H_{a_1}), \ldots, m(H_{a_n})),$$

$$P_\varphi^-(a_1, \ldots, a_n) \Leftrightarrow a_1, \ldots, a_n \in M \;\&\; \neg\varphi(m(H_{a_1}), \ldots, m(H_{a_n})).$$

Then both P_φ^+, P_φ^- are in Γ.

Proof. The preceding lemma 9 gives the result for the case φ is $x_1 = x_2$ (or $x_1 = z$ with a constant z), since e.g.

$$P_=^\pm(a, b) \Leftrightarrow P^+(a, \langle\emptyset\rangle, b, \langle\emptyset\rangle).$$

For the $\in$-relation we can take

$$P_\in^+(a, b) \Leftrightarrow a, b \in M \;\&\; (\exists x)\{P^+(a, \langle\emptyset\rangle, b, \langle x\rangle)\},$$

$$P_\in^-(a, b) \Leftrightarrow a, b \in M \;\&\; (\forall x)\{\langle x\rangle \in \breve{H}_b \Rightarrow P^-(a, \langle\emptyset\rangle, b, \langle x\rangle)\}.$$

It is easy to verify that the class of formulas which satisfy the lemma is closed under $\neg$, &, $\vee$ and the restricted quantifiers. For example, if

$$\varphi(y, z) \equiv (\forall x \in y)\psi(x, y, z),$$

we can take

$$P_\varphi^+(b, c) \Leftrightarrow b, c \in M \;\&\; (\forall x)[\langle x\rangle \in \breve{H}_b \Rightarrow (\exists d)[d \in M \;\&\; P^+(b, \langle x\rangle, d, \langle\emptyset\rangle) \;\&\; P_\psi^+(d, b, c)]].$$

To complete an inductive proof of the lemma, we need only treat the prime formula $R(x, y, z)$. Choose first a Δ_0 formula $\varphi(z)$ such that

$$\varphi(z) \Leftrightarrow z \text{ is an ordinal} > 1.$$

By the part of the lemma we have proved, both P_φ^+, P_φ^- are in Γ. Also it is immediate that whenever $c \in M$ and $m(H_c)$ is an ordinal $\xi > 1$, then

$$\xi = rank\{(\langle s_1, s_2\rangle, \langle t\rangle) : \langle s_1, s_2\rangle, \langle t\rangle \in H_c \;\&\; m(H_c, \langle s_1, s_2\rangle) \in m(H_c, \langle t\rangle)\}.$$

Using the inductive relations P^+, P^- of lemma 9 it is easy to find an inductive and a coinductive relation which agree whenever $c \in M$ and define the binary relation in the braces, so by the Good Parametrization Theorem 9C.8 there is a fixed function $j(c)$ in Δ such that whenever $c \in M$ and $m(H_c) = \xi$ is an ordinal > 1, then

$$(13) \qquad j(c) \in I^2, \qquad \xi = rank(H^2_{j(c)}).$$

Also if $a \in M$ and $\sigma(a) > 1$, then

$$\sigma(a) = rank\{(s, t) : s <_\sigma^* t <_\sigma^* a\}.$$

Using the definition of a Γ-norm and 9C.8, we can find a function $k(a)$ in Δ such that

$$a \in M \Rightarrow [k(a) \in I^2 \;\&\; H^2_{k(a)} = \{(s, t): s <^*_\sigma t <^*_\sigma a\}].$$

Notice that if $a \in M$ and $\sigma(a) \leqslant 1$, then $H^2_{k(a)} = \emptyset$. From this and the remarks above, it follows that if $a, c \in M$ and $m(H_c) = \xi > 1$, then

$$\sigma(a) \leqslant \xi \Leftrightarrow rank(H^2_{k(a)}) \leqslant rank(H^2_{j(c)}), \tag{14}$$

where $k(a)$, $j(c)$ are in Δ. Now the relation

$$X \leqslant Y \Leftrightarrow X, Y \text{ are wellfounded } \& \; rank(X) \leqslant rank(Y)$$

is inductive and hence Γ on Δ. Thus, *there is a fixed relation $Q(a, c)$ in Γ such that whenever $a, c \in M$ and $m(H_c)$ is an ordinal > 1, then*

$$\sigma(a) \leqslant m(H_c) \Leftrightarrow Q(a, c). \tag{15}$$

Now

$$\begin{aligned} P^+_R(a, b, c) \Leftrightarrow\; & a, b, c \in M \;\&\; P^+_\varphi(c) \\ & \& \; (\exists x)(\exists u)\{x \in M \;\&\; m(H_a) = x \\ & \qquad \& \; u \in H_a \;\&\; m(H_b) = u \\ & \qquad \& \; \sigma(x) \leqslant m(H_c)\} \\ \Leftrightarrow\; & a, b, c \in M \;\&\; P^+_\varphi(c) \\ & \& \; (\exists x)(\exists u)\{x \in M \;\&\; P^\pm_=(a, i(x)) \\ & \qquad \& \; u \in H_a \;\&\; P^\pm_=(b, i(u)) \\ & \qquad \& \; Q(x, c)\}, \end{aligned}$$

so that P^+_R is in Γ.

A similar computation reduces the problem of showing that $P^-_R(a, b, c)$ is in Γ to finding some $Q'(a, c)$ in Γ such that whenever $a, c \in M$ and $m(H_c)$ is an ordinal > 1, then

$$a \notin M \vee m(H_c) < \sigma(a) \Leftrightarrow Q'(a, c). \tag{16}$$

If $m(H_c)$ is an ordinal, then $m(H_c) < o(\Delta)$, since $m(H_c)$ is the rank of a wellfounded relation in Δ. On the other hand, $\sigma: M \twoheadrightarrow o(\Delta)$, since M is not in Δ by lemma 5. Hence we can set

$$Q'(a, c) \Leftrightarrow (\exists y)\{y \in M \;\&\; rank(H^2_{j(c)}) \leqslant rank(H^2_{k(y)}) \;\&\; y <^*_\sigma a\}$$

and (16) will hold. That Q' is in Γ follows easily as before.

From lemma 10 follows easily the main part of the theorem.

Lemma 11. *$\mathscr{M}$ is admissible relative to R.*

Proof. We have already shown that $\mathscr{M}$ is transitive and closed under pairing and union, so it remains to check Δ_0-Separation and Δ_0-Collection relative to R.

The Good Parametrization Theorem 9C.8 implies that there is a function (a, s) in $\varDelta$ such that whenever $a \in I^1$, then $j(a, s) \in I^1$ and

$$H_{j(a,s)} = \{u: \langle s\rangle^\frown u \in H_a\};$$

hence

$$(17) \qquad a \in M \Rightarrow \{(\forall s)[j(a, s) \in M] \,\&\, m(H_a) = \{m(H_{j(a,s)}): \langle s\rangle \in H_a\}\}.$$

To prove Δ_0-Separation, suppose $z = m(H_a)$ is a set in $\mathscr{M}$ and $\varphi(x)$ is $\Delta_0(R)$. We must show that

$$w = \{x \in z: \varphi(x)\} \in \mathscr{M}.$$

Let $P_\varphi^+(b)$, $P_\varphi^-(b)$ be the relations in Γ associated with $\varphi(x)$ by lemma 10 and put

$$\begin{aligned} X &= \{j(a, s): \langle s\rangle \in H_a \,\&\, P_\varphi^+(j(a, s))\} \\ &= \{j(a, s): \langle s\rangle \in H_a \,\&\, \neg P_\varphi^-(j(a, s))\}. \end{aligned}$$

Clearly $X \subseteq M$, X is in $\varDelta$ and

$$\{m(H_c): c \in X\} = \{x \in z: \varphi(x)\} = w,$$

so that $w \in \mathscr{M}$ by lemma 6.

To prove Δ_0-Collection, suppose $z = m(H_a)$ is a set in $\mathscr{M}$, $\varphi(x, y)$ is $\Delta_0(R)$ and

$$(\forall x \in z)(\exists y)\varphi(x, y).$$

Letting again $P_\varphi^+(b, c)$ be the relation in Γ associated with $\varphi(x, y)$, by lemma 10 we know that

$$(\forall s)\{\langle s\rangle \in H_a \Rightarrow (\exists c)P_\varphi^+(j(a, s), c)\}.$$

By the $\varDelta$ Selection Theorem 9C.2, there is a $\varDelta$ relation $Q(s, c)$ such that

$$Q(s, c) \Rightarrow P_\varphi^+(j(a, s), c),$$

$$(\forall s)[\langle s\rangle \in H_a \Rightarrow (\exists c)Q(s, c)].$$

Put

$$X = \{c: (\exists s)[\langle s\rangle \in H_a \,\&\, Q(s, c)]\}.$$

X is obviously in $\varDelta$ and

$$\begin{aligned} c \in X &\Rightarrow (\exists s)Q(s, c) \\ &\Rightarrow (\exists s)P_\varphi^+(j(a, s), c) \\ &\Rightarrow c \in M. \end{aligned}$$

Hence by lemma 6,

$$w = \{m(H_c)\colon c \in X\} \in \mathscr{M}.$$

It is now trivial to check that

$$(\forall x \in z)(\exists y \in w)\varphi(x, y),$$

which completes the proof of Δ_0-Collection and the lemma.

Now lemmas 11 and 3 imply immediately:

Lemma 12. $\mathscr{M} = \mathscr{M}(\Delta)$.

We can collect these lemmas into a proof of the theorem.

Proof of part (c). Assume first that P is $\Sigma_1(R)$ on $\mathscr{M}$,

$$P(x_1, \ldots, x_n) \Leftrightarrow (\exists y)\varphi(y, x_1, \ldots, x_n),$$

with φ some $\Delta_0(R)$ formula. Using lemmas 8 and 10,

$$P(x_1, \ldots, x_n) \Leftrightarrow (\exists a)\{a \in M \ \&\ P_\varphi^+(a, i(x_1), \ldots, i(x_n))\},$$

so that P is in Γ.

To prove the converse, notice first that M is $\Sigma_1(R)$ since

$$a \in M \Leftrightarrow (\exists \xi)R(a, \langle \emptyset \rangle, \xi).$$

By lemma 5, M is Γ-complete, so given $P(\bar{x})$ in Γ, there is a function $f(\bar{x})$ in Δ such that

$$P(\bar{x}) \Leftrightarrow f(\bar{x}) \in M.$$

Choose $z \in \mathscr{M}$ by lemma 3 so that

$$(\bar{x}, y) \in z \Leftrightarrow f(\bar{x}) = y;$$

then

$$P(\bar{x}) \Leftrightarrow (\exists y)[(\bar{x}, y) \in z \ \&\ y \in M],$$

so that P is $\Sigma_1(R)$.

Proof of part (a). $\mathscr{M}$ is admissible by lemma 11. That $o(\mathscr{M}) = o(\Delta)$ is obvious since every ordinal in $\mathscr{M}$ is the rank of some wellfounded relation in Δ and conversely, every wellfounded relation in Δ is a member of $\mathscr{M}$ and hence has rank an ordinal of $\mathscr{M}$. The last assertion follows immediately from part (c) and lemma 3.

Proof of part (b). $\mathscr{M}$ is admissible relative to R by lemma 11. The proof of resolvability and projectibility follows by an application of 9D.3.

Notice first that the map

$$H(\xi, a) = \{(u^\frown\langle s\rangle, u) \colon \mathscr{R}(a, u^\frown\langle s\rangle, \xi)\}$$

is $\Delta_0(R)$ and

$$H(\xi, a) = \begin{cases} \{(u^\frown\langle s\rangle, u) \colon u^\frown\langle s\rangle \in H_a\} & \text{if } a \in M \;\&\; \xi > 1 \;\&\; \sigma(a) \leqslant \xi, \\ \emptyset & \text{otherwise.} \end{cases}$$

The map

$$(18) \qquad G(w, f, x) = \{f(t) \colon t \in \mathit{Domain}(f) \;\&\; (t, x) \in w\}$$

is Δ_0, hence by 9D.3 there is a Δ_1 function $F(w, x)$ such that whenever w is a wellfounded relation, $x \in \mathit{Field}(w)$,

$$(19) \qquad F(w, x) = \{F(w, t) \colon (t, x) \in w\}.$$

Put

$$\mu(\xi, a) = F(H(\xi, a), \langle\emptyset\rangle);$$

now μ is $\Delta_1(R)$ and

$$(20) \qquad [\xi > 1 \;\&\; a \in M \;\&\; \sigma(a) \leqslant \xi] \Rightarrow \mu(\xi, a) = m(H_a).$$

We can get a $\Delta_1(R)$ resolution of $\mathscr{M}$ by setting

$$\tau(\xi) = \{\mu(\eta, a) \colon \eta < \xi, a \in A\}.$$

For a $\Delta_1(R)$ projection of $\mathscr{M}$ on A, put

$$(21) \qquad a \in D \Leftrightarrow a \in M \;\&\; (\forall b)[b <_\sigma^* a \Rightarrow m(H_b) \neq m(H_a)]$$

and define

$$\pi \colon D \twoheadrightarrow \mathscr{M}$$

by

$$\pi(a) = m(H_a).$$

It is easy to check that

$$\begin{aligned} \pi(a) = y \Leftrightarrow (\exists\xi)\{&R(a, \langle\emptyset\rangle, \xi) \;\&\; \mu(\xi, a) = y \\ &\& \; (\forall\eta < \xi)(\forall b \in A)[R(b, \langle\emptyset\rangle, \eta) \Rightarrow \mu(\eta, b) \neq y]\} \\ \Leftrightarrow (\forall\eta)(\forall b \in A)\{&[R(b, \langle\emptyset\rangle, \eta) \;\&\; \mu(\eta, b) = y] \\ &\Rightarrow (\exists\xi \leqslant \eta)[R(a, \langle\emptyset\rangle, \xi) \;\&\; \mu(\xi, a) = y]\}, \end{aligned}$$

so that the graph of π is $\Delta_1(R)$. ⊣

We call a structure $\langle \mathcal{M}(\Delta), R_1, \ldots, R_l \rangle$ a *companion* of the Spector class Γ, if the sequence of relations

$$\bar{R} = R_1, \ldots, R_l$$

on $\mathcal{M}(\Delta)$ satisfies (b), (c) of Theorem 9E.1, in particular for $P \subseteq A^n$,

$$P \text{ is in } \Gamma \Leftrightarrow P \text{ is } \Sigma_1(\bar{R}) \text{ on } \mathcal{M}(\Delta).$$

The next result shows that the companion is essentially unique. First a lemma.

9E.2. LEMMA. *Let $\mathcal{M}$ be admissible relative to both $\bar{R} = R_1, \ldots, R_l$ and $\bar{R}' = R'_1, \ldots, R'_{l'}$, let A be a set in $\mathcal{M}$ such that the following hold:*

(a) *$\mathcal{M}$ is projectible on A relative to $\bar{R}$.*

(b) *There is a coding scheme on A which is $\Delta_1(\bar{R})$.*

(c) *If $P \subseteq A^n$, then P is Σ_1 $(\bar{R})$ if and only if P is $\Sigma_1(\bar{R}')$.*

Then for every $P \subseteq \mathcal{M}^n$, P is $\Sigma_1(\bar{R})$ if and only if P is $\Sigma_1(\bar{R}')$.

PROOF. Let

$$\pi : D \twoheadrightarrow \mathcal{M} \qquad (D \subseteq A)$$

be a $\Delta_1(\bar{R})$ projection of $\mathcal{M}$ on A. If $b \in D$, then

$$a \in D \;\&\; \pi(a) \in \pi(b) \Leftrightarrow (\exists y)[\pi(b) = y \;\&\; (\exists x \in y)[\pi(a) = x]]$$
$$\Leftrightarrow (\forall y)[(\pi(b) = y) \Rightarrow (\exists x \in y)[\pi(a) = x]].$$

Using this, it is easy to verify that for each $b \in D$, the wellfounded tree

(22) $\quad H(b) = \{\langle x_1, \ldots, x_n \rangle : x_1, \ldots, x_n \in D \;\&\; \pi(b) \ni \pi(x_1) \ni \ldots \ni \pi(x_n)\}$

is in $\mathcal{M}$. Moreover, the relation

(23) $\quad W(b) = \{(\langle x_1, \ldots, x_n, y \rangle, \langle x_1, \ldots, x_n \rangle) : \langle x_1, \ldots, x_n, y \rangle \in H(b)\}$

is also in $\mathcal{M}$, and in fact the relations

$$W^+(b, u, v) \Leftrightarrow b \in D \;\&\; (u, v) \in W(b),$$

$$W^-(b, u, v) \Leftrightarrow b \in D \;\&\; (u, v) \notin W(b)$$

are $\Sigma_1(\bar{R})$, hence they are $\Sigma_1(\bar{R}')$.

Let $F(w, x)$ be the function defined by (18) and (19) in the proof of 9E.1. Since for each $b \in D$ the tree $H(b)$ determines the set $\pi(b)$, we have

$$b \in D \Rightarrow \pi(b) = F(W(b), \langle \emptyset \rangle).$$

Moreover, $F(w, x)$ is Δ_1 on $\mathcal{M}$, i.e. definable by Σ_1, Π_1 formulas in which $\bar{R}$ does not occur.

Since D, W^+, W^- are all $\Sigma_1(\bar{R})$ and hence $\Sigma_1(\bar{R}')$, the equivalence

$$\pi(b) = y \Leftrightarrow b \in D \ \& \ (\exists w)\{W(b) = w \ \& \ F(w, \langle\emptyset\rangle) = y\}$$

implies that the graph of π is $\Sigma_1(\bar{R}')$.

If $P(x_1, \ldots, x_n)$ is $\Sigma_1(\bar{R})$, put

$$P^\pi(a_1, \ldots, a_n) \Leftrightarrow a_1, \ldots, a_n \in D \ \& \ P(\pi(a_1), \ldots, \pi(a_n));$$

then P^π is $\Sigma_1(\bar{R})$, hence $\Sigma_1(\bar{R}')$ and

$$P(x_1, \ldots, x_n) \Leftrightarrow (\exists a_1, \ldots, a_n)[\pi(a_1) = x_1 \ \& \ldots \& \ \pi(a_n) = x_n \ \& \ P^\pi(a_1, \ldots, a_n)],$$

so that P is $\Sigma_1(\bar{R}')$.

The same argument shows that every relation P which is $\Sigma_1(\bar{R}')$ is also $\Sigma_1(\bar{R})$, so that the proof is complete. ⊣

9E.3. THEOREM. *Let Γ be a Spector class on the infinite transitive set A, let $\bar{R}$, $\bar{R}'$ be relations on $\mathcal{M} = \mathcal{M}(\Delta)$ so that both $\langle \mathcal{M}, \bar{R}\rangle$, $\langle \mathcal{M}, \bar{R}'\rangle$ are companions of Γ. Then a relation P on $\mathcal{M}$ is $\Sigma_1(\bar{R})$ if and only if P is $\Sigma_1(\bar{R}')$.*

PROOF is immediate from the definitions and the lemma. ⊣

There is also the following converse to Theorem 9E.1.

9E.4. THEOREM. *Let $\mathcal{M}$ be admissible and resolvable with respect to $\bar{R} = R_1, \ldots, R_l$, let $A \in \mathcal{M}$ be transitive and assume that $\mathcal{M}$ is projectible on A relative to $\bar{R}$ and that there is a one-to-one pairing function*

$$f: A \times A \rightarrow A$$

which is in $\mathcal{M}$. Let Γ be the collection of all relations on A which are $\Sigma_1(\bar{R})$. Then Γ is a Spector class, $\mathcal{M}(\Delta) = \mathcal{M}$ and $\langle \mathcal{M}, \bar{R}\rangle$ is a companion of Γ.

In particular, if $\mathcal{M}$ is admissible, resolvable and projectible with respect to $\bar{R}$, then $\langle \mathcal{M}, \bar{R}\rangle$ is the companion of some Spector class.

PROOF. Choose $a \neq b$ in A and put

$$a_0 = f(a, a), \ a_1 = f(a_0, b), \ a_2 = f(a_1, b), \ldots, a_{n+1} = f(a_n, b), \ldots.$$

It is easy to check that the sequence $a_0, a_1, a_2, \ldots$ with its natural ordering gives a copy of ω in Δ, and from this and f we can easily construct a coding scheme in Γ. The other closure properties of Γ are trivial.

To prove the parametrization property, let

$$\pi: D \twoheadrightarrow \mathcal{M} \qquad (D \subseteq A)$$

be a $\Delta_1(\bar{R})$ projection of $\mathcal{M}$ on A, for each n choose a $\Sigma_1(\bar{R})$ relation $S^n \subseteq \mathcal{M}^{n+1}$ which parametrizes the n-ary relations on $\mathcal{M}$ and put

$$U^n(a, x_1, \ldots, x_n) \Leftrightarrow a \in D \;\&\; S^n(\pi(a), x_1, \ldots, x_n).$$

To prove that Γ is normed let

$$\tau\colon o(\mathcal{M}) \to \mathcal{M}$$

be a $\Delta_1(\bar{R})$ resolution of $\mathcal{M}$. If $P \subseteq A^n$ is in Γ, then there is a $\Delta_0(\bar{R})$ relation $Q(y, \bar{x})$ such that

$$P(\bar{x}) \Leftrightarrow (\exists z)Q(z, \bar{x}).$$

Put

$$\sigma(\bar{x}) = \textit{least } \xi \textit{ such that } (\exists z \in \tau(\xi))Q(z, \bar{x}).$$

Now σ is a Γ-norm on P, since

$$\bar{x} \leqslant_\sigma^* \bar{y} \Leftrightarrow (\exists \xi)[(\exists z \in \tau(\xi))Q(z, \bar{x}) \;\&\; (\forall \eta < \xi)(\forall z \in \tau(\eta))\neg Q(z, \bar{y})],$$

$$\bar{x} <_\sigma^* \bar{y} \Leftrightarrow (\exists \xi)[(\exists z \in \tau(\xi))Q(z, \bar{x}) \;\&\; (\forall \eta \leqslant \xi)(\forall z \in \tau(\eta))\neg Q(z, \bar{y})].$$

It is now sufficient to show that $\mathcal{M}(\varDelta) = \mathcal{M}$, and since $\mathcal{M}$ is admissible and $\varDelta \subseteq \mathcal{M}$, it is enough to verify $\mathcal{M} \subseteq \mathcal{M}(\varDelta)$, i.e. for all $\mathcal{N}$,

$$\mathcal{N} \textit{ admissible } \& \; \varDelta \subseteq \mathcal{N} \Rightarrow \mathcal{M} \subseteq \mathcal{N}.$$

Given $\pi(b) \in \mathcal{M}$ with $b \in D$, define $H(b)$ and $W(b)$ by (22) and (23) in the proof of 9E.2. As we argued in that proof, $H(b)$, $W(b)$ are in $\varDelta$, hence in $\mathcal{N}$ and

$$\pi(b) = F(W(b), \langle\emptyset\rangle),$$

where $F(w, x)$ is Δ_1 on $\mathcal{N}$, so that $\pi(b) \in \mathcal{N}$.

To prove the last assertion, choose z such that $\mathcal{M}$ is projectible on z relative to $\bar{R}$, choose $A \supseteq z$ such that A is transitive and closed under the ordinary settheoretic pairing and apply the main claim of the theorem. ⊣

The representation theorem 9E.1 applies only to a Spector class Γ on a domain A which is a transitive, infinite set. It should be pointed out that this restriction is not essential; but if we want to extend the result to arbitrary Spector classes, we must allow for admissible sets which may have arbitrary objects (nonsets) as members. Such a theory of *admissible sets with urelements* has been developed recently by Barwise [1973]. Since the proofs in this section do not use any deep properties of admissible sets, they should extend to admissible sets with urelements. This would establish a correspondence between arbitrary Spector classes and admissible sets with urelements which are projectible and resolvable.

9F. The next admissible set

We now establish the characterization of the class of inductive relations on a nice transitive set, the main result of Barwise–Gandy–Moschovakis [1971]. First a lemma.

9F.1. LEMMA. *Let A be a transitive infinite set, let $R_1, \ldots, R_l$ be relations on A, let $\mathcal{M}$ be an admissible set such that $A, R_1, \ldots, R_l \in \mathcal{M}$, put*

$$\mathfrak{A} = \langle A, \in \restriction A, R_1, \ldots, R_l \rangle.$$

Then

$$\kappa^{\mathfrak{A}} \leqslant o(\mathcal{M})$$

and every relation P which is inductive on $\mathfrak{A}$ is Σ_1 on $\mathcal{M}$.

PROOF. Let $\varphi(\bar{x}, S)$ be any positive formula in the language of $\mathfrak{A}$. We can consider $\varphi(\bar{x}, S)$ as a Δ_0 formula in $\mathcal{M}$, where all the quantifiers are restricted to A and the relations $R_1, \ldots, R_l$ are members of $\mathcal{M}$. Then the function

$$G(f, \xi) = \{\bar{x} \in A^n : \varphi(\bar{x}, \bigcup \{f(\eta) : \eta \in Domain(f)\})\}$$

is Δ_0 on $\mathcal{M}$, so that by 9D.5 the function F which is determined by the recursion

$$F(\xi) = G(F \restriction \xi, \xi) = \{\bar{x} : \varphi(\bar{x}, \textstyle\bigcup_{\lambda < \xi} F(\eta))\}$$

is Δ_1 on $\mathcal{M}$. Obviously,

$$F(\xi) = I_\varphi^\xi,$$

so in particular each stage I_φ^ξ of the induction determined by φ is in $\mathcal{M}$, as long as ξ is an ordinal of $\mathcal{M}$. Moreover there is a Δ_0 relation $Q(z, \xi, \bar{x})$ such that

(1) $$\bar{x} \in I_\varphi^\xi \Leftrightarrow (\exists z) Q(z, \xi, \bar{x}).$$

Let

$$\kappa = o(\mathcal{M})$$

be the ordinal of $\mathcal{M}$ and assume that for a fixed tuple $\bar{x} \in A^n$, $\bar{x} \in I_\varphi^\kappa$, i.e. the sentence ψ satisfying

$$\psi \Leftrightarrow \varphi(\bar{x}, \textstyle\bigcup_{\xi < \kappa} I_\varphi^\xi)$$

is true. The sentence ψ is obtained from $\varphi(\bar{x}, S)$ by replacing each occurrence of

(2) $\bar{y} \in S$ *by* $(\exists \xi)(\exists z) Q(z, \xi, y)$.

Clearly ψ is Σ on $\mathcal{M}$ and the only unrestricted quantifiers in ψ are those introduced by (2). By the Σ-Reflection Principle 9D.2, there is some $w \in \mathcal{M}$

such that $\psi^{(w)}$ is true. Now $\psi^{(w)}$ is obtained from $\varphi(\bar{x}, S)$ by replacing each occurrence of

(3) $\quad \bar{y} \in S$ *by* $(\exists \xi \in w)(\exists z \in w)Q(z, \xi, \bar{y})$.

Let λ be any ordinal in $\mathcal{M}$ which is not in w and form χ by replacing each occurrence of

(4) $\quad \bar{y} \in S$ *by* $(\exists z)Q(z, \lambda, \bar{y})$

in $\varphi(\bar{x}, S)$. Since

$$\xi < \lambda \Rightarrow I_\varphi^\xi \subseteq I_\varphi^\lambda,$$

we have

$$(\exists \xi \in w)(\exists z \in w)Q(z, \xi, y) \Rightarrow (\exists z)Q(z, \lambda, \bar{y})$$

and since S occurs positively in φ we know that χ is true. But

$$\chi \Leftrightarrow \varphi(\bar{x}, \textstyle\bigcup_{\xi<\lambda} I_\varphi^\xi),$$

hence $\bar{x} \in I_\varphi^\lambda$. Thus we have shown

$$\bar{x} \in I_\varphi^\kappa \Rightarrow \bar{x} \in I_\varphi^{<\kappa},$$

i.e. the closure ordinal $\|\varphi\|$ of φ is $\leqslant \kappa = o(\mathcal{M})$. Also

$$\begin{aligned} \bar{x} \in I_\varphi &\Leftrightarrow (\exists \xi)\bar{x} \in I_\varphi^\xi \\ &\Leftrightarrow (\exists \xi < \kappa)(\exists z \in \mathcal{M})Q(z, \xi, \bar{x}) \end{aligned}$$

and the fixed point I_φ is Σ_1 on $\mathcal{M}$.

It follows immediately that every inductive relation on $\mathfrak{A}$ is Σ_1 on $\mathcal{M}$. ⊣

9F.2. THEOREM. *Let A be a transitive, infinite set, let $R_1, \ldots, R_l$ be relations on A such that the structure*

$$\mathfrak{A} = \langle A, \in \restriction A, R_1, \ldots, R_l\rangle$$

admits a hyperelementary coding scheme, put

$$\mathfrak{A}^+ = \bigcap \{\mathcal{M} : \mathcal{M} \text{ is admissible}, A, R_1, \ldots, R_l \in \mathcal{M}\}.$$

Then $\mathfrak{A}^+$ is admissible and resolvable, $\mathfrak{A}^+$ is projectible on A and

$$o(\mathfrak{A}^+) = \kappa^{\mathfrak{A}}.$$

Moreover, if $P \subseteq A^n$, then

(5) $\quad$ *P is hyperelementary on $\mathfrak{A} \Leftrightarrow P \in \mathfrak{A}^+$,*

(6) $\quad$ *P is inductive on $\mathfrak{A} \Leftrightarrow P$ is Σ_1 on $\mathfrak{A}^+$.*

In particular, the theorem applies if A is closed under pairing or if $A = \lambda$ is an infinite ordinal.

PROOF. Let Γ be the collection of all inductive relations on $\mathfrak{A}$. This is a Spector class, so choose R on

$$\mathcal{M} = \mathcal{M}(\Delta) \tag{7}$$

so that $\langle \mathcal{M}, R\rangle$ is a companion of Γ by Theorem 9F.1.

It is immediate from the definitions that

$$\mathfrak{A}^+ \subseteq \mathcal{M}.$$

On the other hand, if $X \subseteq A$ is in Δ, then X is Δ_1 on every admissible set $\mathcal{N}$ such that $A, R_1, \ldots, R_l \in \mathcal{N}$ by 9F.1, hence X is a member of every such admissible $\mathcal{N}$ by Δ_1-Separation, hence $X \in \mathfrak{A}^+$. Thus

$$\mathcal{M} = \mathfrak{A}^+. \tag{8}$$

We now apply Lemma 9E.2. Since $\mathfrak{A}^+$ is admissible relative to R and also relative to the empty list of relations, since $\mathfrak{A}^+$ is projectible on A relative to R and since for $P \subseteq A^n$

$$P \textit{ is } \Sigma_1(R) \Leftrightarrow P \textit{ is in } \Gamma \Leftrightarrow P \textit{ is } \Sigma_1$$

by 9E.1 and 9F.1, Lemma 9E.2 implies that for every relation P on $\mathfrak{A}^+$,

$$P \textit{ is } \Sigma_1(R) \Leftrightarrow P \textit{ is } \Sigma_1. \tag{9}$$

The theorem follows immediately from (9) and 9F.1.

The last assertion, that $\mathfrak{A}$ admits a hyperelementary coding scheme if A is closed under pairing or if A is an ordinal follows by Exercises 9.4 and 2.2. ⊣

We could avoid proving Lemma 9F.1 and give a slightly easier proof of this theorem if we were willing to use the results of Chapter 7, in particular the Spector–Gandy Theorem 7D.2 and the characterization of $\mathcal{HE}$ as the smallest model of Δ^1_1-Comprehension 7F.1. But Lemma 9F.1 is interesting in its own right and the technique we used in proving 9F.2 can be used to compute the companion of many interesting Spector classes, especially in the theory of nonmonotone inductive definability. It would take us far afield from the subject of this book to do this here. Instead we look briefly at the illuminating example of positive inductive definability in the language $\mathcal{L}(\mathrm{Q})$.

Let A be a transitive infinite set, $R_1, \ldots, R_l$ relations on A and let $\mathrm{Q} \subseteq$ *Power*(A) be a nontrivial monotone quantifier on A, as in (2) of Section 3E. We proved in Theorem 9B.1 that the class Γ of all relations which are positive $\mathcal{L}^{\mathfrak{A}}(\mathrm{Q})$-inductive on the structure

$$\mathfrak{A} = \langle A, \in \restriction A, R_1, \ldots, R_l\rangle$$

is a Spector class. The problem arises of giving a concrete description of the companion of this Γ in terms of the quantifier Q.

Suppose $\mathcal{M}$ is a transitive set, $A \in \mathcal{M}$, Q is a nontrivial monotone quantifier on A and $P_1, \ldots, P_k$ are relations on $\mathcal{M}$. We enlarge the language of the structure $\langle \mathcal{M}, \in \restriction \mathcal{M}, P_1, \ldots, P_k \rangle$ by the restricted quantifiers $(\exists x \in y)$, $(\forall x \in y)$, $(\mathrm{Q}x \in A)$, $(\mathrm{Q}^{\cup}x \in A)$, so that, for example, if φ is a formula, so are

$$(\mathrm{Q}x \in A)\varphi, \qquad (\mathrm{Q}^{\cup}x \in A)\varphi.$$

These are interpreted in the natural way,

$$(\mathrm{Q}x \in A)\varphi \textit{ is true} \Leftrightarrow \{x \in A : \varphi\} \in \mathrm{Q},$$

$$(\mathrm{Q}^{\cup}x \in A)\varphi \textit{ is true} \Leftrightarrow \neg(\mathrm{Q}x \in A)\neg\varphi \textit{ is true}$$

$$\Leftrightarrow \{x \in A : \neg\varphi\} \notin \mathrm{Q}.$$

A formula φ is $\Delta_0(\mathrm{Q}; P_1, \ldots, P_k)$ if it is built up from the prime formulas by the propositional connectives and the restricted quantifiers. The classes $\Sigma_1(\mathrm{Q}; P_1, \ldots, P_k)$, $\Pi_1(\mathrm{Q}; P_1, \ldots, P_k)$ and $\Delta_1(\mathrm{Q}; P_1, \ldots, P_k)$ relations are defined in the obvious way.

We call $\mathcal{M}$ *Q-admissible relative to* $P_1, \ldots, P_k$ if it is closed under pairing and union and if it satisfies the schemas of $\Delta_0(\mathrm{Q})$-*Separation* and $\Delta_0(\mathrm{Q})$-*Collection* relative to $P_1, \ldots, P_k$, i.e. all formulas of the form

$(\Delta_0(\mathrm{Q})$-*Sep*$)$ $\quad (\exists w)(\forall x)\{x \in w \Leftrightarrow [x \in z \;\&\; \varphi(x)]\}$,

$(\Delta_0(\mathrm{Q})$-*Coll*$)$ $\quad (\forall x \in z)(\exists y)\varphi(x, y) \Rightarrow (\exists w)(\forall x \in z)(\exists y \in w)\varphi(x, y)$,

where $\varphi(x)$ and $\varphi(x, y)$ are arbitrary $\Delta_0(\mathrm{Q}; P_1, \ldots, P_k)$ formulas.

It turns out that we need a stronger property than Q-admissibility. We call $\mathcal{M}$ *strongly Q-admissible relative to* $P_1, \ldots, P_k$ if it is Q-admissible and if it also satisfies the following two schemata of *strong* $\Delta_0(\mathrm{Q})$-*Collection relative to* $P_1, \ldots, P_k$,

$$(\mathrm{Q}x \in A)(\exists y)\varphi(x, y) \Rightarrow (\exists w)(\mathrm{Q}x \in A)(\exists y \in w)\varphi(x, y),$$

$$(\mathrm{Q}^{\cup}x \in A)(\exists y)\varphi(x, y) \Rightarrow (\exists w)(\mathrm{Q}^{\cup}x \in A)(\exists y \in w)\varphi(x, y),$$

where of course $\varphi(x, y)$ ranges over arbitrary $\Delta_0(\mathrm{Q}; P_1, \ldots, P_k)$ formulas.

9F.3. THEOREM. *Let A be a transitive, infinite set, let $R_1, \ldots, R$ be relations on A such that the structure*

$$\mathfrak{A} = \langle A, \in \restriction A, R_1, \ldots, R_l \rangle$$

admits a hyperelementary coding scheme, let Q *be a nontrivial monotone quantifier on A, put*

$$\mathfrak{A}^+(\mathrm{Q}) = \bigcap \{\mathcal{M} : A, R_1, \ldots, R_l \in \mathcal{M} \;\&\; \mathcal{M} \textit{ is strongly Q-admissible}\}.$$

Then $\mathfrak{A}^+(\mathrm{Q})$ *is strongly* Q-*admissible*,

$$o(\mathfrak{A}^+(\mathrm{Q})) = \textit{supremum}\{\|\varphi\|: \varphi \textit{ is a positive formula in } \mathscr{L}^{\mathfrak{A}}(\mathrm{Q})\}$$

and $\mathfrak{A}^+(\mathrm{Q})$ *admits a projection on* A *and a resolution which are* $\Delta_1(\mathrm{Q})$.

Moreover, if $P \subseteq A^n$, *then*

$$P \textit{ is hyper-}\mathscr{L}^{\mathfrak{A}}(\mathrm{Q})\textit{-definable} \Leftrightarrow P \in \mathfrak{A}^+(\mathrm{Q}),$$

$$P \textit{ is positive } \mathscr{L}^{\mathfrak{A}}(\mathrm{Q})\textit{-inductive} \Leftrightarrow P \textit{ is } \Sigma_1(\mathrm{Q}) \textit{ on } \mathfrak{A}^+(\mathrm{Q}).$$

PROOF is quite straightforward and we shall outline it in Exercises 9.8–9.14. ⊣

The notions of Q-admissibility and strong Q-admissibility are not immediately transparent, but special cases of them have been studied. We will see in the exercises that in the case of the *Suslin quantifier* these notions (essentially) coincide with the so-called β and *strong* β properties of admissible sets. Theorem 9F.3 then applies and gives an inductive definability characterization of *the next strongly* β *set*.

Exercises for Chapter 9

For each $n \geqslant 0$, define the set T^n of *objects of type n over* ω by the induction

$$T^0 = \omega,$$

$$T^{n+1} = \textit{all unary functions on } T^n \textit{ to } \omega.$$

It is sometimes convenient to use variables $\alpha^n, \beta^n, F^n, \ldots$ over T^n. Following Kleene, we let ${}^{n+2}E$ be the object of type $n+2$ which represents quantification over T^n,

$${}^{n+2}E(\alpha^{n+1}) = \begin{cases} 0 & \textit{if } (\exists\beta^n)[\alpha^{n+1}(\beta^n) = 0], \\ 1 & \textit{if } (\forall\beta^n)[\alpha^{n+1}(\beta^n) \neq 0]. \end{cases}$$

The first exercise is simply an observation, trivial to those familiar with recursion theory on higher types, particularly Kleene [1959b], Gandy [1962], Moschovakis [1967], Platek [1966], Grilliot [1967].

9.1. Let $n \geqslant 2$ and $j = 0$ or $j+3 \leqslant n$, let F^n be a fixed object of type n, put

$$\Gamma = \textit{all relations } P \textit{ on } T^j \textit{ which are semirecursive in } {}^nE, F^n \textit{ and some } \alpha^j \in T^j.$$

Prove that Γ is a Spector class. Prove that the condition on j is necessary by arguing that for $j = 1$, $n = 3$ the result fails. ⊣

Let Q be a nontrivial monotone unary quantifier on A and let Γ be a class of relations on A. We naturally call Γ *closed under* Q if for $P \subseteq A^{n+1}$ in Γ,

$$Q^A P = \{\bar{x}: (Qy)P(y, \bar{x})\} \in \Gamma.$$

9.2. Let Γ be a Spector class on A closed under the nontrivial monotone quantifier Q and its dual $Q^{\cup}$. Prove that if $Q_1, \ldots, Q_m$ are in Γ and R is positive $\mathscr{L}^A(Q)$-inductive in $Q_1, \ldots, Q_m$, then R is in Γ. Infer a corollary analogous to 9A.3. ⊣

9.3. Let Γ be a Spector class on A closed under the nontrivial monotone quantifier Q, let $P \subseteq A^{n+1}$ be in Γ and assume that

$$(Qx)(\exists\bar{y})P(x, \bar{y}).$$

Prove that there is some $P^* \subseteq P$ in Δ, such that

$$(Qx)(\exists\bar{y})P^*(x, \bar{y}).$$

HINT: If P is in Δ, the result is trivial, so assume that P is not in Δ. Fix a Γ-norm $\sigma: P \twoheadrightarrow o(\Delta)$ and put

$$R(x, \bar{y}) \Leftrightarrow (Q^{\cup}u)(\forall\bar{v})\neg[(u, \bar{v}) \leqslant_\sigma^* (x, \bar{y})].$$

Clearly R is in $\neg\Gamma$. Prove that $R \subsetneqq P$ and then show that we can take

$$P^*(x, \bar{y}) \Leftrightarrow (x, \bar{y}) \leqslant_\sigma^* (s, \bar{t}),$$

where $(s, \bar{t})$ is any fixed pair in $P-R$. ⊣

9.4. Prove that if A is a transitive set closed under pairing, then $\langle A, \in \restriction A\rangle$ admits a hyperelementary coding scheme. ⊣

9.5. Prove the Parametrization Theorem for Σ_1 relations on an admissible set, 9D.6.

HINT: Assign codes to the $\Delta_0(R_1, \ldots, R_l)$ formulas so that the relation

$$Sat_0(a, x) \Leftrightarrow a \textit{ is the code of some } \Delta_0(R_1, \ldots, R_l) \textit{ formula } \varphi$$
$$\& \; Seq(x) \;\&\; (x)_1, \ldots, (x)_{lh(x)}, 0, 0, \ldots \models \varphi$$

is $\Delta_1(R_1, \ldots, R_l)$ on $\mathscr{M}$. ⊣

9.6. Prove that for every infinite ordinal λ, $\lambda^{(\prime)} = \kappa^{\langle\lambda, \leqslant\rangle}$ is the next admissible ordinal, i.e. the smallest admissible ordinal greater than λ. ⊣

9.7. Prove that if $\mathfrak{A}$ is acceptable, then $\kappa^{\mathfrak{A}}$ is admissible. ⊣

In the next seven problems 9.8–9.14 we outline a proof of Theorem 9F.3. For these problems, fix

$$\mathfrak{A} = \langle A, \in \restriction A, R_1, \ldots, R_l \rangle,$$

where A is a transitive infinite set and let Q be a nontrivial monotone quantifier on A and

$$\Gamma = \textit{all positive } \mathscr{L}^{\mathfrak{A}}(\mathrm{Q})\textit{-inductive relations on } A.$$

9.8. Let $\varphi(\bar{x}, S)$ be an S-positive formula in $\mathscr{L}^{\mathfrak{A}}(\mathrm{Q})$, let $\mathscr{M}$ be a Q-admissible set such that $A, R_1, \ldots, R_l \in \mathscr{M}$. Prove that the map

$$\xi \mapsto I_\varphi^\xi$$

is $\Delta_1(\mathrm{Q})$ on $\mathscr{M}$, and in particular $\mathscr{M}$ is closed under this map. ⊣

9.9. Formulate and prove the $\Sigma(\mathrm{Q})$-*Reflection Principle* for strongly Q-admissible sets, in the fashion of Theorem 9D.2. ⊣

9.10. Prove that if $\mathscr{M}$ is strongly Q-admissible and $A, R_1, \ldots, R_l \in \mathscr{M}$, then for every S-positive formula φ in $\mathscr{L}^{\mathfrak{A}}(\mathrm{Q})$, $\|\varphi\| \leqslant o(\mathscr{M})$ and every positive $\mathscr{L}^{\mathfrak{A}}(\mathrm{Q})$-inductive relation on A is $\Sigma_1(\mathrm{Q})$ on $\mathscr{M}$. ⊣

9.11. Let $\langle \mathscr{M}, R \rangle$ be the companion of Γ. Prove that every $\Sigma_1(\mathrm{Q}; R)$ relation on $\mathscr{M}$ is $\Sigma_1(R)$.

HINT: Let $\pi: D \twoheadrightarrow \mathscr{M}$ be a $\Delta_1(R)$ projection of $\mathscr{M}$ on A. Prove that for every $\Delta_0(\mathrm{Q}; R)$ formula $\varphi(x_1, \ldots, x_n)$, the relation

$$\varphi^\pi(y_1, \ldots, y_n) \Leftrightarrow y_1, \ldots, y_n \in D \;\&\; \varphi(\pi(y_1), \ldots, \pi(y_n))$$

on A is in Γ. ⊣

9.12. Prove that if $\langle \mathscr{M}, R \rangle$ is the companion of Γ, then $\mathscr{M}$ is strongly Q-admissible relative to R. ⊣

HINT: Use Exercises 9.11 and 9.3.

9.13. Verify the version of 9E.2 for strong Q-admissibility. ⊣

9.14. Prove Theorem 9F.3. ⊣

HINT: Imitate the proof of 9F.2.

The *Suslin quantifier* S is the dual of the classical operation $\mathscr{A}$ and has received some attention in the literature, see Enderton [1967] and especially

Aczel [1970], where inductive definability in the language $\mathscr{L}(\mathsf{Q})$ on $\mathbb{N}$ is studied extensively. Relative to a fixed coding scheme on A, we put

$$\mathsf{S} = \{X \subseteq A: \{(\forall s_1)(\forall s_2)\ldots\}(\exists n)[\langle s_1, \ldots, s_n\rangle \in X]\}.$$

It is immediate that this is a nontrivial monotone quantifier on A.

We say that an admissible set $\mathcal{M}$ *has the β property* or *is β*, if the relation of wellfoundedness is absolute for $\mathcal{M}$. This means that for every binary relation $\prec$ which is a member of $\mathcal{M}$, if $\prec$ is not wellfounded, then there is some $z \in \mathcal{M}$, $z \subseteq Field(\prec)$, $z \neq \emptyset$, such that z has no $\prec$-least element.

We say that an admissible set $\mathcal{M}$ *has the strong β property* or *is strongly β*, if for every $w \in \mathcal{M}$ and for every binary relation $\prec \subseteq w$, if $\prec$ is Σ_1 on $\mathcal{M}$ and if $\prec$ is not wellfounded, then there is some $z \in \mathcal{M}$, $z \subseteq Field(\prec)$, $z \neq \emptyset$ such that z has no $\prec$-least element. This is simply expressed by the following schema in the second order language over $\mathcal{M}$, where $\varphi(x, y)$ ranges over all Σ_1 formulas:

$$\begin{aligned}&\{(\forall x)(\forall y)[\varphi(x, y) \Rightarrow x, y \in w]\\ &\quad \& (\exists S)\{(\exists y)S(y) \;\&\; (\forall y)[S(y) \Rightarrow (\exists x)[S(x) \;\&\; \varphi(x, y)]]\}\}\\ &\qquad\qquad \Rightarrow (\exists z)\{(\exists y)(y \in z) \;\&\; (\forall y)[y \in z \Rightarrow (\exists x)[(x \in z) \;\&\; \varphi(x, y)]]\}.\end{aligned}$$

In this formulation the strong β property can be considered a *reflection principle*.

The next two results are due to P. Aczel for $A = \omega$.

9.15. Let $\mathcal{M}$ be an admissible set, let A be a transitive infinite member of $\mathcal{M}$ which admits a coding scheme that is Δ_1 on $\mathcal{M}$, let S be the Suslin quantifier on A relative to that coding scheme. Prove the following four propositions:

(i) If $\mathcal{M}$ is β, then $\mathcal{M}$ is S-admissible and every $\Sigma_1(\mathsf{S})$ relation on $\mathcal{M}$ is Σ_1.

(ii) If $\mathcal{M}$ is S-admissible and projectible on A by a $\Delta_1(\mathsf{S})$ function, then $\mathcal{M}$ is β.

(iii) If $\mathcal{M}$ is strongly β, then $\mathcal{M}$ is strongly S-admissible.

(iv) If $\mathcal{M}$ is strongly S-admissible and projectible on A by a $\Delta_1(\mathsf{S})$ function, then $\mathcal{M}$ is strongly β. ⊣

9.16. In the notation of 9F.3 with $\mathsf{Q} = \mathsf{S}$, prove that $\mathfrak{A}^+(\mathsf{S})$ is the smallest strongly β set which contains $A, R_1, \ldots, R_l$ as elements. ⊣

REFERENCES

S. AANDERAA

[1973] Inductive definitions and their closure ordinals, in: J. E. Fenstad and P. Hinman, eds., *Generalized Recursion Theory* (North-Holland, Amsterdam, 1973).

P. ACZEL

[1970] Representability in some systems of second order arithmetic, *Israel J. Math.* **8** (1970) 309–328.

P. ACZEL and W. RICHTER

[1972] Inductive definitions and analogues of large cardinals, *Conf. in Mathematical Logic, London*, 1970, Springer Lecture Notes in Math. **255** (1972) 1–9.

[1973] Inductive definitions and reflecting properties of admissible ordinals, in: J. E. Fenstad and P. Hinman, eds., *Generalized Recursion Theory* (North-Holland, Amsterdam, 1973).

K. J. BARWISE

[1972] Absolute logics and $\mathscr{L}_{\infty,\omega}$, *Ann. Math. Logic* **4** (1972) 309–340.

[1973] Admissible sets over models of set theory, in: J. E. Fenstad and P. Hinman, eds., *Generalized Recursion Theory* (North-Holland, Amsterdam, 1973).

K. J. BARWISE, R. O. GANDY and Y. N. MOSCHOVAKIS

[1971] The next admissible set, *J. Symbolic Logic* **36** (1971) 108–120.

C. C. CHANG and Y. N. MOSCHOVAKIS

[1970] The Suslin–Kleene theorem for V_κ with cofinality(κ) = ω, *Pacific J. Math.* **35** (1970) 565–569.

M. DAVIS

[1950] *On the theory of recursive unsolvability*, Doctoral Dissertation, Princeton University (1950).

H. B. ENDERTON

[1967] An infinitistic rule of proof, *J. Symbolic Logic* **32** (1967) 447–451.

S. FEFERMAN

[1965] Some applications of forcing and generic sets, *Fund. Math.* **56** (1965) 325–345.

[1971] Review of Moschovakis [1969c], *Math. Rev.* **42** (1971). MR #5791 1053–1054.

D. GALE and F. M. STEWART

[1953] Infinite games of perfect information, *Ann. Math. Studies* **28** (1953) 245–266.

R. O. GANDY

[1960] Proof of Mostowski's conjecture, *Bull. Acad. Polon. Sci. Sér. Sci. Math. Astron. Phys.* **8** (1960) 571–575.

[1962] General recursive functionals of finite type and hierarchies of functions, in: *Proc. Logic Colloq. Clermont Ferrand* (1962) 5–24.

R. GANDY, G. KREISEL and W. TAIT

[1960] Set existence, *Bull. Acad. Polon. Sci. Sér. Sci. Math. Astron. Phys.* **8** (1960) 577–582.

K. GÖDEL

[1940] *The consistency of the axiom of choice and the generalized continuum hypothesis with the axioms of set theory*, 4th printing (Princeton, N.J., 1958).

T. J. GRILLOT

[1967] *Recursive functions of finite higher types*, Doctoral Dissertation, Duke University, Durham, N.C. (1967).

[1971] Inductive definitions and computability, *Trans. Amer. Math. Soc.* **158** (1971) 309–317.

[1972] Omitting types: applications to recursion theory, *J. Symbolic Logic* **37** (1972) 81–89.

L. HENKIN

[1961] Some remarks on infinitely long formulas, in: *Infinitistic methods* (Pergamon, New York, 1961) 167–183.

I. KAPLANSKY

[1954] *Infinite abelian groups* (University of Michigan Press, 1954).

A. S. KECHRIS

[1972] *Projective ordinals and countable analytical sets*, Doctoral Dissertation, Univ. of California, Los Angeles, Calif. (1972).

H. J. KEISLER

[1965] Finite approximations of infinitely long formulas, in: J. W. Addison et al., eds., *The Theory of Models* (North-Holland, Amsterdam, 1965) 158–169.

[1970] Logic with the quantifier "there exist uncountably many", *Ann. Math. Logic* **1** (1970) 1–93.

[1971] *Model theory for infinitary logic* (North-Holland, Amsterdam, 1971).

S. C. KLEENE

[1944] On the forms of the predicates in the theory of constructive ordinals, *Amer. J. Math.* **66** (1944) 41–58.

[1952] *Introduction to metamathematics* (Van Nostrand, Princeton, N.J., 1952).

[1955a] On the forms of the predicates in the theory of constructive ordinals (second paper), *Amer. J. Math.* **77** (1955) 405–428.

[1955b] Arithmetical predicates and function quantifiers, *Trans. Amer. Math. Soc.* **79** (1955) 312–340.

[1955c] Hierarchies of number theoretic predicates *Bull. Amer. Math. Soc.* **61** (1955) 193–213.
[1959a] Quantification of number theoretic functions, *Compositio Math.* **14** (1959) 23–40.
[1959b] Recursive functionals and quantifiers of finite types I, *Trans. Amer. Math. Soc.* **91** (1959) 1–52.

G. KREISEL
[1961] Set-theoretic problems suggested by the notion of potential totality, in: *Infinitistic methods* (Pergamon, New York, 1961) 103–140.
[1962] The axiom of choice and the class of hyperarithmetic functions, *Indag. Math.* **24** (1962) 307–319.

S. KRIPKE
[1964] Transfinite recursion on admissible ordinals I, II (abstracts), *J. Symbolic Logic* **29** (1964) 161–162.

K. KURATOWSKI
[1966] *Topology*, Vol. 1 (Academic Press, New York, 1966).

A. LEVY
[1965] A hierarchy of formulas in set theory, *Mem. Amer. Math. Soc.* **57** (1965).

R. MANSFIELD
[1970] Perfect subsets of definable sets of real numbers, *Pacific J. Math.* **35** (1970) 451–457.

Y. N. MOSCHOVAKIS
[1967] Hyperanalytic predicates, *Trans. Amer. Math. Soc.* **129** (1967) 249–282.
[1969a] Abstract first order computability I, *Trans. Amer. Math. Soc.* **138** (1969) 427–464.
[1969b] Abstract first order computability II, *Trans. Amer. Math. Soc.* **138** (1969) 465–504.
[1969c] Abstract computability and invariant definability, *J. Symbolic Logic* **34** (1969) 605–633.
[1970] The Suslin–Kleene theorem for countable structures, *Duke Math. J.* **37** (1970) 341–352.
[1971a] The game quantifier, *Proc. Amer. Math. Soc.* **31** (1971) 245–250.
[1971b] Axioms for computation theories—first draft, in: R. O. Gandy and C. E. M. Yates, eds., *Logic Colloquium* '69 (North-Holland, Amsterdam, 1971) 199–255.

A. MOSTOWSKI
[1951] A classification of logical systems, *Studia Phil.* **4** (1951) 237–274.

R. PLATEK
[1966] *Foundations of recursion theory*, Doctoral Dissertation, Stanford Univ., Stanford, Calif., (1966).

W. RICHTER
[1971] Recursively Mahlo ordinals and inductive definitions, in: R. O. Gandy and C. E. M. Yates, eds., *Logic Colloquium* '69 (North-Holland, Amsterdam, 1971) 273–288.

H. Rogers, Jr.

[1967] *Theory of recursive functions and effective computability* (McGraw-Hill, New York, 1967).

J. Shoenfield

[1967] *Mathematical logic* (Addison-Wesley, Reading, Mass., 1967).

C. Spector

[1955] Recursive wellorderings, *J. Symbolic Logic* **20** (1955) 151–163.

[1960] Hyperarithmetical quantifiers, *Fund. Math.* **48** (1960) 313–320.

[1961] Inductively defined sets of natural numbers, in: *Infinitistic methods* (Pergamon, New York, 1961) 97–102.

INDEX

INDEX OF SYMBOLS

A CATALOG OF SELECTED

DOVER BOOKS

IN SCIENCE AND MATHEMATICS

Astronomy

BURNHAM'S CELESTIAL HANDBOOK, Robert Burnham, Jr. Thorough guide to the stars beyond our solar system. Exhaustive treatment. Alphabetical by constellation: Andromeda to Cetus in Vol. 1; Chamaeleon to Orion in Vol. 2; and Pavo to Vulpecula in Vol. 3. Hundreds of illustrations. Index in Vol. 3. 2,000pp. 6⅛ x 9¼.
Vol. I: 0-486-23567-X
Vol. II: 0-486-23568-8
Vol. III: 0-486-23673-0

EXPLORING THE MOON THROUGH BINOCULARS AND SMALL TELESCOPES, Ernest H. Cherrington, Jr. Informative, profusely illustrated guide to locating and identifying craters, rills, seas, mountains, other lunar features. Newly revised and updated with special section of new photos. Over 100 photos and diagrams. 240pp. 8¼ x 11. 0-486-24491-1

THE EXTRATERRESTRIAL LIFE DEBATE, 1750–1900, Michael J. Crowe. First detailed, scholarly study in English of the many ideas that developed from 1750 to 1900 regarding the existence of intelligent extraterrestrial life. Examines ideas of Kant, Herschel, Voltaire, Percival Lowell, many other scientists and thinkers. 16 illustrations. 704pp. 5⅜ x 8½. 0-486-40675-X

THEORIES OF THE WORLD FROM ANTIQUITY TO THE COPERNICAN REVOLUTION, Michael J. Crowe. Newly revised edition of an accessible, enlightening book recreates the change from an earth-centered to a sun-centered conception of the solar system. 242pp. 5⅜ x 8½. 0-486-41444-2

A HISTORY OF ASTRONOMY, A. Pannekoek. Well-balanced, carefully reasoned study covers such topics as Ptolemaic theory, work of Copernicus, Kepler, Newton, Eddington's work on stars, much more. Illustrated. References. 521pp. 5⅜ x 8½.
0-486-65994-1

A COMPLETE MANUAL OF AMATEUR ASTRONOMY: TOOLS AND TECHNIQUES FOR ASTRONOMICAL OBSERVATIONS, P. Clay Sherrod with Thomas L. Koed. Concise, highly readable book discusses: selecting, setting up and maintaining a telescope; amateur studies of the sun; lunar topography and occultations; observations of Mars, Jupiter, Saturn, the minor planets and the stars; an introduction to photoelectric photometry; more. 1981 ed. 124 figures. 25 halftones. 37 tables. 335pp. 6½ x 9¼. 0-486-40675-X

AMATEUR ASTRONOMER'S HANDBOOK, J. B. Sidgwick. Timeless, comprehensive coverage of telescopes, mirrors, lenses, mountings, telescope drives, micrometers, spectroscopes, more. 189 illustrations. 576pp. 5⅝ x 8¼. (Available in U.S. only.) 0-486-24034-7

STARS AND RELATIVITY, Ya. B. Zel'dovich and I. D. Novikov. Vol. 1 of *Relativistic Astrophysics* by famed Russian scientists. General relativity, properties of matter under astrophysical conditions, stars, and stellar systems. Deep physical insights, clear presentation. 1971 edition. References. 544pp. 5⅝ x 8¼. 0-486-69424-0

Chemistry

THE SCEPTICAL CHYMIST: THE CLASSIC 1661 TEXT, Robert Boyle. Boyle defines the term "element," asserting that all natural phenomena can be explained by the motion and organization of primary particles. 1911 ed. viii+232pp. 5⅜ x 8½. 0-486-42825-7

RADIOACTIVE SUBSTANCES, Marie Curie. Here is the celebrated scientist's doctoral thesis, the prelude to her receipt of the 1903 Nobel Prize. Curie discusses establishing atomic character of radioactivity found in compounds of uranium and thorium; extraction from pitchblende of polonium and radium; isolation of pure radium chloride; determination of atomic weight of radium; plus electric, photographic, luminous, heat, color effects of radioactivity. ii+94pp. 5⅜ x 8½. 0-486-42550-9

CHEMICAL MAGIC, Leonard A. Ford. Second Edition, Revised by E. Winston Grundmeier. Over 100 unusual stunts demonstrating cold fire, dust explosions, much more. Text explains scientific principles and stresses safety precautions. 128pp. 5⅜ x 8½. 0-486-67628-5

THE DEVELOPMENT OF MODERN CHEMISTRY, Aaron J. Ihde. Authoritative history of chemistry from ancient Greek theory to 20th-century innovation. Covers major chemists and their discoveries. 209 illustrations. 14 tables. Bibliographies. Indices. Appendices. 851pp. 5⅜ x 8½. 0-486-64235-6

CATALYSIS IN CHEMISTRY AND ENZYMOLOGY, William P. Jencks. Exceptionally clear coverage of mechanisms for catalysis, forces in aqueous solution, carbonyl- and acyl-group reactions, practical kinetics, more. 864pp. 5⅜ x 8½. 0-486-65460-5

ELEMENTS OF CHEMISTRY, Antoine Lavoisier. Monumental classic by founder of modern chemistry in remarkable reprint of rare 1790 Kerr translation. A must for every student of chemistry or the history of science. 539pp. 5⅜ x 8½. 0-486-64624-6

THE HISTORICAL BACKGROUND OF CHEMISTRY, Henry M. Leicester. Evolution of ideas, not individual biography. Concentrates on formulation of a coherent set of chemical laws. 260pp. 5⅜ x 8½. 0-486-61053-5

A SHORT HISTORY OF CHEMISTRY, J. R. Partington. Classic exposition explores origins of chemistry, alchemy, early medical chemistry, nature of atmosphere, theory of valency, laws and structure of atomic theory, much more. 428pp. 5⅜ x 8½. (Available in U.S. only.) 0-486-65977-1

GENERAL CHEMISTRY, Linus Pauling. Revised 3rd edition of classic first-year text by Nobel laureate. Atomic and molecular structure, quantum mechanics, statistical mechanics, thermodynamics correlated with descriptive chemistry. Problems. 992pp. 5⅜ x 8½. 0-486-65622-5

FROM ALCHEMY TO CHEMISTRY, John Read. Broad, humanistic treatment focuses on great figures of chemistry and ideas that revolutionized the science. 50 illustrations. 240pp. 5⅜ x 8½. 0-486-28690-8

Engineering

DE RE METALLICA, Georgius Agricola. The famous Hoover translation of greatest treatise on technological chemistry, engineering, geology, mining of early modern times (1556). All 289 original woodcuts. 638pp. 6¾ x 11. 0-486-60006-8

FUNDAMENTALS OF ASTRODYNAMICS, Roger Bate et al. Modern approach developed by U.S. Air Force Academy. Designed as a first course. Problems, exercises. Numerous illustrations. 455pp. 5⅜ x 8½. 0-486-60061-0

DYNAMICS OF FLUIDS IN POROUS MEDIA, Jacob Bear. For advanced students of ground water hydrology, soil mechanics and physics, drainage and irrigation engineering and more. 335 illustrations. Exercises, with answers. 784pp. 6⅛ x 9¼. 0-486-65675-6

THEORY OF VISCOELASTICITY (Second Edition), Richard M. Christensen. Complete consistent description of the linear theory of the viscoelastic behavior of materials. Problem-solving techniques discussed. 1982 edition. 29 figures. xiv+364pp. 6⅛ x 9¼. 0-486-42880-X

MECHANICS, J. P. Den Hartog. A classic introductory text or refresher. Hundreds of applications and design problems illuminate fundamentals of trusses, loaded beams and cables, etc. 334 answered problems. 462pp. 5⅜ x 8½. 0-486-60754-2

MECHANICAL VIBRATIONS, J. P. Den Hartog. Classic textbook offers lucid explanations and illustrative models, applying theories of vibrations to a variety of practical industrial engineering problems. Numerous figures. 233 problems, solutions. Appendix. Index. Preface. 436pp. 5⅜ x 8½. 0-486-64785-4

STRENGTH OF MATERIALS, J. P. Den Hartog. Full, clear treatment of basic material (tension, torsion, bending, etc.) plus advanced material on engineering methods, applications. 350 answered problems. 323pp. 5⅜ x 8½. 0-486-60755-0

A HISTORY OF MECHANICS, René Dugas. Monumental study of mechanical principles from antiquity to quantum mechanics. Contributions of ancient Greeks, Galileo, Leonardo, Kepler, Lagrange, many others. 671pp. 5⅜ x 8½. 0-486-65632-2

STABILITY THEORY AND ITS APPLICATIONS TO STRUCTURAL MECHANICS, Clive L. Dym. Self-contained text focuses on Koiter postbuckling analyses, with mathematical notions of stability of motion. Basing minimum energy principles for static stability upon dynamic concepts of stability of motion, it develops asymptotic buckling and postbuckling analyses from potential energy considerations, with applications to columns, plates, and arches. 1974 ed. 208pp. 5⅜ x 8½. 0-486-42541-X

METAL FATIGUE, N. E. Frost, K. J. Marsh, and L. P. Pook. Definitive, clearly written, and well-illustrated volume addresses all aspects of the subject, from the historical development of understanding metal fatigue to vital concepts of the cyclic stress that causes a crack to grow. Includes 7 appendixes. 544pp. 5⅜ x 8½. 0-486-40927-9

Mathematics

FUNCTIONAL ANALYSIS (Second Corrected Edition), George Bachman and Lawrence Narici. Excellent treatment of subject geared toward students with background in linear algebra, advanced calculus, physics and engineering. Text covers introduction to inner-product spaces, normed, metric spaces, and topological spaces; complete orthonormal sets, the Hahn-Banach Theorem and its consequences, and many other related subjects. 1966 ed. 544pp. 6⅛ x 9¼. 0-486-40251-7

ASYMPTOTIC EXPANSIONS OF INTEGRALS, Norman Bleistein & Richard A. Handelsman. Best introduction to important field with applications in a variety of scientific disciplines. New preface. Problems. Diagrams. Tables. Bibliography. Index. 448pp. 5⅜ x 8½. 0-486-65082-0

VECTOR AND TENSOR ANALYSIS WITH APPLICATIONS, A. I. Borisenko and I. E. Tarapov. Concise introduction. Worked-out problems, solutions, exercises. 257pp. 5⅜ x 8¼. 0-486-63833-2

AN INTRODUCTION TO ORDINARY DIFFERENTIAL EQUATIONS, Earl A. Coddington. A thorough and systematic first course in elementary differential equations for undergraduates in mathematics and science, with many exercises and problems (with answers). Index. 304pp. 5⅜ x 8½. 0-486-65942-9

FOURIER SERIES AND ORTHOGONAL FUNCTIONS, Harry F. Davis. An incisive text combining theory and practical example to introduce Fourier series, orthogonal functions and applications of the Fourier method to boundary-value problems. 570 exercises. Answers and notes. 416pp. 5⅜ x 8½. 0-486-65973-9

COMPUTABILITY AND UNSOLVABILITY, Martin Davis. Classic graduate-level introduction to theory of computability, usually referred to as theory of recurrent functions. New preface and appendix. 288pp. 5⅜ x 8½. 0-486-61471-9

ASYMPTOTIC METHODS IN ANALYSIS, N. G. de Bruijn. An inexpensive, comprehensive guide to asymptotic methods–the pioneering work that teaches by explaining worked examples in detail. Index. 224pp. 5⅜ x 8½ 0-486-64221-6

APPLIED COMPLEX VARIABLES, John W. Dettman. Step-by-step coverage of fundamentals of analytic function theory–plus lucid exposition of five important applications: Potential Theory; Ordinary Differential Equations; Fourier Transforms; Laplace Transforms; Asymptotic Expansions. 66 figures. Exercises at chapter ends. 512pp. 5⅜ x 8½. 0-486-64670-X

INTRODUCTION TO LINEAR ALGEBRA AND DIFFERENTIAL EQUATIONS, John W. Dettman. Excellent text covers complex numbers, determinants, orthonormal bases, Laplace transforms, much more. Exercises with solutions. Undergraduate level. 416pp. 5⅜ x 8½. 0-486-65191-6

RIEMANN'S ZETA FUNCTION, H. M. Edwards. Superb, high-level study of landmark 1859 publication entitled "On the Number of Primes Less Than a Given Magnitude" traces developments in mathematical theory that it inspired. xiv+315pp. 5⅜ x 8½. 0-486-41740-9

Math–Decision Theory, Statistics, Probability

ELEMENTARY DECISION THEORY, Herman Chernoff and Lincoln E. Moses. Clear introduction to statistics and statistical theory covers data processing, probability and random variables, testing hypotheses, much more. Exercises. 364pp. 5⅜ x 8½. 0-486-65218-1

STATISTICS MANUAL, Edwin L. Crow et al. Comprehensive, practical collection of classical and modern methods prepared by U.S. Naval Ordnance Test Station. Stress on use. Basics of statistics assumed. 288pp. 5⅜ x 8½. 0-486-60599-X

SOME THEORY OF SAMPLING, William Edwards Deming. Analysis of the problems, theory and design of sampling techniques for social scientists, industrial managers and others who find statistics important at work. 61 tables. 90 figures. xvii +602pp. 5⅜ x 8½. 0-486-64684-X

LINEAR PROGRAMMING AND ECONOMIC ANALYSIS, Robert Dorfman, Paul A. Samuelson and Robert M. Solow. First comprehensive treatment of linear programming in standard economic analysis. Game theory, modern welfare economics, Leontief input-output, more. 525pp. 5⅜ x 8½. 0-486-65491-5

PROBABILITY: AN INTRODUCTION, Samuel Goldberg. Excellent basic text covers set theory, probability theory for finite sample spaces, binomial theorem, much more. 360 problems. Bibliographies. 322pp. 5⅜ x 8½. 0-486-65252-1

GAMES AND DECISIONS: INTRODUCTION AND CRITICAL SURVEY, R. Duncan Luce and Howard Raiffa. Superb nontechnical introduction to game theory, primarily applied to social sciences. Utility theory, zero-sum games, n-person games, decision-making, much more. Bibliography. 509pp. 5⅜ x 8½. 0-486-65943-7

INTRODUCTION TO THE THEORY OF GAMES, J. C. C. McKinsey. This comprehensive overview of the mathematical theory of games illustrates applications to situations involving conflicts of interest, including economic, social, political, and military contexts. Appropriate for advanced undergraduate and graduate courses; advanced calculus a prerequisite. 1952 ed. x+372pp. 5⅜ x 8½. 0-486-42811-7

FIFTY CHALLENGING PROBLEMS IN PROBABILITY WITH SOLUTIONS, Frederick Mosteller. Remarkable puzzlers, graded in difficulty, illustrate elementary and advanced aspects of probability. Detailed solutions. 88pp. 5⅜ x 8½. 65355-2

PROBABILITY THEORY: A CONCISE COURSE, Y. A. Rozanov. Highly readable, self-contained introduction covers combination of events, dependent events, Bernoulli trials, etc. 148pp. 5⅜ x 8¼. 0-486-63544-9

STATISTICAL METHOD FROM THE VIEWPOINT OF QUALITY CONTROL, Walter A. Shewhart. Important text explains regulation of variables, uses of statistical control to achieve quality control in industry, agriculture, other areas. 192pp. 5⅜ x 8½. 0-486-65232-7

Physics

OPTICAL RESONANCE AND TWO-LEVEL ATOMS, L. Allen and J. H. Eberly. Clear, comprehensive introduction to basic principles behind all quantum optical resonance phenomena. 53 illustrations. Preface. Index. 256pp. 5⅜ x 8½. 0-486-65533-4

QUANTUM THEORY, David Bohm. This advanced undergraduate-level text presents the quantum theory in terms of qualitative and imaginative concepts, followed by specific applications worked out in mathematical detail. Preface. Index. 655pp. 5⅜ x 8½. 0-486-65969-0

ATOMIC PHYSICS (8th EDITION), Max Born. Nobel laureate's lucid treatment of kinetic theory of gases, elementary particles, nuclear atom, wave-corpuscles, atomic structure and spectral lines, much more. Over 40 appendices, bibliography. 495pp. 5⅜ x 8½. 0-486-65984-4

A SOPHISTICATE'S PRIMER OF RELATIVITY, P. W. Bridgman. Geared toward readers already acquainted with special relativity, this book transcends the view of theory as a working tool to answer natural questions: What is a frame of reference? What is a "law of nature"? What is the role of the "observer"? Extensive treatment, written in terms accessible to those without a scientific background. 1983 ed. xlviii+172pp. 5⅜ x 8½. 0-486-42549-5

AN INTRODUCTION TO HAMILTONIAN OPTICS, H. A. Buchdahl. Detailed account of the Hamiltonian treatment of aberration theory in geometrical optics. Many classes of optical systems defined in terms of the symmetries they possess. Problems with detailed solutions. 1970 edition. xv + 360pp. 5⅜ x 8½. 0-486-67597-1

PRIMER OF QUANTUM MECHANICS, Marvin Chester. Introductory text examines the classical quantum bead on a track: its state and representations; operator eigenvalues; harmonic oscillator and bound bead in a symmetric force field; and bead in a spherical shell. Other topics include spin, matrices, and the structure of quantum mechanics; the simplest atom; indistinguishable particles; and stationary-state perturbation theory. 1992 ed. xiv+314pp. 6⅛ x 9¼. 0-486-42878-8

LECTURES ON QUANTUM MECHANICS, Paul A. M. Dirac. Four concise, brilliant lectures on mathematical methods in quantum mechanics from Nobel Prize-winning quantum pioneer build on idea of visualizing quantum theory through the use of classical mechanics. 96pp. 5⅜ x 8½. 0-486-41713-1

THIRTY YEARS THAT SHOOK PHYSICS: THE STORY OF QUANTUM THEORY, George Gamow. Lucid, accessible introduction to influential theory of energy and matter. Careful explanations of Dirac's anti-particles, Bohr's model of the atom, much more. 12 plates. Numerous drawings. 240pp. 5⅜ x 8½. 0-486-24895-X

ELECTRONIC STRUCTURE AND THE PROPERTIES OF SOLIDS: THE PHYSICS OF THE CHEMICAL BOND, Walter A. Harrison. Innovative text offers basic understanding of the electronic structure of covalent and ionic solids, simple metals, transition metals and their compounds. Problems. 1980 edition. 582pp. 6⅛ x 9¼. 0-486-66021-4